Design of
New Materials

Design of New Materials

Edited by
D. L. Cocke
and
A. Clearfield

Texas A&M University
College Station, Texas

Plenum Press • New York and London

Library of Congress Cataloging in Publication Data

Texas A & M University. Industry–University Cooperative Chemistry Program. Symposium
  (4th: 1986)
  Design of new materials.

  Sponsored by Celanese Chemical Company and others.
  Bibliography: p.
  Includes index.
  1. Materials—Congresses. I. Cocke, D. L. (David L.) II. Clearfield, Abraham. III. Celanese
Chemical Company. IV. Title.
TA401.3.T47   1986   620.1'1   87-3131
ISBN-13: 978-1-4615-9503-8   e-ISBN-13: 978-1-4615-9501-4
DOI: 10.1007/978-1-4615-9501-4

Proceedings of the Fourth Annual Industry–University Cooperative Chemistry Program Symposium,
held March 24–26, 1986, in College Station, Texas

© 1987 Plenum Press, New York
Softcover reprint of the hardcover 1st edition 1987
A Division of Plenum Publishing Corporation
233 Spring Street, New York, N.Y. 10013

PREFACE

Dr. George P. Thomon, Nobel Laureate in Physics said, "We have labelled civilizations by the main materials which they have used: The Stone Age, the Bronze Age and the Iron Age . . . a civilization is both developed and limited by the materials at its disposal. Today, man lives on the boundary between the Iron Age and a New Materials Age." The ever more stringent requirements for materials to accomplish specific functions and withstand extreme conditions, as dictated by the needs of industry and defense, continue to spur ever more intensive research in Materials Science.

According to the recent report "Trends and Opportunities in Materials Research" a vital goal of materials research is to design synthesize and fabricate in high yield, new materials with properties that can be predicted, varied and controlled. In the past this has been a fairly empirical process, but as we gain more comprehensive understanding of the behavior of matter on an atomic and molecular scale this goal becomes ever more attainable. An important recent trend is the increasing sophistication and power of theoretical approaches. Aided by the development of computers and versatile numerical techniques, as well as concepts from statistical mechanics, theorists are beginning to confront the complexity of real materials. Important advances are expected through a concentrated attack on model systems in which the theorist, experimental scientist and engineer all work together towards designing new materials and controlling their properties.

We have tried to bring together in this small conference a blend of experimentalists and theorists in diverse areas of materials science. We wished to highlight both the progress made in frontier areas of synthesis and some new ideas and theoretical concepts aimed at a better understanding of materials. Whether we have succeeded is for the reader to judge, but certainly this volume highlights recent progress in a number of interesting areas and the approaches being pursued by leaders in their respective fields.

In preparing this volume we have relied extensively on the managerial and editorial skills of Elizabeth Porter who supervised the many details that go into a project of this complexity. We are also grateful to Ms. Dawne Welch for her assistance in text preparation. Her skillful help greatly eased our task.

Abraham Clearfield

Professor of Chemistry

Program Co-Chair, 1986 IUCCP
Research Symposium

David L. Cocke

Associate Professor of Chemistry

Program Co-Chair, 1986 IUCCP
Research Symposium

CONTENTS

New Perspectives on Materials Design
    David L. Cocke   1

Modifications of Molecular Size and Structure During the
    Hydrolytic Polycondensation of Metal Alkoxides
    Bulent E. Yoldas   13

Design of Microstructures in Sol-Gel Processed Silicates
    Lisa C. Klein   39

Inorganic Macromolecules and the Search for New Electroactive
    and Structural Materials
    Harry R. Allcock   67

The Preceramic Polymer Route to Silicon-Containing Ceramics
    Dietmar Seyferth and Yuan-Fu Yu   79

Synthesis of Ceramic Powders and Thin Films from Laser
    Heated Gases
    John S. Haggerty   95

New Approaches to the Design of Materials Via Preparative
    Inorganic Chemistry
    Abraham Clearfield   121

Materials Design by Means of Discharge Plasmas
    S. Veprek   135

Advanced Ceramic Materials and Processes
    Roy W. Rice   169

Research on Hydrogenated Amorphous Silicon
    K. Weiser   195

Structure and Optical Properties of Amorphous Semiconductors
    Richard Zallen   217

Surface Modification by Rapid Solidification Laser and
    Electron Beam Processing
    B. H. Kear and P. R. Strutt   229

Materials-By-Design: Prospects and Promise
    James J. Eberhardt   257

Crystallographic Engineering
    R. E. Newnham   275

Structure - Property Relationships in Metallic and
  Oxide Glasses
    Philip H. Gaskell   291

Infrared and Raman Studies of Si-Chalcogenide Glasses
    M. Tenhover, R. S. Henderson, M. A. Hazle,
        D. Lukco and R. K. Grasselli   329

Compound Index   357

Index   363

NEW PERSPECTIVES ON MATERIALS DESIGN

David L. Cocke

Department of Chemistry
Texas A&M University
College Station, Texas 77843

INTRODUCTION

In order to have a good perspective on "materials design", it is necessary to define the concept. Materials design can be viewed as the best application of available information and understanding for the prediction and synthesis of materials with desired properties. Here, understanding is differentiated from information and is used to imply correlations usable to formulate workable concepts. We know what functions a material should have or with better understanding what properties a material should have to provide a given function. However, the means to predict and then prepare the material with the desired properties for a given function or process is now generally lacking.

The need and hope for materials design has been around for some time - but mainly confined to individual disciplines. For example, Evans (1) has pointed out in 1952 that crystal chemistry would enable the prediction and synthesis of chemical compounds having any desired combination of desired properties. From these early aspirations sprang the materials science area of structure-property relationships (2). In 1962, von Hipple (3) called for allying science and engineering for the design of materials with prescribed properties. In metallurgy, Tien and Ansell (4) have produced a monogram on alloy and microstructure design. In catalysis, Trimm has laid the ground work for heterogeneous catalyst design in his recent reviews (5) and book (6).

Why, then are we not further along in designing materials? There would appear to be several reasons. First, the task is an enormous undertaking and requires the collaboration of many disciplines. Second, although each recent generation of disciplinary scientists has marveled at the advances in his area and its potential application to materials design, the complexity of the task has rarely been appreciated and the advances have been insufficient. Third, education and training in the skills and philosophy of materials design have been practically nonexistent. Forth, the long-term nature of materials design and the relatively short turn-around time for trial and error approaches have limited industrial attempts. However, significant progress is being made in the drug industry (7) which might serve as a guide and an inspiration for true materials design.

Today, we are faced with challenges from abroad that are adversely affecting not only our standard of living, but our future ability to perform the necessary science due to an accompanying slide in our economic base. An example of this in the area of advanced ceramics (8) is shown in Figure 1. With the prospects of a 30 billion/year market by the year 2010, one can clearly see the relative efforts of Japan, England, West Germany and the USA. About 1000 USA scientists are involved while more than 2000 Japanese researchers are involved. Given the strong coordination of university, government and industry in Japan, it can be expected that the impact may be much larger for Japan. This consequently calls for a closer look at the "Japanese model"(9).

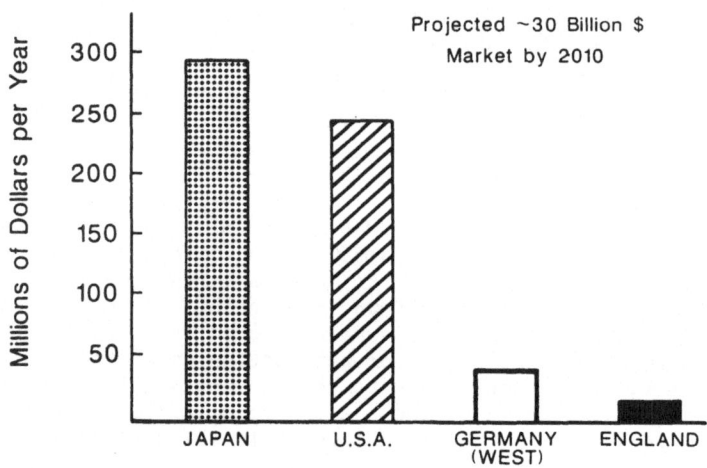

Fig. 1. Relative commitment to advanced ceramics research.

As pointed out by Kevin Marsden (10), European and Japanese efforts are being directed toward collaborative, interdisciplinary programs for "creating and developing new materials." The Japanese Government subsidizes about 25% of their national R & D budget of 23 billion dollars. They also give generous tax credits including accelerated depreciation on R & D facilities as well as low interest loans. Only 2% of Japan's research capital is in the military sector. This makes Japan's civilian research effort approximately equal the U.S. civilian research effort! The Japanese Ministry of International Trade and Industry (M.I.T.I.) promotes collaborative research involving amorphous materials, advanced ceramics, new alloys, semiconductors, intercalation compounds, and polymers. These areas have been identified as important future materials growth areas. All of these areas except polymers are covered in this book.

At this point, it is appropriate to establish what is the relationship between materials and the U.S. economy (11). The Bureau of Mines has estimated that extracted raw materials contribute approximately 8% to our total national economy of about $3T(GNP). Combining this with synthetic materials and their products will raise this by several times. The Federal Government in 1985 had an R&D budget of about $51 billion of which $1 B was allocated for materials research. A new National Center for Advanced Materials (CAM) at Lawrence Berkely National Laboratories is being established at a multimillion dollar/year cost.

The need for materials design is clear. The resources being allocated for materials research are substantial. The question to be answered is - should some of these resources be directed at materials design.

COMPONENTS OF MATERIALS DESIGN

Materials design is extremely multidisciplinary. Figure 2 illustrates the interplay needed between experiment, theory, modeling, statistics and computation which are the five major advancement areas relevant to Materials Design.

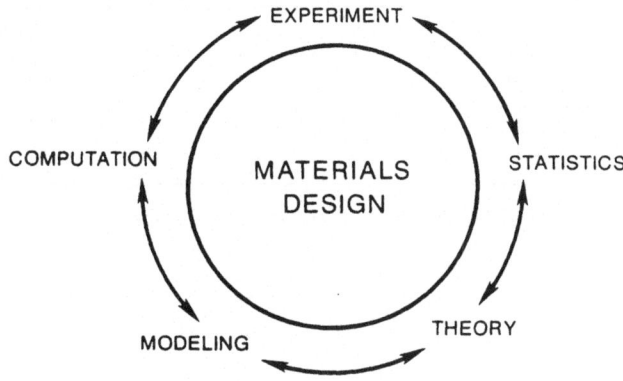

COMPUTATION
  PC's to Super Computers
EXPERIMENT
  Preparation - Characterization - Testing
THEORY
  Empirical to Quantum
MODELING
  Hierarchy of Models
STATISTICS
  Pattern Recognition

Fig. 2. Five major components of materials design. The interplay needed between experiment, theory, modeling, statistics and computation.

The design process will involve interdisciplinary action between materials scientists, chemists, physicists, mathematicians, statisticians, engineers and computational experts. Possible interplay between the design process and advancing areas is illustrated in Figure 3.

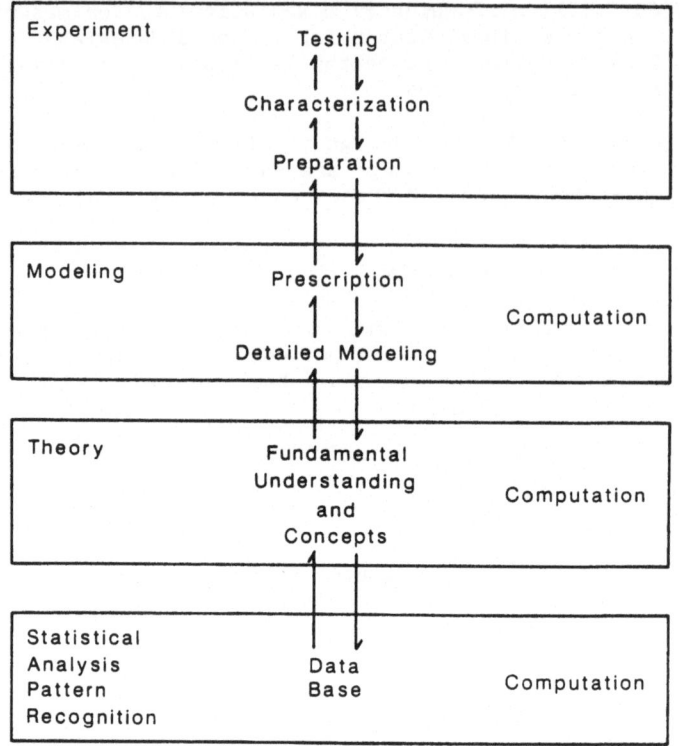

Fig. 3. Interplay between the design process and advancing
areas.

The foundation of materials design is the "available" data base as indicated in the previous definition. This is a critical aspect since lack of information can prevent materials design. The data base must be in interaction with every step in the process. Missing data must be obtained via direct experimentation which is a relatively slow process or it must be predicted via semiempirical or theoretical models. Statistical analysis of the data by pattern recognition will maximize the correlation of data and aid in identification of the most relevant data that is needed in the design process. This latter process is extremely important since it will focus attention and the R&D effort to the most fertile aspects of the problem. This will certainly save materials and human resources and result in increased competitiveness.

Theory provides fundamental understanding and develops concepts which are the vehicles of materials design. Through fundamental understanding and concepts, realistic models ranging from fundamental electronic structure to structure-property relationships and even applied processes can be formulated.

Modeling will thus involve development of a hierarchy of models. Successful modeling will provide the framework for prescribing the materials.

Finally, the experimental approaches of preparation, characterization and testing will conclude a materials design cycle. Rapid iterative interaction between preparation, characterization and testing is common today and a mainstay of materials development. Materials design calls for integration of these steps in a process which promises better understanding and better focused expenditure of effort.

The increasing demand placed on materials today is challenging the innovation of preparative, characterization and testing scientists and engineers. The emphasis of this book is on the experimental advances impacting materials design with a strong coverage of preparative technology.

The topics highlight some of the break-through technologies with great promise in materials design. In sol-gel technology, new precursor chemistry and growth conditions are allowing control of topology from dense spheres to random fractals. Control of pore size, pore volume and bulk properties are keys to catalysis, sensoring and photonic advances. New routes to advanced ceramics such as polymers are already paying dividends. Advances in preparative methods for noncrystalline materials are providing exciting opportunities in fundamental, as well as practical applications. Advances in materials modification and preparation via non-equilibrium chemical environments such as plasmas are producing outstanding results. Finally, advances in structure-property relationships hold enormous promise where fundamental understanding is being achieved. The remainder of this report will examine the advances in the other areas and hopefully place the preparative advances in perspective.

SELECTED ADVANCES

For materials design to become a reality, significant progress will need to be made in all the areas shown in the scheme in Figure 3. While highlighting selected advances, it is appropriate to review in a general way the progress in the five major advancement areas.

DATA BASE   Statistical analysis and in particular pattern recognition has shown important progress. The general state-of-the-art can be found in recent reviews (12,13). Cocke and Bleeker (14) have used ARTHUR (15) to design corrosion inhibitors for acid media. Here pattern recognition on a data base of over 100 compounds tested for their inhibiting power resulted in identification of the major physical chemical properties of the compounds most relevant to their inhibiting ability. In heterogeneous catalysis, Strouf et al.(16) have used pattern recognition to predict hydrogenation activity of new materials. The use of pattern recognition in drug design again stands as a model for study by materials designers.

THEORY   Enormous advances in all forms of theory from semi-empirical to ab initio have promoted a study of the potential of theory and modeling in materials design. The details of this study can be found in the book edited by Eberhardt, Allen, and Cocke (17). The study found that in spite of the enormous work to be done, it was not too early to begin a push in this direction. The most important advance in semiempirical theory in materials has been the successful application of the atomic cell model of Miedema (18-25). It has been used to predict numerous properties of metals and alloys such as heats of solution, surface energies, heats of adsorption, etc. Obviously these data are directly useful to a materials design effort. More fundamental theories are providing concepts which are quite helpful. The recent work of

Goddard (26) on molybdenum oxide catalysts is a good example of concepts from theory. Materials scientists have long been using Pauling's five rules (27). Structural field maps have also proven quite valuable to materials research. The recent theoretical basis placed on these by Burdett (28) and others (29) has shown that the interjection of theoretical considerations significantly improve usefulness of these "standard" tools. Figure 4 shows such a structural field map based on pseudopotential radii.

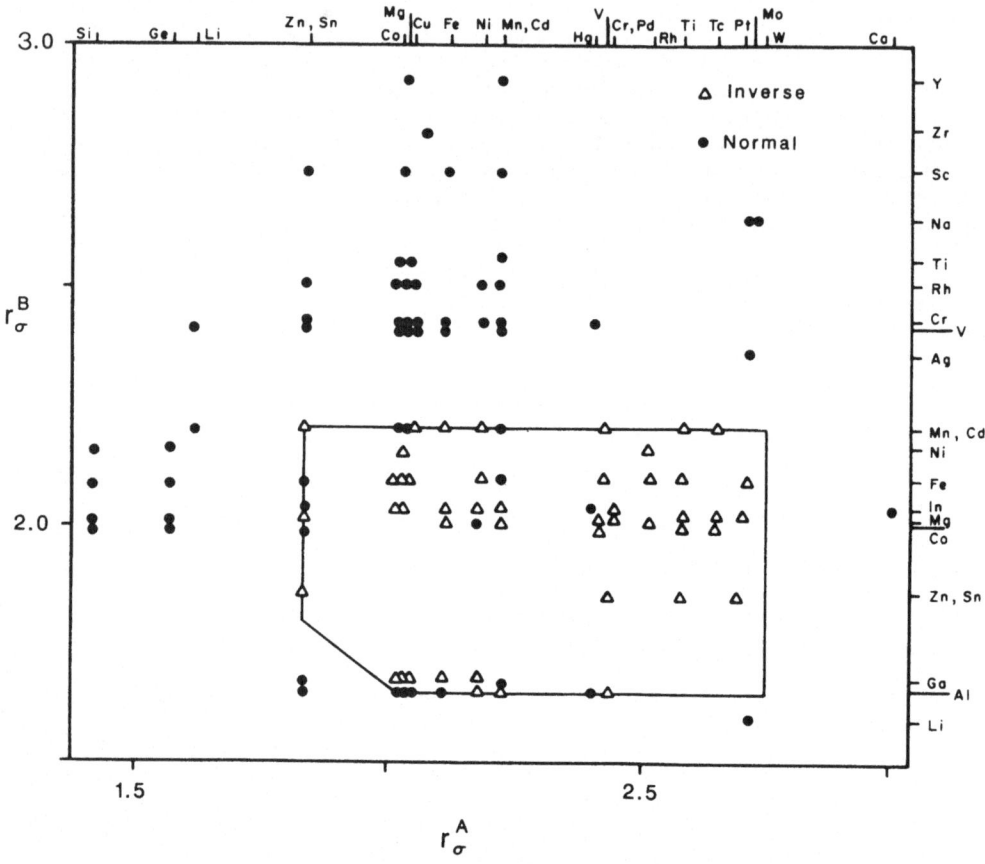

Fig. 4. A structural field map based on pseudopotential radii.

MODELING   There is general agreement that realistic models at many levels will be needed for materials design. The models will overlap extending from fundamental electronic theory to real engineering processes. A schematic of this hierarchy of models is shown in Figure 5. A specific example of modeling is the prediction of surface segregation induced by a reactive environment. Here the thermodynamic approach (32) using Miedema's model for heats of solution and surface energies produces the results shown in figure 6. This information is very important in predicting the surface composition and subsequent properties under reaction conditions. It can be seen that palladium segregates under both oxidizing and reducing conditions to the surface of a Pt-Pd alloy. This is a plausible explanation why Pt-Pd alloys only show palladium catalytic activity.

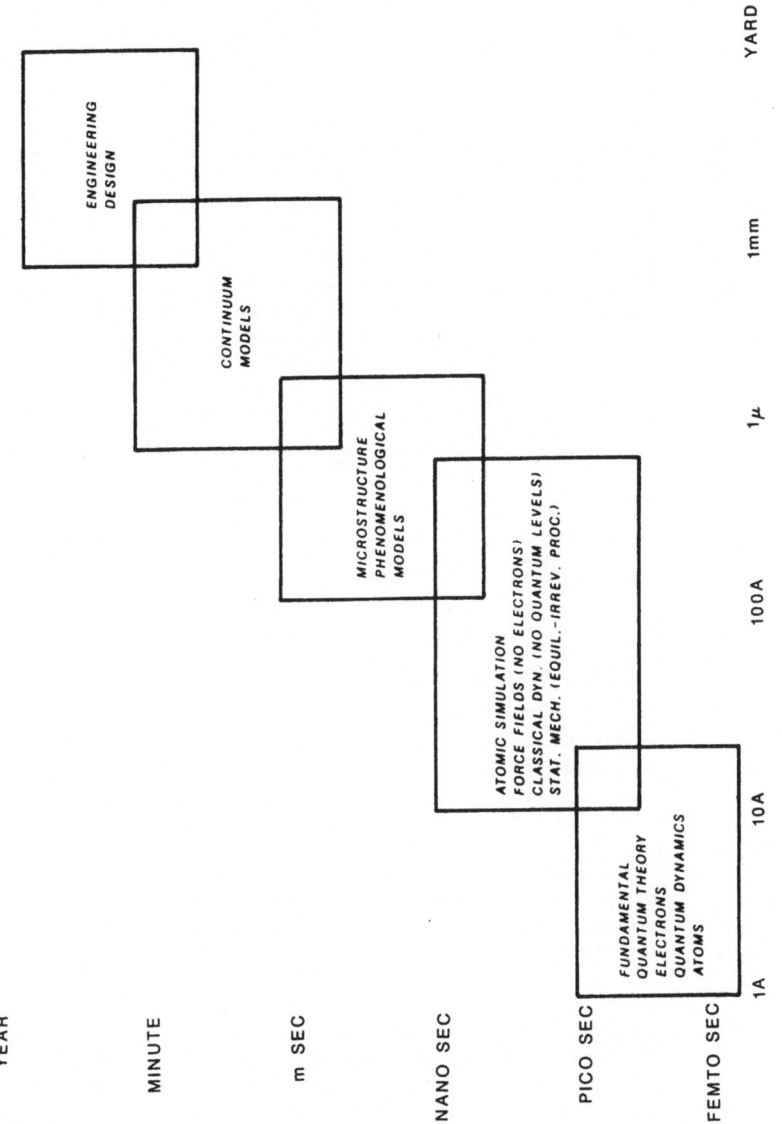

SIZE SCALE

Figure 5. Hierarchy of models.

7

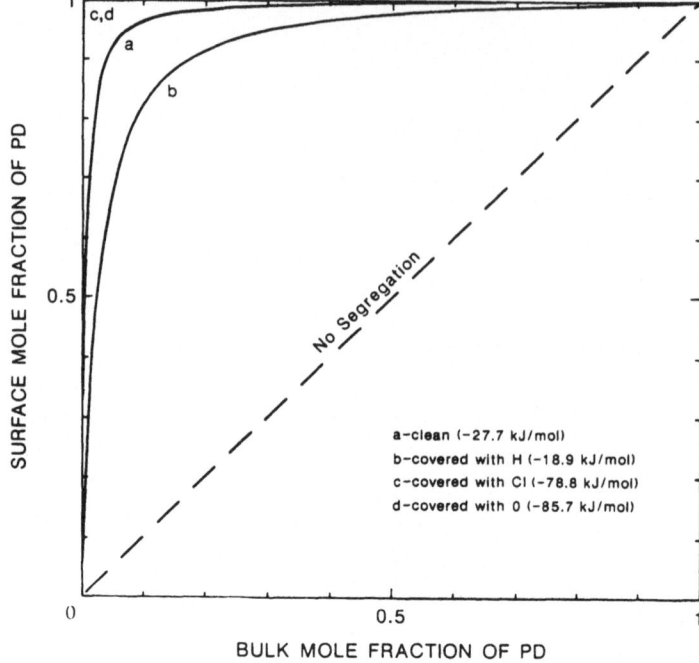

Figure 6. Surface segregation or clean and covered surface of PD-PT. Thermodynamic approach using Miedema's model.

EXPERIMENT   The experimental aspects of materials design can be divided into three areas: preparation, characterization, and testing. The preparation of materials is a rapidly developing interdisciplinary science with contributions from chemistry, physics, and engineering. The "break-through" technologies with great promise for materials design are sol-gel, new routes to advanced ceramics, amorphous materials, materials modification and preparation via non-equilibrium chemical environments (plasmas), and structure-property relationships. These are the main topics of this text and are covered in detail in the following chapters.

Characterization instrumentation and techniques have been the premier advancement area in materials design. This has not been confined to any given characterization area although surface techniques have been in the spotlight. Table 1 lists some of the outstanding recent advances in characterization instrumentation. The arrangement of the methods from ideal materials in pristine environments to real materials in process conditions shows the range of advances needed in materials design. Great strides are being made with ideal samples. For materials design, bulk and surface structure must be examined under process conditions as well.

Perhaps most noteworthy has been the recent development of the scanning tunneling microscope (33-38). Here is an instrument that is capable of seeing single atoms on the surface and can operate in vacuum or possibly in realistic process conditions such as high pressure or under fluids (39). This technique may be able to bridge the range of conditions needed for materials design.

Table 1. Selected Instrumental Advances

Low Energy Electron Diffraction	LEED
Angle Resolved Photoemission Spectroscopy	ARPS
Electron Energy Loss Spectroscopy	EELS
X-ray photoelectron spectroscopy	XPS
Extended X-ray Absorption Fine Structure	EXAFS
Magic Angle Spinning NMR (Solid State NMR)	MAS NMR
Conversion Electron Mossbauer Spectroscopy	CEMS

Techniques using high energy sources have been used to examine materials under process conditions. Synchrotron radiation has provided the radiation to examine catalysts during reactions (40) as well as producing data on the primary structure of materials and surfaces - the latter under vacuum conditions. Mossbauer spectroscopy (41) has provided a similar ability and has been used to determine the structure hydrotreating catalysts during reaction.

Extraction of thermodynamic data from electron spectroscopy (42) offers exciting new possibilities in materials design. This has been made possible using the equivalent core approximation and Born-Haber cycles. This may provide data such as heats of adsorption and heats of segregation that are difficult to measure by other means.

CONCLUSIONS

Although materials design is in part a futuristic concept now definitely the time to start toward this achievable goal. The payoff will be great and the sustained progress occurring in the five major areas will allow it to come to fruition.

REFERENCES

1. Evans, R. C., "An Introduction to Crystal Chemistry," Cambridge University Press, 1952.

2. Newnham, R. E., "Structure-Property Relationships," Springer-Verlag: Berlin, 1975.

3. Von Hipple, A., _Science_ (1962) _138_, 91.

4. Tien, J. K. and Ansell, G. S., "Alloy and Microstructural Design"; Academic Press: N.Y., 1976.

5. Trimm, D. L., "Designing Catalysts" _Chem. Tech._ (1979) 572.

6. Trimm, D. L., "Design of Industrial Catalysts," Elsevier: New York, 1980.

7. Kowalski, B. R. and Benaler, C. F., *J. Amer. Chem. Soc.* (1974) **96**, 916.

8. Global Report, *Advanced Materials and Processes* (1986) **2**, 7.

9. English, L. K., *Materials Engineering* (1985) **102**, 38.

10. Marsden, K., *J. Metals* (2986) **38**, 45.

11. Promisel, N. E., *Metall. Trans.* (1985) **16B**, 5.

12. Varmuza, K., "Pattern Recognition in Chemistry," Springer-Verlag: Berlin.

13. Ramiller, N., *American Laboratory* (1984) **6**, 78.

14. Cocke, D. L. and Bleeker, F.; unpublished results.

15. Koskinen, J. R. and Kowalski, B. R., *J. Chem. Inf. Comput. Sci.* (1975) **15**, 119.

16. Strout, O., Fusek, J. and Kuchynka, K., *Coll. Czec. Chem. Comm.* (1981) **46**, 75, 2328.

17. Eberhardt, J. J., Allen, R. E. and Cocke, D. L.. eds., Theory and Modeling in Materials Design" in press.

18. Miedema, A. R.; DeBoer, F. R. and Boom, R., *Physica* (1981) **103B**, 67.

19. Miedema, A. R., *Z. Metallkde.* (1979) **70**, 345.

20. Miedema, A. R. and Broeder, F. J. A., *Z. Metallkde.* (1979) **70**, 14.

21. Miedema, A. R. "The Heat of Formation of Alloys" (1976) **26**, 217.

22. Miedema, A. R. and Dorleijn, J. W. F., *Surface Science* (1980) **95**, 447.

23. Miedema, A. R., *Bull. Euro. Phys. Soc.* (1980) **11**.

24. Boom., R., DeBoer, F. R. and Miedema, A. R., *J. Less Common Metals* (1976) **45**, 237.

25. Miedema, A. R.; DeBoer, F. R.; Boom, R. and Dorleijn, J. W. F., *Calphad* (1977) **1**, 341.

26. Goddard III, W. A., *Science* (1985) **227**, 917.

27. Pauling, L., *J. Amer. Chem. Soc.* (1929) **51**, 1010.

28. Burdett, J. K.; Price, G. D. and Price, S. L., *Phys. Rev. B* (1981) **24**, 24.

29. Chelikowsky, J. R. and Phillips, J. C., *Phys. Rev. B* (1978) **17**, 2453.

30. Zanger, A., in "Structure and Bonding in Crystals," Naurotsky,, A. and O'Keeffe, M.; Academic Press: New York, 1981.

31. Block, A.N. and Schatterman, G. C. in "Structure and Bonding in Crystals," Naurotsky, A and O'Keeffe, M.; Academic Press: New York, 1981.

32. Yoon, C. and Cocke, D. L.,; to be published.

33. Binnig, G. and Rohrer, H., Helv. Phys. Acta (1982) 55, 726.

34. Binnig, G., Rohrer, H., Gerber, Ch. and Weibel, E., Phys. Rev. Lett. (1982) 50, 57.

35. Binnig, G., Rohrer, H., Gerber, Ch. and Weibel, E., Phys. Rev. Lett. (1983) 50, 120.

36. Binnig, G., Rohrer, H., Salvan, F., Gewrber, Ch. and Baro, A., Surface Sci. (1985) 152, 17.

37. Becker, R. S., Golovchenko, J. and Swartzentraber, B.S., Phys. Rev. Lett. (1985) 54, 2678.

38. Moreland, J. and Hansma, P. K., Rev. Sci. Instr., (1985); to be published.

39. Drake, B., Sonnenfeld, R., Schneir, J. and Hansma, P. K., Rev. Sci. Instr. (1986) 57, 441.

40. Topsoe, H., Clausen, B. S., Candia, R., Wivel, C. and Morup, S., J. Catal. (1981) 433.

41. Topsoe, H. Clausen, B. S., Candia, R., Wivel, C. and Morup, S., Bull. Soc. Chim. Belg. (1981) 90, 1189.

42. Martensson, N. and Johansson, B., Solid State Comm. (1979) 32, 791.

MODIFICATIONS OF MOLECULAR SIZE AND STRUCTURE DURING THE HYDROLYTIC POLYCONDENSATION OF METAL ALKOXIDES

Bulent E. Yoldas

PPG Industries, Inc.
Glass Research and Development
P. O. Box 11472
Pittsburgh, Pennsylvania 15238

ABSTRACT

Inorganic materials, such as ceramics and glasses, are traditionally formed by thermal processes. Application of heat has been such an integral part of their formation that the scientific definition of the very materials often refers to this fact, e.g., product of fusion, etc. Recently, it has been demonstrated that inorganic networks of ceramics and glasses can also be formed by low temperature chemical polymerization. In the latter cases high temperature reactions which limit the glass formation can be avoided. More importantly, molecular-structural variations can be introduced into the polymeric structure, allowing property modifications without compositional alterations. These nonequilibrium conditions are found to be quite stable even in relatively high temperatures.

In this paper, hydrolytic polycondensation of metal alkoxides and alkoxysilanes is reviewed within the context of forming inorganic networks and modifying such networks. Since the paper is intended to be an overview, it contains extensive quotations from this author's previous works in the field.

INTRODUCTION

Use of metal-organic compounds as precursors to ceramic and glass formation is relatively recent, since synthesis of these materials dates only from the last century. The initial attraction of these materials as ceramic precursors was largely based on their high purity and reactivity (1), submicron particle size (2), and capability of being utilized to deposit thin films (3). With the exception of thin films deposited on substrates, the inorganic materials obtained previously from these compounds have been in a particulate form, requiring subsequent melting or sintering to attain coherency. Further advances in ceramic processing were made by forming monolithic glass and ceramic materials. Formation of a monolithic transparent alumina produced from a sol obtained from aluminum alkoxides was disclosed in the early 1970's (4), followed by several articles on the monolithic glass formation by chemical polymerization (5-6). These publications shifted the focus from the use of colloids and metal-organic compounds merely as raw materials to the formation of a monolithic glass and ceramic materials by chemical polymerization. Since then, the number of publications relating to the retainment of monolithic gel state (7-9), of the formation of films (10-13) and the fibers (14), as well as the general condensation and glass formation (15-21), have appeared.

One of the most recent developments in this field has been the modification of material properties by the introduction of molecular-structural variations (21-23).

HYDROLYTIC POLYCONDENSATION OF ALKOXIDES

Alkoxides of group III and group V metals, which constitute a large portion of glass and ceramic forming systems, are readily synthesized (24-26) and commercially available at a reasonable cost. These compounds do not occur naturally. Although alkoxy derivatives of silicon were investigated as early as 1846 (27), only in this century has the alkoxide chemistry become widely known. Metal alkoxides have the general formula $M(OR)_n$ where M is a metal having a valence m, and R is an alkyl group C_kH_{2k+1} (n, often used for valence, is used for another purpose in this paper). They can either be considered derivatives of alcohol, R(OH), or derivatives of hydroxides $M(OH)_n$. Due to the electronegativity of oxygen, the M-O-C bonds are often highly polarized. The nature of the M-O bond, i.e., covalency, varies not only with the electronegativity of the central metal, M, but also with the inductive effect of the alkyl group. Detailed chemistry of these compounds has been covered by D. C. Bradley and his coworkers in an excellent book (26). We will only deal with the reactions leading to the formation of inorganic polymers, i.e., oxide glasses and ceramics.

Metal alkoxides may be converted to oxides either by thermal degradation:

$$M(OR)_m \longrightarrow MO_{m/2} + m/2\ ROH + \text{olefins} \tag{1}$$

or by hydrolysis and thermal dehydration.

It is this latter process that presents interesting new capabilities in design and modifications of inorganic materials. In most chemistry books, hydrolysis of metal organic compounds is often represented by one of the following equations:

$$M(OR)_m + m/2 \; H_2O \longrightarrow MO_{m/2} + mR(OH) \tag{2}$$

$$M(OR)_m + mH_2O \longrightarrow M(OH)_m + mR(OH) \tag{3}$$

$$M(OR)_m + (m-x)H_2O \longrightarrow M(OH)_{m-x}(OR)_x + (m-x)R(OH) \tag{4}$$

All of these equations are incorrect, not because they are oversimplifications, but because they do not represent the chemical make-up of the end product which requires that the polymerization process must be taken into account. For example, the hydrolysis product of metal alkoxides can never be a pure oxide as equation (2) implies (because of the terminal bonds). The reactions (3) and (4) ignore the presence of bridging oxygens and imply that only monomers are formed.

Hydrolytic condensation reactions of metal alkoxides take place in several steps, often involving competing reactions between hydrolysis and polymerization. The process leading to the oxide network formation requires at least two steps:

I. Hydrolysis

$$(RO)_{m-1}M-OR + HOH \longrightarrow (RO)_{m-1}M-OH + R(OH) \tag{5}$$

What happens after this necessary initial reaction depends on the chemical encounters and nature of the hydrolysis system. Encounters of the partially hydrolyzed species with water molecules may lead to further hydrolysis, creating a higher degree of hydrolyzed monomeric species, i.e., $(RO)_{m-2}M(OH)_2$, $(RO)_{m-3}M(OH)_3 \ldots M(OH)_m$. It is more likely however, that as soon as the first partially hydrolyzed species are formed a second type of reaction will be initiated creating bridging oxygens, i.e., polymerization.

II. Polymerization

$$\equiv M-OH + RO-M\equiv \longrightarrow \; \equiv M-O-M\equiv + ROH, \text{ or} \tag{6}$$

$$\equiv M-OH + HO-M\equiv \longrightarrow \; \equiv M-O-M\equiv + H_2O \tag{7}$$

The above reactions form the inorganic skeleton of the condensates.

Progression of the polymerization requires the presence of hydroxyl bonds. These reactions arrange the various partially hydrolyzed species into macromolecules having an oxide backbone framed by hydroxyl and organic groups. The relative concentrations of "OH" and "OR" groups in the terminal bonds depend on the availability of water, aging time, nature of the

alkoxide, and host liquor, etc. The chemical composition of such a conden-
sate continuously changes and can only be represented by taking into account
its polymeric nature. For example, the hydrolytic polycondensation product
of an alkoxysilane can only be represented as:

$$Si_nO_{2n-(\frac{x+y}{2})}(OH)_x(OR)_y \tag{8}$$

where "n" is the number of silicon atoms polymerized in the polymer molecule
and "x" and "y" are the number of terminal groups, "OH" and "OR". The
oxide content of the polymer becomes a function of size and morphology. In
silica and titania systems for example, condensates having oxide contents
varying from less than 70% by weight to over 90% can be formed (Figure 1).

RELATIONSHIP BETWEEN OXIDE CONTENT AND POLYMER SIZE AND MORPHOLOGY

Once the hydrolytic polycondensation is performed, the resultant
materials can be converted to oxide systems by pyrolytic degradation at
around 300-500°C. The relationship between the oxide content and the
polymer size and morphology has been addressed in two previous papers
(21,28). Here it is sufficient to show how the chemical make-up of a
spherically expanding silanol polymer varies during the hydrolytic
polycondensation. In such a polymer unit, the volume can be written in
terms of total silicon atoms, n.

$$V = \frac{4}{3}\pi r^3 = n \qquad or \qquad r = (3n/4\pi)^{1/3} \tag{9}$$

Those silicon atoms located on the surface will contain an OH bond whose
number will be found by expressing surface area in terms of n from equation
(9)

$$\text{number of OH} = 4\pi r^2 = 4\pi(3n/r\pi)^{2/3} = (36\pi n^2)^{1/3} \tag{10}$$

Since the number of bridging oxygens is twice the number of silicon
atoms minus one-half of the number of OH bonds, the number of bridging
oxygens is:

$$BO = 2n - \frac{(36\pi n^2)^{1/3}}{2} \tag{11}$$

The chemical make-up of the spherical polymer will be

$$Si_nO_{2n-(9/2\pi n^2)^{1/3}}(OH)_{(36\pi n^2)^{1/3}} \tag{12}$$

The oxide content of this polymer will be:

$$\text{Oxide content} = \frac{60}{60+9(\frac{36\pi}{n})^{1/3}} \qquad (Q_\infty \rightarrow 100\%) \tag{13}$$

It is apparent that if the nature of the terminal bonds is known, or
when certain assumptions can be made, the molecular size can be calculated
directly from the oxide content of these polymers.

Figure 2 shows, in the silanol system, how the chemical composition and oxide content would vary for linear and planar polymerization as a function of molecular size, n. The rate of change of the oxide content for each polymer as a function of size is significantly different for different polymerizations. To achieve 99% of the maximum oxide content requires

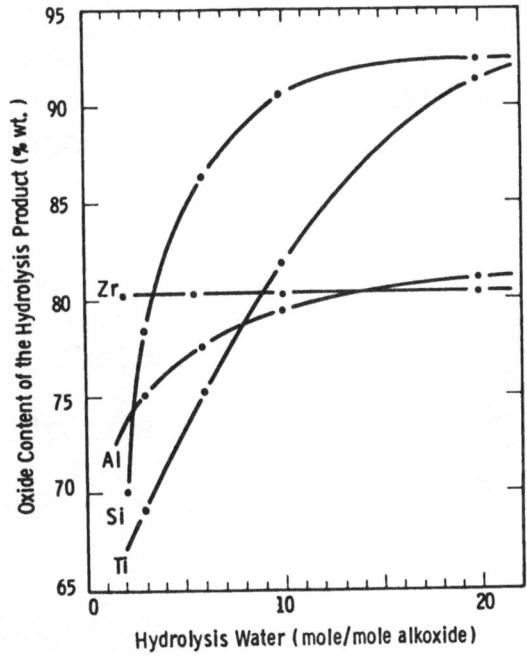

Fig. 1. Oxide content of polycondensates varies as a function of hydrolysis water $Al(OC_4H_9)_3$, $Zr(OC_4H_9)$, $Ti(OC_2H_5)_4$ and $Si(OC_2H_5)_4$ were hydrolytically polycondensed in their alcohols at 5% eq. oxide condensation.

polymerization of 18 silicons in the case of linear polymerization, $2x10^3$ silicons in the case of planar and $4x10^5$ in the case of spherical polymerization (referred to as critical size n_c). Not only the oxide content, but more significantly the overall network connectivity (average number of

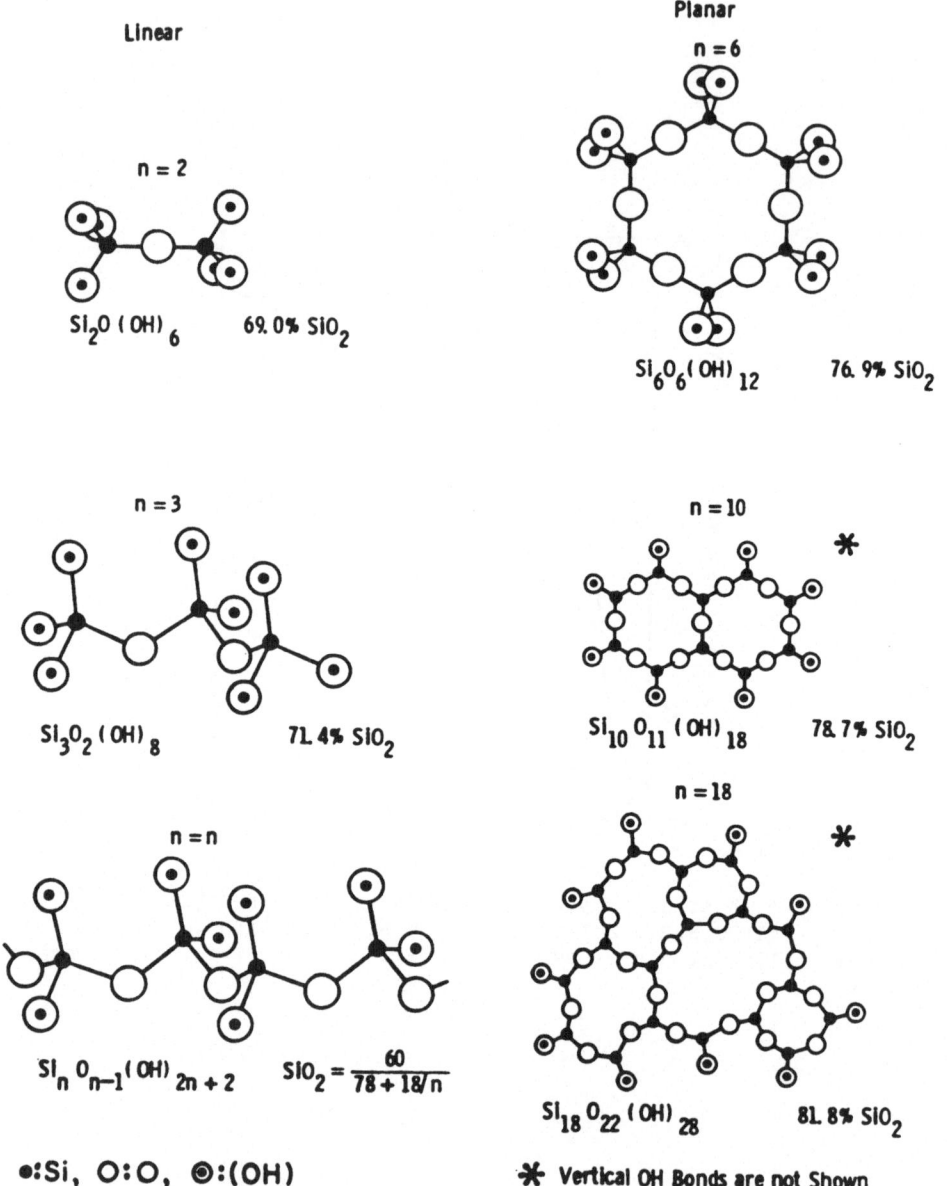

Linear

n = 2

$Si_2O(OH)_6$ 69.0% SiO_2

n = 3

$Si_3O_2(OH)_8$ 71.4% SiO_2

n = n

$Si_nO_{n-1}(OH)_{2n+2}$ $SiO_2 = \dfrac{60}{78 + 18/n}$

Planar

n = 6

$Si_6O_6(OH)_{12}$ 76.9% SiO_2

n = 10   ✱

$Si_{10}O_{11}(OH)_{18}$ 78.7% SiO_2

n = 18   ✱

$Si_{18}O_{22}(OH)_{28}$ 81.8% SiO_2

●:Si, O:O, ◉:(OH)   ✱ Vertical OH Bonds are not Shown

Fig. 2.   Schematic representation of how molecular size and polymerization type affect chemical make-up and oxide content of silanol polymers.

bridging oxygens per silicon) is also determined by the molecular size, n. These parameters are summarized in Table 1. This last parameter not only affects the behavior of the gels but also affects the properties of glasses produced from these polymers (21). The properties such as melting point, viscosity and strength are closely related to network connectivity.

MODIFICATION OF THE CONDENSATION PROCESS

A most significant development in this field has been the ability to affect the substructure of ceramic materials by controlling the kinetics of the chemical reaction that form their network structure. These structural variations significantly alter the behavior of these materials and modify their basic properties. These modifications include even such high temperature behaviors of ceramic materials as sintering, crystallization, structural transformations, and properties such as melting point and viscosity (21,28).

The principal parameters that introduce structural variations in polymer-gel networks are listed as follows (21):

1. Kinetics of hydrolysis and polymerization reactions, which are affected by: Selection of starting compounds and host medium; Water/alkoxide ratio; Molecular separation by dilution; Catalytic effects; and Reaction temperatures.

2. Order of reactions in multicomponent systems.

3. Post-mixing effects: Aging; Heat treatment (time, temperature, atmosphere).

Under any conditions, of course, the hydrolysis product will contain mixtures of various types of polymeric structures whose distribution will be determined by statistical interactions. Nevertheless, as a main reaction product, a particular type of polymer may be obtained by the manipulation of the above parameters. These parameters have been discussed in Reference 21 and in the following sections as they apply to the silica system.

The starting compound and host medium have an important influence on the hydrolysis product. The polymerization reaction represented by equations (6) and (7) requires diffusion of the partially hydrolyzed species, which is in turn affected by the size of the alkyl groups attached to these species. For a given metal alkoxide, $M(OR)_m$, the rate of hydrolysis is affected by the steric effect of the alkyl groups, $C_k H_{kx+1}$, as well. For example, the hydrolysis rates of $Al(OC_4H_9)_3^s$ and $Al(OC_3H_7)_3^i$ are different; consequently, their condensates have a different constitution (29). The alkoxides with lower alkyl groups generally produce larger polymer units; consequently, the oxide components of their gels are higher.

Table 1. Chemical Makeup, Oxide Content and Connectivity of Silica Polymers as a Function of Molecular Size (n).

Polymer Type	Polymer Formula	Connectivity (C_n)	Polymer Oxide Content	Maximum Oxide Content (%) $(n \to \infty)$	Critical Size*
Linear	$Si_n O_{n-1}(OH)_{2n+2}$	$2 - (\frac{2}{n})$ $(C_\infty \to 2)$	$\dfrac{60}{78 + \dfrac{18}{n}}$	76.92	18
Planar**	$Si_n O_{1.5n-(\pi n)^{1/2}}(OH)_{n+2(\pi n)^{1/2}}$	$3 - (\frac{4\pi}{n})^{1/2}$ $(C_\infty \to 3)$	$\dfrac{60}{69 + 18(\frac{\pi}{n})^{1/2}}$	86.96	$\approx 2 \times 10^3$
3 Dimen.†	$Si_n O_{2n-(9/2\pi n^2)^{1/3}}(OH)_{(36\pi n^2)^{1/3}}$	$4 - (\frac{36\pi}{n})^{1/3}$ $(C_\infty \to 4)$	$\dfrac{60}{60 + 9(\frac{36\pi}{n})^{1/3}}$	100	4×10^5

* Size, above which oxide content is stable within 1%

** Circular Expansion

† Spherical Expansion

The host medium also affects the diffusion rates of the reacting species, that is, H_2O and the partially hydrolyzed species. For this reason hydrolysis products produced in lower alcohols have a higher equivalent oxide content, indicating a greater extent of polymerization. For example, when $Ti(OC_2H_5)_4$ is hydrolyzed in ethanol (C_2H_5OH) and dried at 100°C, the oxide content of the gel by weight is ≈83-84%. This figure drops to about 73% when the hydrolysis is performed in butanol, $C_4H_9(OH)$[S].

The polymer structure also depends on the functionality of the building blocks. For example monofunctional and difunctional units, i.e., R_3SiOR and $R_2Si(OR)_2$, form only linear polymers. Whereas the use of compounds higher than difunctional, e.g., $RSi(OR)_3$, leads to branching in the polymer.

The fact that one starts from an organic compound of a metal and ends up with an oxide structure of that metal, without going through high temperatures, is often sufficient to bring about rather unusual structural conditions. For example, aluminum ions are six-coordinated in its oxide and four-coordinated in its alkoxides. When Al_2O_3 is formed from alkoxides, it is forced to remain in four-coordination. This is indicated by IR spectroscopy and electron spectroscopy (23). The Al remains in its coordination until heated above 1200°C, at which temperature it has enough activation energy to convert to its stable six-coordinated state.

Similarly when ZrO_2 is produced from metal alkoxides, the initial crystalline phase formed is the cubic phase which is the high temperature phase of ZrO_2 under normal circumstances.

FORMATION OF CLEAR POLYMER SOLUTIONS AND GELLING

The hydrolytic polycondensation of metal alkoxides under normal conditions leads to the formation and precipitation of particulate materials except for silicon and a few other alkoxides. This is perhaps one of the reasons why the hydrolysis of silicon alkoxides has been investigated more than any other alkoxide. In that system, formation of clear solutions naturally occurs. Such precipitate formation and self-condensation can be prevented by careful control of molecular interactions during the hydrolysis, such that certain amounts of (OR) groups are left in the molecular structure (6). This is done first by controlling the amount of water and dilution of the system, and second, by the presence of a critical amount of certain acids. These solutions are useful in depositing optical oxide coatings of precise thickness, as well as forming monolithic glass and ceramic materials (10,11,30).

In a titania system, condensation of particulates is avoided by introducing at least 0.014 mol of HNO_3 or HCl acid. If the acid is introduced
into the water-alcohol solution before mixing with the alkoxide, no cloudiness ever occurs; therefore, this is the preferred method of introducing the
acid. Acid can be added any time after the mixing occurs and will cause the
cloudy slurry to turn into a clear solution. However, if more than 0.3
moles of acid are used per mole of alkoxide, the stability of the solution
is reduced and it will turn cloudy after several days to several months,
depending on the acid concentration (31).

One must also consider the concentration and water/alkoxide ratio
criteria in these clear solutions. If the water/alkoxide ratio is less than
1.7, the solution does not deposit clear continuous films. If the water/-
alkoxide ratio is much above 2, the solution tends to gel to a clear single
phase within hours or days, depending on the concentration. Usually 3%
equivalent TiO_2 by weight, produced by 2 mole water hydrolysis with about
0.15 mole of HNO_3 per mole of alkoxide produces clear stable solutions in
this system (31). The conditions that produce clear zirconia solutions are
given in Reference 32. In these systems, gelling occurs both by
electrolytic interaction and by formation of polymeric networks through
chemical reactions. These are the fundamental differences between the
occurrence of gelling by electrolytic interaction of colloidal particles and
the occurrence of gelling by chemical polymerization (33).

POLYCONDENSATION OF ALKOXYSILANES, MODIFICATIONS OF MOLECULAR SIZE AND SIZE
DISTRIBUTION

General hydrolytic polycondensation of alkoxysilanes has been investigated by others (34-40). We recently investigated the effects of specific
parameters on the molecular size, and size distribution, by using nuclear
magnetic resonance (NMR) and size-exclusion liquid chromatography (SEC)
(41).

Hydrolysis of silicon alkoxides is generally carried out in an alcohol,
which is a solvent for the mutually immiscible liquids $Si(OR)_4$ and H_2O. In
the following experiments, silicon tetraethoxide $Si(OC_2H_5)_4$ was hydrolyzed

in ethanol. Ethanol was used as the host liquor to eliminate the additional complexities of ester exchange. Figure 3 shows the miscibility range of $Si(OC_2H_5)_4$ and water as a function of water/alkoxide ratio in ethanol.

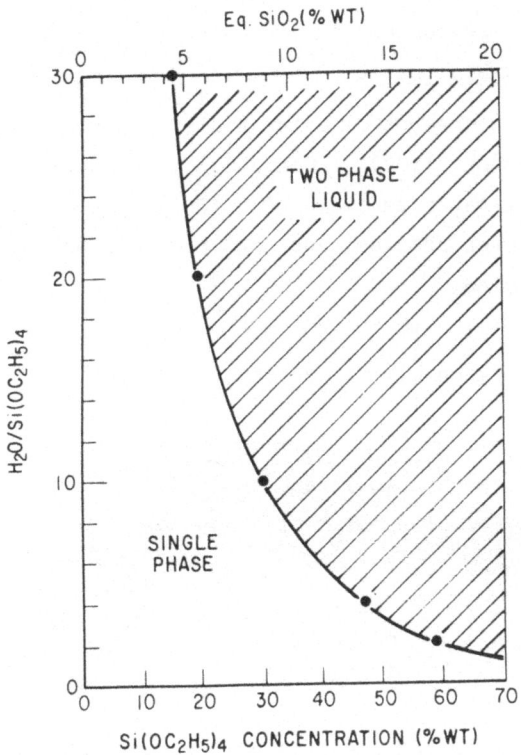

Fig. 3. Miscibility as a function of $H_2O/Si(OC_2H_5)_4$ ratio and $Si(OC_2H_5)_4$ concentration in the $Si(OC_2H_5)_4$-H_2O-C_2H_5OH ternary.

100 mg HNO_3 per mole of $Si(OC_2H_5)_4$ was used as a catalyst, except when the effect of HNO_3 acid was investigated specifically. The solutions were mixed at room temperature and immediately heated to 60°C; the solutions were held in closed containers at that temperature for 24 hours prior to the molecular size distribution and morphology related tests. The molecular size distribution was studied by size-exclusion liquid chromatography.

SEC measurements were carried out on a DuPont Model 850 liquid chromatograph at room temperature. Solutions were diluted to 1% in tetrahydrofuran inhibited by 0.025% butylated hydroxytoluene as the mobile phase and tested at flow rates of 1.5 ml per minute through a 10 and 100 nm styrene divinylbenzene copolymer packed column under 56 bars pressure. The deviations of the refractive index from the baseline of 1.4065" were detected by a Waters Associate Model 401 refractometer and computer integrated.

Effect of $H_2O/Si(OC_2H_5)_4$ Ratio

The chemical requirement of water falls between 2 and 4 moles for a complete hydrolysis of $Si(OR)_4$. Formation of a monomer $Si(OH)_4$ species requires 4 moles of water per mole of $Si(OR)_4$. However, this precludes any polymerization between species, and thus requires an infinite separation of the hydrolyzed species:

$$Si(OR)_4 + 4\ H_2O \rightarrow OH-\overset{\displaystyle OH}{\underset{\displaystyle OH}{\overset{|}{\underset{|}{Si}}}}-OH + 4\ R(OH) \qquad (14)$$

On the other hand, the chemical requirement of water would be 2 moles per mole $Si(OR)_4$ if an infinite polymerization takes place:

$$n\ Si(OR)_4 + 2n\ H_2O \rightarrow \left[-O-\overset{\displaystyle -O-}{\underset{\displaystyle -O-}{\overset{|}{\underset{|}{Si}}}}-O-\right]_n + 4n\ ROH \qquad (15)$$

This reaction precludes the presence of any surface ($n\rightarrow\infty$) and the presence of hydroxyl groups. Reality falls between these two extreme cases, but is much closer to the latter, since the molecular weights of these polycondensed materials are often in the thousands. The water requirement, from a purely chemical point of view, becomes slightly in excess of 2 moles, but its actual effect goes much beyond 4 moles of water as apparent from Figure 1, for example.

Figure 4 shows NMR results for the chemical shift anisotropy of ^{29}Si in two solutions where $Si(OC_2H_5)_4$ is hydrolyzed with 2 and 4 moles of water in ethanol under 1% equivalent SiO_2 concentration. In the top, (A), the water concentration is 2 moles and in the bottom, (B), 4 moles. When the hydrolysis is performed with 2 moles of H_2O the highest NMR peak occurs at -89.6 ppm; when the hydrolysis is performed with 4 moles of H_2O, there are at least three strong NMR peaks higher than this value, at -93.6, -95.8 and -102.3 ppm. The broad peak occurring at around -110 ppm in these figures is due to the fused silica sample holder (42,43). As the average number of

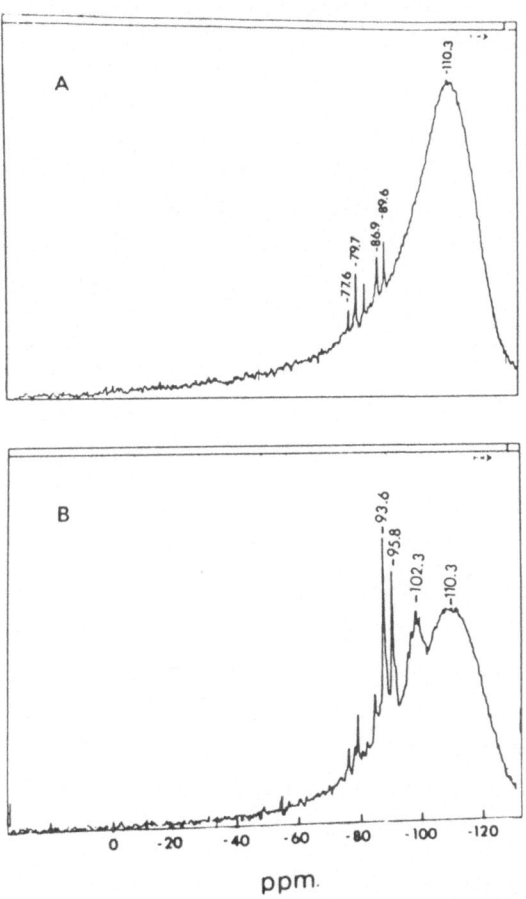

Fig. 4. A higher degree oxide network formation is indicated by silicon[29] NMR spectra when $Si(OC_2H_5)_4$ is hydrolyzed with 4 moles of H_2O (B), than when it is hydrolyzed with 2 moles H_2O (A). Hydrolysis is performed in ethanol at 1% eq. SiO_2 concentration.

bridging oxygens increases, the NMR peaks shift toward a higher field. Table 2 gives the relationship between the ^{29}Si NMR peaks and the number of bridging oxygens in hydroxysiloxanes. This increasing oxide network formation with an increasing amount of hydrolysis by water is also confirmed by the higher oxide content of the polycondensates which are formed under high water conditions, as shown in Table 3. Table 3 also shows the effect of the hydrolysis medium on the hydrolytic condensation of $Si(OC_2H_5)_4$ in various alcohols within 24 hours.

Effect of Molecular Separation

Since hydrolytic polycondensation reactions involve chemical encounter rates and the diffusion process, the molecular separation of the interacting species and their mobilities in the host liquor play a fundamental role in shaping the molecular structure and molecular size distribution.

In a compound, the average intermolecular distance ℓ_o, can be calculated from the molecular volume, V_m, or from the molecular weight, W_m:

$$\ell_o = (\frac{V_m}{N_o})^{1/3} = (\frac{W_m}{d \cdot N_o})^{1/3} \tag{16}$$

where N_o is Avogadro's Number and d is the density. For example, the average intermolecular distance calculated from equation (16) is 7.2Å for $Si(OC_2H_5)_4$ and 3.2Å for H_2O at room temperature.

This intermolecular distance can be increased by diluting the compound in an inert solvent. In the diluted state the intermolecular distance, ℓ_d, is given by:

$$\ell_d = (\frac{V_d}{V_o})^{1/3} \cdot \ell_o \tag{17}$$

where V_d is the diluted volume and V_o is the undiluted volume. We define the relative molecular separation of species, MS, as the ratio of molecular spacings in diluted and undiluted states:

$$MS = \frac{\ell_d}{\ell_o} = (\frac{V_d}{V_o})^{1/3} = (\frac{V_o + V_s}{V_o})^{1/3} \tag{18}$$

where V_s is the volume of the solvent used to dilute V_o to a volume V_d. Since the term inside the parentheses in equation (18) is the inverse of the

Table 2

EFFECT OF HYDROXYL BONDS
ON THE ^{29}Si NMR

BRIDGING OXYGEN	STRUCTURE	^{29}Si NMR

I. B.O. = 2
$(\equiv Si-O)_2 Si(OH)_2$

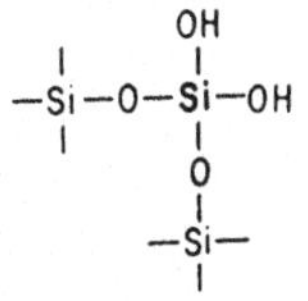

$$-Si-O-\overset{\overset{\displaystyle OH}{|}}{\underset{\underset{\displaystyle -Si-}{\overset{|}{O}}}{Si}}-OH$$

- 90.6 ppm

II. B.O. = 3
$(\equiv Si-O)_3 Si(OH)$

$$-Si-O-\overset{\overset{\displaystyle OH}{|}}{\underset{\underset{\displaystyle -Si-}{\overset{|}{O}}}{Si}}-O-Si-$$

- 99.8 ppm

III. B.O. = 4
$(\equiv Si-O)_4 Si$

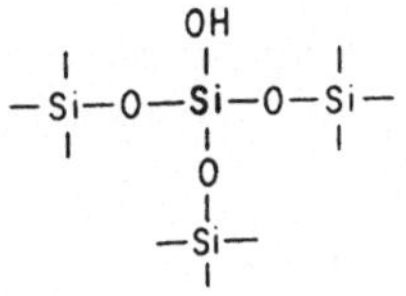

$$-Si-O-\overset{\overset{\displaystyle -Si-}{\overset{|}{O}}}{\underset{\underset{\displaystyle -Si-}{\overset{|}{O}}}{Si}}-O-Si-$$

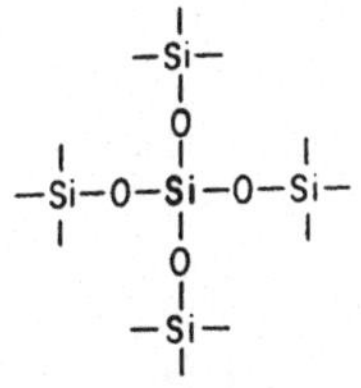

- 109.3 ppm

Table 3. Effect of Reaction Medium on the Oxide Content of Polymers
Obtained by 2 and 16 Moles H_2O Hydrolysis of $Si(OC_2H_5)_4$.

Water per $SiO(CH_2H_5)_4$	Oxide Content* of Gels (% Wt.) Formed in		
	Methanol	Ethanol	n-Propanol
2 moles	≈81	≈66	≈46
16 moles	≈93	≈89	≈83

*After 24 hours at 60°C.

volume fraction of the substance in the total volume, the molecular spacing
becomes:

$$MS = (\frac{1}{f_{V_o}})^{1/3} \tag{19}$$

where f_{V_o} is the volume fraction.

If the total volume is made up of more than one substance, then the
molecular spacing of each substance i, MS, can be expressed as:

$$MS_i = (\frac{V_t}{V_i})^{1/3} = (\frac{V_1 + V_2 + \ldots V_n}{V_i})^{1/3} = (\frac{1}{f_{V_i}})^{1/3} \tag{20}$$

where V_t is the total volume of the mixture and V_i is the volume of the
species i under consideration.

Figure 5 shows the effect of concentration on the molecular size distribution as indicated by SEC retention time when $Si(OC_2H_5)_4$ is hydrolyzed
with 4 moles H_2O in ethanol. In the top curve (A), the concentration is 1%
weight equivalent SiO_2, and in the bottom one, (B), the concentration is 10%
equivalent SiO_2. It is clear that higher molecular sizes are obtained at
higher concentrations. This is consistent with the increased molecular encounter rates at higher concentrations which build the molecular network.

The liquid in which the hydrolytic polycondensation reactions are
carried out constitutes a mixture of $Si(OC_2H_5)_4$, H_2O and C_2H_5OH. Obviously,
the same molecular spacing of $Si(OC_2H_5)_4$ can be attained at different

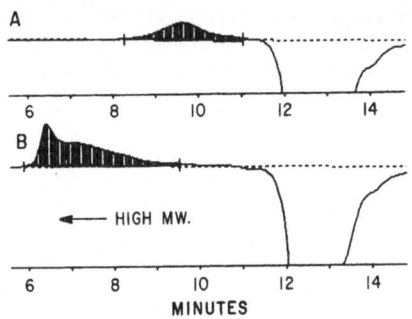

Fig. 5. Effect of the concentration on the molecular size distribution, as
indicated by SEC. $Si(OC_2H_5)_4$ is hydrolyzed with 4 moles of water
in ethanol.
A) The hydrolysis is performed at 1% eq. SiO_2 concentration.
B) Hydrolysis is performed at 10% eq. SiO_2 concentration.

proportional mixings of the remaining two components. Since water is one of
the chemically reacting species, it makes a great deal of difference how the
molecular separation of $Si(OC_2H_5)_4$ is attained. To make things simple, one
can fix the $H_2O/Si(OR)_4$ ratio constant and change the molecular separation
by various amounts of ethanol additions. Table 4 gives the molecular
spacing of $Si(OC_2H_5)_4$ when hydrolyzed with two moles of water in ethanol at
various concentrations. We will consider the effect of $Si(CO_2H_5)_4$ molecular
spacing in a limited water environment and in a water-rich environment where
the division between these two states is necessarily arbitrary.

Effect of the Molecular Spacing of $Si(OC_2H_5)_4$ in a Limited Water
Environment. In limited water conditions, the molecular growth terminates
when the condensation reactions substantially exhaust the hydroxyl groups.
When the water/alkoxide ratio is around 2, essentially all the terminal
bonds should eventually be alkyl groups. In these cases, increased
encounter rates cannot lead to network formation, and the variations in the
oxide contents of these polymers can reasonably be assumed to be due to the
relative variation in their molecular size and morphology. The oxide
content of the organosiloxane polymers formed by 2 moles water hydrolysis of
$Si(OC_2H_5)_4$ in ethanol at various molecular spacings clearly indicates that a

Table 4. Relationship Between Concentration (Expressed as Equivalent Weight Percent SiO_2 in Solution) and Molecular Separation of $Si(OC_2H_5)_4$ when Mixed with 2 Moles Water in Ethanol at Room Temperature.

Eq. SiO_2 (%wt.)	Molecular Spacing	Make-Up of Solution			
		$Si(OC_2H_5)_4$	H_2O	C_2H_5OH	Total Vol. (cc)
28.85	1.00*	208 g	–	–	≈ 223
24.6	1.05	208 g	36 g	–	≈ 259
16.0	1.24	208 g	36 g	131 g	≈ 424
8.0	1.59	208 g	36 g	506 g	≈ 899
4.0	2.02	208 g	36 g	1256 g	≈1849
2.0	2.56	208 g	36 g	2756 g	≈3747
1.0	3.23	208 g	36 g	5756 g	≈7545

*Molecular separation of $Si(CO_2H_5)_4$ in pure $Si(OC_2H_5)_4$ taken as unity (corresponds to slightly over 7Å). All other molecular separation is expressed in terms of this unit.

higher oxide content results in lower molecular spacings, i.e., concentrations, (Figure 3 in Reference 1), consistent with SEC and NMR results.

Effect of the Molecular Separation of $Si(OC_2H_5)_4$ in a Water-Rich Environment. Hydrolytic polycondensation of $Si(OC_2H_5)_4$ can only be done at low concentrations in a water-rich liquor, due to immiscibility of $Si(OC_2H_5)_4$ and H_2O. The maximum $H_2O/Si(OC_2H_5)_4$ ratios achievable as a function of $Si(OC_2H_5)_4$ concentration in ethanol in the miscible state is shown in Figure 3. The figure shows that if the concentration of $Si(OC_2H_5)_4$ in the solution is approximately 60% by weight, the maximum amount of water that can be incorporated is approximately 2 moles, per mole of $Si(OC_2H_5)_4$. This amount of water can be increased to 8 moles at 35% $Si(OC_2H_5)_4$ concentration, 20 moles at 20% $Si(OC_2H_5)_4$ concentration, and over 200 moles at about 3.5% $Si(OC_2H_5)_4$ concentration. Thus, at very low concentrations, hydrolysis of $Si(OC_2H_5)_4$ can be performed in a solution that is overwhelmingly water. For example, at 1% equivalent SiO_2 concentrations, one can hydrolyze 1 mole of $Si(OC_2H_5)_4$ with 200 moles of water, requiring only approximately 20 moles of ethanol for miscibility. After the reactions, the ethanol portion can be evaporated, leaving a liquid which is essentially water.

Figure 6 shows the effect of $Si(OC_2H_5)_4$ molecular separation on the molecular size distribution when $Si(OC_2H_5)_4$ is hydrolyzed at 1% equivalent SiO_2 (A) and at 5% equivalent SiO_2 concentrations in water-rich liquor (200 and 16 moles H_2O per mole of $Si(OC_2H_5)_4$, respectively). In a water-rich environment, the molecular weight is higher and the molecular size distribution is much wider than the case under limited water hydrolysis. In the water-rich environment the concentration generally also makes a bigger difference than the case under limited water conditions. Table 5 summarizes consequences of these condensation conditions.

Effect of Concentrating Solutions

The effect of concentrating the already formed organosiloxane solutions also leads to an increased chemical encounter rate between the existing polymeric molecules. The increased encounter rates, however, cannot lead to chemical reactions when the terminal bonds are alkyl groups and the liquor is devoid of water. This is essentially the case in the limited water hydrolysis. Under these conditions, concentrating a solution makes little difference in the molecular size distribution (see Figure 7). On the other hand, when the terminal bonds contain a substantial number of hydroxyl groups, further network formation becomes possible upon concentration (see Figure 8) by reactions:

$$P_1-OH + HO-P_2 \longrightarrow P_3 + H_2O \tag{21}$$
$$P_1-OH + RO-P_2 \longrightarrow P_3 + R(OH) \tag{22}$$

where P_1, P_2 and P_3 are large molecules. This often causes a bimodal

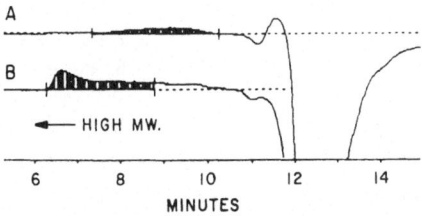

Fig. 6. Effect of concentration on the molecular weight distribution when $Si(OC_2H_5)_4$ is hydrolyzed in water-rich ethanol liquor.
A) 1% eq. SiO_2 concentration: 200 moles/mole water.
B) 5% eq. SiO_2 concentration: 16 moles/mole water.

Table 5. Effect of Molecular Spacing of $Si(OC_2H_5)_4$ on the Hydrolytic Polycondensation in Ethanol.

Molecular Spacing of $Si(OC_2H_5)_4$	Hydrolysis Water per $Si(OC_2H_5)_4$	Consequences
	A. Limited Water (2-4 Moles)	Molecular size limited. Relatively inactive polymers. Not subject to significant change during aging and concentration. Carbon deposition during the pyrolysis. Low oxide content. Very poor tendency to remain monolithic.
I. High MS. (>2.0) (Conc. <4% eq. SiO_2)	B. High Water (20-200 Moles)	Active polymers. Molecular size changes significantly during concentration and aging. Very little, if any, tendency to form carbon. Very high oxide content. Good tendency to remain monolithic during drying and pyrolysis.
	A. Limited Water (2-4 moles)	Molecular size is higher and distribution is wider than the case I.A. Tendency of carbon formation is less and oxide content is higher. Molecules are rather inactive especially at the lower end of water.
II. Low MS. ($\approx$1.2-1.6) (Conc. 8-17% eq. SiO_2)	B. High Water	$Si(OC_2H_5)_4$ cannot be hydrolyzed with very high water at high concentrations in ethanol. (See Figure 4.) In between cases, polymer characteristics fall between IB and IIA.

molecular weight distribution since these "recombination" reactions multiply the molecular weight, whereas the "growth" reactions consist of smooth continuous addition to the molecular weight.

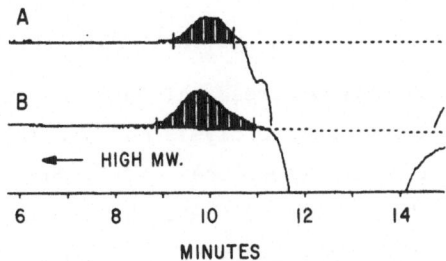

Fig. 7. No significant molecular size change occurs during the
concentration of a siloxane solution formed under limited water
conditions.
A) $Si(OC_2H_5)_4$ is hydrolyzed with 4 moles of water in ethanol at 1%
eq. SiO_2 concentration.
B) Above solution is concentrated to 5% eq. SiO_2 by evaporation.

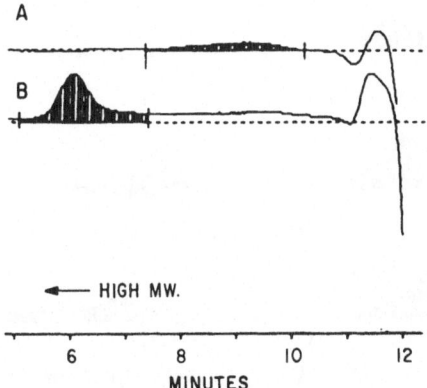

Fig. 8. Significant molecular size expansion occurs during the
concentration of water-rich siloxane solutions. The change in the
molecular size distribution occurring during the concentration of
1% eq. SiO_2 solution to 5% eq. SiO_2 concentration level by
evaporation is shown above.

The effect of the concentration of species during the hydrolytic polycondensation and the effect of concentrating the already formed organosiloxane solutions on the molecular make-up for the limited water and water-
rich conditions is summarized in Figure 9.

MATERIAL CONSEQUENCES OF MOLECULAR-STRUCTURAL MODIFICATIONS

In addition to the parameters already discussed, the condensation temperature, the catalyst, aging and hydrolysis medium, all affect the hydrolysis polymerization. Kinetics, therefore, modify chemical composition and
structure of the condensates. Some of these structural and chemical variations show their effect during the pyrolysis. For example, Figure 10 shows
the carbon formation tendencies of polyorganosilanols produced by the hydrolytic polycondensation of $Si(OC_2H_5)_4$ with 2.2 moles of water in ethanol at

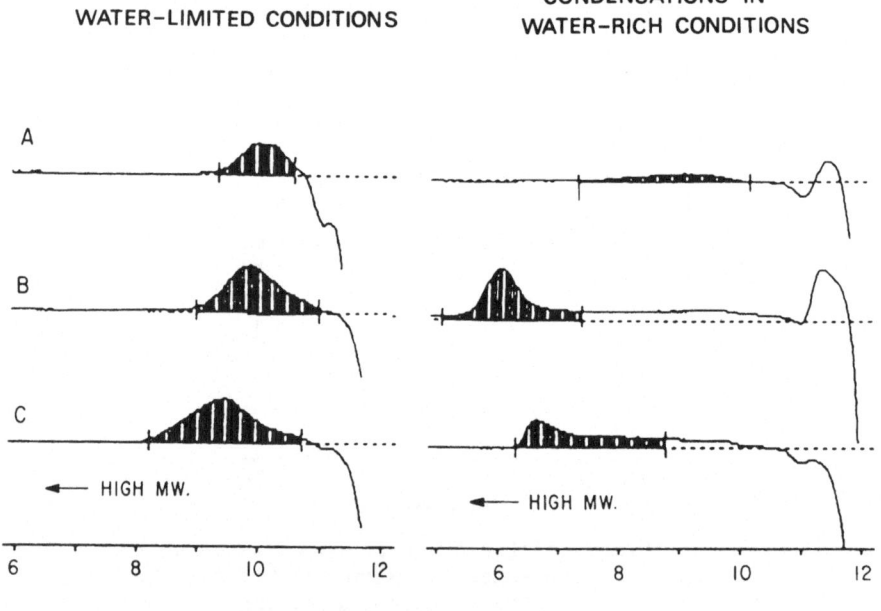

Fig. 9. Modification of molecular size distribution by concentration under
limited water and water-rich conditions.
A) $Si(OC_2H_5)_4$ is hydrolyzed with 4 moles of water in ethanol (on
the left) and 200 moles of water (on the right) at 1% eq. SiO_2
concentration.
B) The solutions are concentrated to 5% eq. SiO_2 by evaporation.
C) When the solutions are prepared at 5% eq. SiO_2 initially.

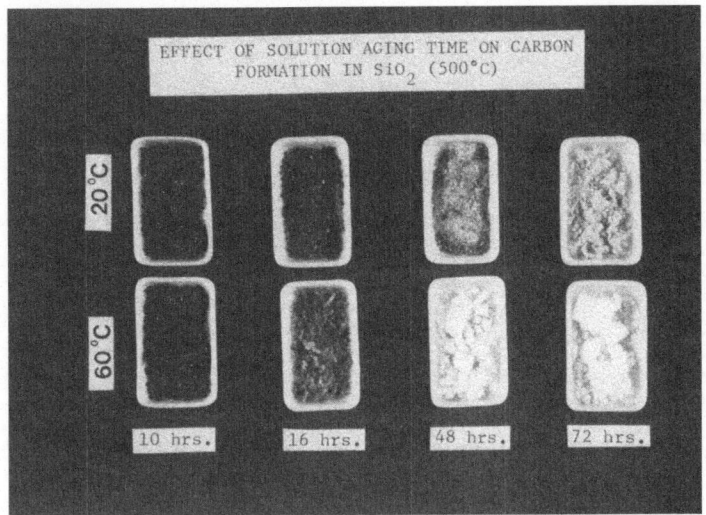

Fig. 10. Carbon formation tendencies of polyorganosilanol condensates during the conversion to oxide decreases with increasing hydrolysis temperature and time. (Other parameters also play a role in this process.)

$\approx 15\%$ equivalent SiO_2 concentration as a function of condensation temperature and aging of the liquor.

From these polymer solutions clear, monolithic oxide glasses can be produced (5,6). Figure 11 shows a 4 cm diameter SiO_2 sample produced by the

Fig. 11. A 4 cm diameter SiO_2 sample is obtained by hydrolysis and polymerization of $Si(OC_2H_5)_4$. Heat treated to 500°C.

gelling of such a solution followed by heat treatment to 500°C. What is extremely interesting is that the effect of these molecular-structural variations induced into the initial gel networks is maintained in the glass and ceramic materials derived from gels, allowing property modifications of these materials without compositional variations (21,28). The nature and extent of modifications attainable in these systems have great scientific and technological significance in the design of high technology materials.

ACKNOWLEDGEMENT

The author wishes to thank Kathleen A. Sobotka for her valuable assistance in the assembly and preparation of this paper.

REFERENCES

1. Roy, R. J. Amer. Ceram. Soc., (1959) 39 (4), 145.
2. Mazdiyasni, K. S.; Linch, C. T.; Smith, J. S. J. Amer. Ceram. Soc., (1965) 48 (7), 372.
3. Schroder, H. Optica Acta, (1962) 9 (3), 249.
4. Yoldas, B. E. J. Amer. Ceram. Soc., Amer. Ceram. Soc. Bull., (1975) 54 (3), 286.
5. Idem, J. Mater. Sci., (1977) 12, 1203.
6. Idem, ibid, (1979) 14, 1843.
7. Zarzycki, J.; Prassas, M.; Phalippou, J. J. Mat. Sci., (1982) 17, 3371.
8. Tewari, P. H.; Hunt, A. J.; Lofftus, K. D. Materials Letters, (1985) 3 (9), 363.
9. Hench, L. L.; Wang, S. H.; Park, S. C. SPIE Advanced Optical Materials, (1984) 505, 90.
10. Yoldas, B. E.; O'Keeffe, T. W. Applied Optics, (1971) 18 (18), 3133.
11. Yoldas, B. E. ibid, (1982) 21 (16), 2960.
12. Brinker, C. J.; Mukherjee, S. P. Thin Solid Films, (1981) 77, 141.
13. Yoldas, B. E.; Partlow, D. P. Thin Solid Films, (1985) 129, 1.
14. Sakka, S; Kamiya, K. J. Non-Cryst. Solids, (1980) 42, 403.
15. Artaki, I.; Sinka, S.; Irwin, A. D.; Jonas, J. J. Non-Cryst. Solids, (1985) 72, 391.
16. Brinker, C. J.; Scherer, G. W.; Roth, E. P. J. Non-Cryst. Solids, (1985) 72, 345.
17. Yamani, M.; Aso, S.; Okano, S.; Sakaino, T. J. Mat. Sci., (1979) 14, 607.
18. Klein, L. C.; Garvey, G. J. J. Non-Cryst. Solids, (1982) 48, 97.
19. Mukherjee, S. P. J. Non-Cryst. Solids, (1980) 41, 477.
20. Mackenzie, J. D. J. Non-Cryst. Solids, (1982) 48, 1.
21. Yoldas, B. E. J. Non-Cryst. Solids, (1982) 51, 105.
22. Idem, ibid, (1984) 63, 145.
23. Idem, MRS Symp. Proc., (1984) Vol. 24, 291 .
24. Speer, R. J. J. Organic Chem., (1949) 14, 655.
25. Adkins, H.; Cox, J. J. Am. Chem. Soc., (1938) 60, 1151.
26. "Metal Alkoxides," Bradley, D. C.; Mehrotra, R. C.; Gaur, D. P.; Eds.; Academic Press: New York, NY, 1978.
27. Ebelman, J. J.; Bouquet, M. Ann. Chim. Phys., (1946) 17, 54.
28. Yoldas, B. E. J. Amer. Ceram. Soc., (1982) 65 (8), 387.
29. Idem, J. Appl. Chem. Biotech., (1973) 23, 803.
30. Idem, U.S. Patent 4,346,131 (1982).
31. Idem, J. Mater. Sci., (1986) 21, 1087.
32. Idem, ibid, (1986) 21, 1080.
33. Partlow, D. P.; Yoldas, B. E. J. Non-Cryst. Solids, (1981) 46, 153.

34. Iler, R. K. "The Chemistry of Silica," Wiley: New York, NY, 1974.

35. Stolen, R. M.; Walrafen, G. E. J. Chem. Phys., (1976) 64, 2623.

36. Schmidt, H.; Scholze, H; Kaiser, A. J. Non-Cryst. Solids, (1984) 63, 1.

37. Aelion, R.; Loebel, A; Eirich, F. J. Am. Chem. Soc., (1950) 72, 5750.

38. Schwarz, R; Knauf, K. G. Z. anorg. allg. Chem., (1954) 275, 176.

39. Bechtold, M. F.; Vest, R. D.; Plambeck, Jr., L. P. J. Am. Chem. Soc., (1968) 90, 4590.

40. Bechtold, M. F.; Mahler, W.; Schunn, R. A. J. Polym. Soc., (1980) 18, 2823.

41. Yoldas, B. E. J. Polym. Sci. Chem. Edition, accepted for publication.

42. Brown, I. D.; Shannon, R. D. Acta Crystallogr., Sect. A, (1973) 29, 266.

43. Smith, K. A.; Kirkpatrick, R. J.; Oldfield, E.; Henderson, D. M. Am. Mineral, (1983) 68, 1206.

DESIGN OF MICROSTRUCTURES

IN SOL-GEL PROCESSED SILICATES

Lisa C. Klein

Rutgers-The State University of
New Jersey
Ceramics Department
P.O. Box 909
Piscataway, NJ 08854

INTRODUCTION

More and more, oxides and glasses are being prepared by chemical routes (1). In some cases, chemical routes are used to prepare ideal oxide powder compacts (2,3). These compacts, in turn, densify to defect free polycrystalline ceramics (4). In cases dealt with in this paper, chemical routes are used to generate oxides in a porous preform of their final geometry. The geometries of interest are fibers (5), thin films (6) and shapes called monoliths (7,8).

Broadly, these chemical routes are neutralization of salts (9), flocculation of colloids (10,11), and hydrolysis-polycondensation of alkoxides (12). This last chemical route has come to be known as the sol-gel process. Both its practice and applications have been reviewed in the last few years (13,14,15,16).

The sol-gel process for silicates is the subject of this paper. The precursors, hydrolysis and condensation reaction are handled elsewhere in this volume (17). To avoid being repetitious, this paper will assume the reader is familiar with the reactions, the conditions for these reactions and their kinetics. Of course, the early treatment of the alkoxide solution has consequences evident in the microstructure of the gel. Factors such as composition (18,19), solvent (20,21) and catalyst (22,23,24) will be mentioned when they have a direct effect on microstructure. The topics which will be covered are the time-to-gel, drying, textures of dried gels, densification of gels and applications of microporous materials.

THE SOL-GEL PROCESS

The reactants, intermediates and products for the sol-gel process are shown schematically in Figure 1. Tetraethylorthosilicate (TEOS) has been used for this example. This silicon alkoxide is preferred because it reacts with water to give complex silanols rather than silicic acid. Also ethanol the product of hydrolysis is safer to handle than methanol. The chemistry of silica has been treated fully by Iler (25).

Mixtures of TEOS-water-ethanol form true solutions. The reactions hydrolysis and polymerization occur in solution. The solution viscosity remains low for some time until the solution reaches what is called the sol-gel transition (26). For some time after the sol-gel transition, the reactions continue. Eventually, what was a single phase becomes two phases. This is what is called an alcogel, an oxide skeleton which has condensed in the presence of a solvent (27). Accompanying the sol-gel transition is a change from viscous behavior to viscoelastic behavior (28,29). When the alcogel is dried, the solvent phase is replaced by a gas phase. Then the gel behaves elastically.

The sol-gel transition is irreversible and occurs with no change in volume. It is easier to tell that the gel is past the sol-gel transition than to detect the onset of the transition. Many sophisticated scattering experiments have been performed which were intended to capture the transition (19,30,31). Because there was no discontinuity in scattering curves to signal its onset, the transition was not found this way. Instead the most common determination is the time when the solution viscosity increases abruptly. This time is called time-to-gel.

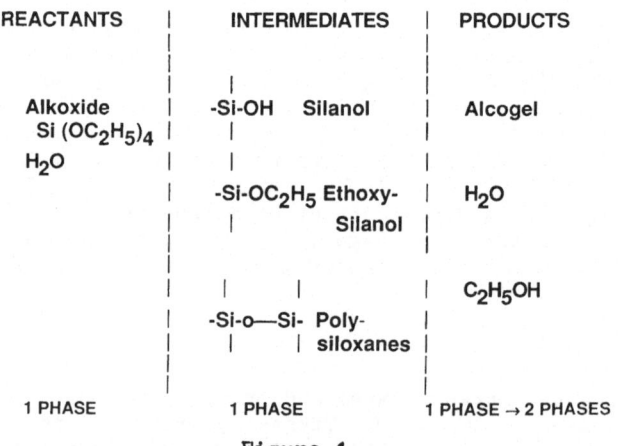

Figure 1.

Sol-Gel Processing: Reactants, Intermediates
and Products for SiO$_2$

Generally, the time to gel is determined by inspection. The time when a solution shows no flow is one such criterion (28). Another description is the time when the gel surface shows a fingerprint when touched lightly (32) or the time when a crack appears when the container is squeezed (8). Often the viscosity of the solution is determined and a reasonable value, perhaps 40 cp in an acid catalyzed system or 1000 cp in a base catalyzed system is chosen. Applying this somewhat arbitrary criterion consistently will at least show trends in time-to-gel.

Solution Viscosity - Typically, solution viscosities are measured with a rotating cylinder (32) or a falling sphere (33). The viscosity is plotted as the log of the viscosity in centipoise vs. real time or, in some cases, vs. reduced time, real time divided by time-to-gel. Using a real time axis, the viscosity is a low value showing shear rate independent behavior for some period of time. The time and manner in which the system loses fluidity depend on composition and catalyst. Low water solutions show a gradual increase in viscosity before losing fluidity, while high water solutions lose fluidity more abruptly from a lower viscosity. This is interpreted to mean that low water solutions produce linear polymers and higher water solutions produce crosslinked polymers or branched clusters. A small volume fraction of branched clusters will restrict flow when the same volume fraction of linear polymers will not. At a higher volume fraction the linear polymers will tangle and the solution will gel.

The Zirconia-Silica System - This system is used to illustrate how the viscosity and time-to-gel are related to microstructure. The technological interest in this system is that zirconia-silica glasses are potentially stronger mechanically and chemically than pure silica glass. In investigations designed to increase the zirconia content of this glass, the problem is that the liquidus temperatures for such compositions are too high for conventional melting equipment. As such, the sol-gel method is required to incorporate greater than 30 mole % ZrO_2 (34,35,36,37,38). Both fibers and thin films have been fabricated.

A series of solutions were prepared with the compositions listed in Table I. TEOS (Dynamit-Nobel) and zirconium isopropoxide (Alfa Chem. ZriP) were used. On the basis of 100 ml TEOS, the quantities indicated in Table I were mixed to obtain between 10 and 50 wt % ZrO_2.

The TEOS is partially hydrolyzed under refluxing in a 3-necked flask. For example, for the composition with 30 wt % ZrO_2, this means 100 ml TEOS dissolved in 125 ml ethanol and reacted with 17 ml water and 10 ml concentrated nitric acid. Another 100 ml ethanol is used to dilute 28 ml ZriP before it is added to the solution. The calculated oxide content for this solution is 10.49 g oxide/100 ml solution.

Table I: ZrO_2-SiO_2 Solutions

Wt % ZrO_2	TEOS	ZriP	Ethanol	Water	Nitric Acid
10	100	9	125	17	10
20	100	16	150	18	10
30	100	28	225	17	10
40	100	40	406	18	10
50	100	62	480	18	20

The calculated mole ratios are listed in Table II. The ratio of water to alkoxide is 2 or less for all 5 solutions. The ratio of nitric acid to alkoxide is about 0.50 except for the solution with 50 wt % ZrO_2 which has a high nitric acid content to prevent precipitation. The reason for prehydrolyzing the TEOS and using low water to alkoxide ratios is to create complex silanols before adding ZriP. The zirconium alkoxide hydrolyzes far more rapidly than TEOS. Sometimes it is questionable that the TEOS and ZriP cohydrolyze. Especially with the high nitric acid levels indicated, a soluble zirconium nitrate may be forming. However, the presence of Zr-O-Si units in the gel has been confirmed with infrared spectroscopy (35).

After refluxing, the reaction flask temperature is raised until the solution starts to boil. As primarily ethanol is removed from the flask, the density of the solution increases.

Table II: Time to gel, Mole Ratios and Oxide
Content for ZrO_2-SiO_2 Solutions

Wt % ZrO_2	Time to gel (hours)	Moles water ----- Moles M(OR)$_4$	Moles HNO$_3$ ------ Moles M(OR)$_4$	Moles ethanol ------ Moles M(OR)$_4$	Oxide Content g/100 ml
10	51	2	0.45	2.88	12.40
20	56	2	0.45	3.25	12.06
30	162	1.75	0.49	5.39	10.49
40	186	1.76	0.39	7.82	7.73
50	325	1.54	0.67	8.09	7.79

The amounts of ethanol removed are recorded so that the oxide content in the concentrated solution can be calculated. In the solution containing 30 wt % ZrO_2, the oxide content increases from 10.49 g oxide/100 ml to 12.52 g oxide/100 ml. These values are listed in Table III. Similar treatments were used for all solutions until the time-to-gel became shorter than experimentally practical for measurement.

The solution viscosity was measured using the cone-plate attachment for the Brookfield viscometer. A fresh bead of the solution is placed on the plate, and the cone-tipped rotor is rotated at 50 rpm. The cone-plate separation is small so that the shear rate experienced everywhere in the sample is the same. The viscosity is read directly on the measuring head, and the viscosity is independent of speed of rotation. The viscosities range from a few centipoise to 10^5 centipoise and the time-to-gel ranges from 40 hours to over 300 hours.

Table III: Effect of Solvent Removal on
30 wt % ZrO_2-70 wt % SiO_2 Solution

Oxide Content (g/100 ml)	Time to gel (hours)	Reduced viscosity at 41 hours
10.49	162	0.11
10.87	125	0.20
11.24	100	0.24
11.74	77	0.60
12.52	41	520

The time dependence of the solution viscosity for the 5 compositions is shown in Figure 2. As the ZrO_2 content is increased and consequently the ethanol level is raised the time-to-gel becomes longer. The solution with 10 wt % ZrO_2 gels in 51 hours and the solution with 50 wt % ZrO_2 gels in 325 hours. It is difficult to say exactly which of the ratios listed in Table II represents the controlling factor but the lowest water/alkoxide ratio and lowest oxide content solution had the longest time-to-gel. At the same time, this solution has a high ethanol/alkoxide ratio and a high nitric acid/alkoxide ratio.

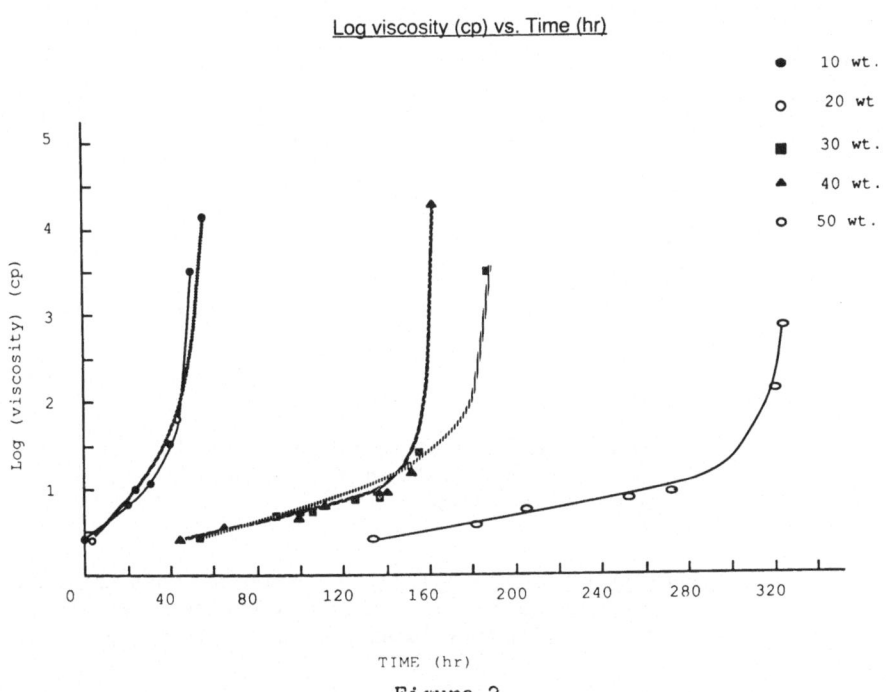

Figure 2.

Time dependence of solution viscosity
for 5 compositions of ZrO_2-SiO_2

The nitric acid/alkoxide ratio is calculated on the basis of total alkoxide, TEOS plus ZriP. If the nitric acid is primarily involved in forming zirconium nitrate groups, which are soluble in water, then the gelling would be postponed. This appears to be the case. The TEOS portion of the solution will bring about gelling despite its low concentration, so in all 5 solutions gelling occurs eventually.

Another way to ensure gelling is to remove solvent after hydrolysis and polycondensation reactions are underway. This brings the polymers in the solution closer together so that entanglement will lead to gelling. This effect is seen in Figure 3. As the oxide content of the solution containing 30 wt % ZrO_2 is increased from 10.49 g oxide/100 ml to 12.52 g oxide/100 ml, the time-to-gel decreases by a factor of 4. The concentrated solution gels in 41 hours.

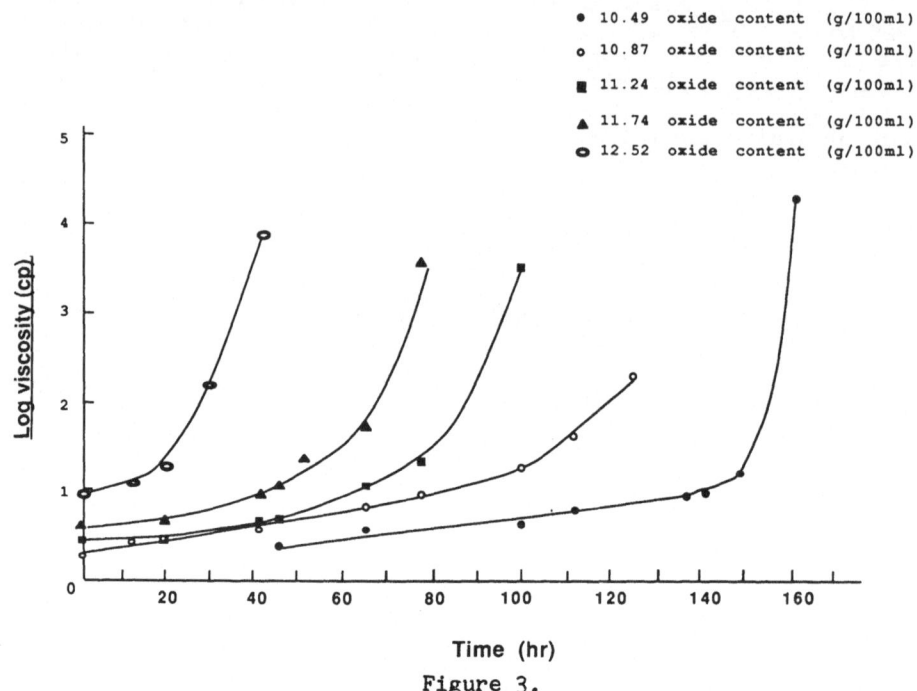

Log viscosity (cp) vs Time (hr)

- ● 10.49 oxide content (g/100ml)
- ○ 10.87 oxide content (g/100ml)
- ■ 11.24 oxide content (g/100ml)
- ▲ 11.74 oxide content (g/100ml)
- ⊖ 12.52 oxide content (g/100ml)

Figure 3.

Time dependence of solution viscosity
for 30 wt % ZrO_2-70 wt % SiO_2 with
5 oxide contents.

        Model: Viscosity of Dilute Solutions - Even the most concentrated
solution shows viscosity behavior of a dilute solution. The way in which a
dilute solution loses its fluidity is an indication of the polymers present
in solution.

        The viscosity of dilute solutions is treated by Sakka and Kamiya
(28). First the measured viscosity is divided by the viscosity of the pure
solvent. This is the relative viscosity. The specific viscosity is defined as the relative viscosity minus one. The specific viscosity is
divided by the oxide content in g/100 ml. This is defined as the reduced
viscosity. The reduced viscosity at infinite dilution is called the intrinsic viscosity. The measured, relative, specific and reduced viscosities are plotted in Figure 4. The time axis is reduced time, time divided
by time-to-gel.

        The results plotted are for the composition with 30 weight % ZrO_2 but
all 5 compositions show identical behavior. In particular, reduced viscosity is an indication of the polymer content of the solution. Linear polymers lead to the type of behavior seen here, a gradual increase in viscosity. Alternately, branched polymers show an abrupt loss of fluidity. Solutions containing linear polymers are suitable for fiber drawing or spinning. All of the solutions prepared in this study showed that a fiber
could be drawn from the surface of the solution by contacting it with a
bait rod. Fibers were 10 cm in length and 20 to 100 microns in diameter.

44

<u>Microstructure</u> <u>of</u> <u>ZrO$_2$-SiO$_2$</u> <u>Gels</u> - Solution viscosity is one way of characterizing molecular structures in the sol-gel process. The viscosities recorded for the ZrO$_2$-SiO$_2$ system are typical of acid-catalyzed, low water solutions. The viscosity-time curve shows the gradual increase that leads to spinnability.

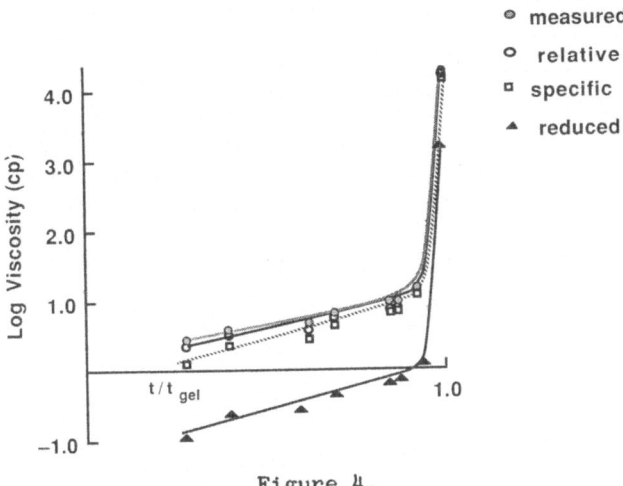

Figure 4.

Variation of viscosity with t/t gel for 30 wt % ZrO$_2$-70 wt % SiO$_2$ with oxide content 10.49 g/100 cm^3 (measured-filled circle, relative-open circle, specific-filled box, reduced-open box).

The same type of measurements performed in a systematic way lead to some general statements about gel microstructures. First, there is a fundamental difference between the microstructure of colloidal gels and that of polymer gels (39). Second, there is a difference between base-catalyzed and acid-catalyzed gels (24). Third, there is a difference between high water and low water gels (40).

Acid-catalyzed gels are characterized by high dry densities and the density decreases with increasing water in the original solution. Base-catalyzed gels are characterized by low dry densities which vary more with amount of base than amount of water. Base-catalyzed gels behave more like particles than acid catalyzed gels, but base-catalyzed gels are not the same as colloidal gels. Base-catalyzed gels are made up of more condensed regions and less condensed regions in an overall polymer framework. Colloidal gels behave like agglomerations of compact spheres. Acid-catalyzed gels have flexible, uniformly crosslinked polymers. The point of this section is that there is no real distinction between molecular structure and microstructure in alkoxide gels.

After the sol-gel transition, the two phases are oxide skeleton and solvent in the pores. The pores are interconnected. The solvent is removed from the pores by one of the drying schemes in Figure 5. Hypercritical evacuation results in aerogels (7,41) and natural evaporation results in xerogels. Xerogels may have water and alcohol in the pores or solvent substituted in various proportions with a drying control chemical additive (DCCA) (21).

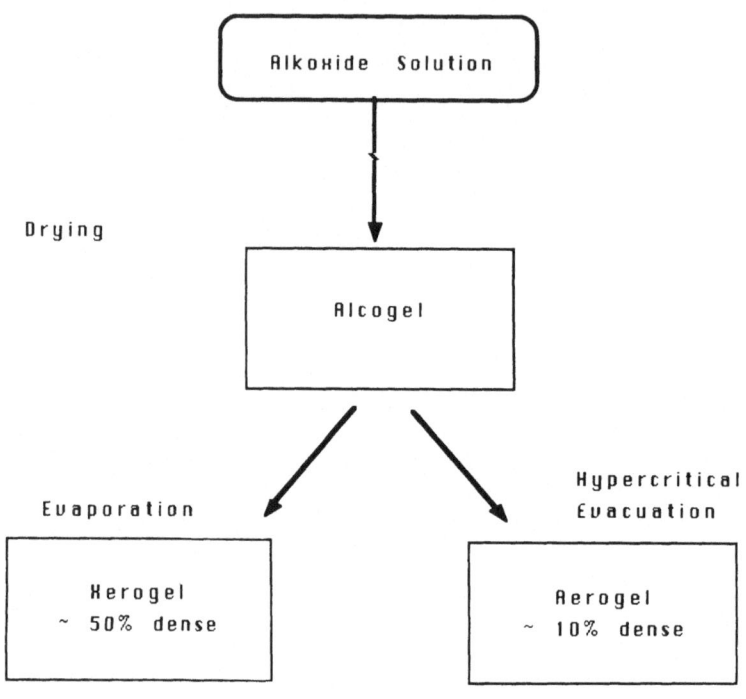

Figure 5.

Drying schemes

The removal of water and solvent from interconnected pores in gels is not unlike the processes involved in conventional drying of ceramics before firing (42), but the magnitudes of the weight loss and volume reduction surpass those usually encountered in conventional ceramics. A typical shrinkage curve is shown in Figure 6. An S-shaped curve is traced after the gel begins to part from the walls of the mold. When the oxide skeleton has sufficient time to age it is able to withstand drying stresses. These stresses can be eliminated by hypercritical evacuation (7,41) or absorbed with a DCCA (21) but the gels resulting in these cases have low density and the same dimensions as the original mold. In aerogels, the pore size is on the scale of 10-50 nm, approaching sizes of those pores in samples prepared by colloidal techniques (43,44). In xerogels, the pore size is on the scale of 2-5 nm (40). Aerogels dry much faster than xerogels. This is significant in forming monolithic shapes without cracks. In fibers and thin films, drying is less of a problem.

The dry condition of a gel is difficult to define. When there is no further weight loss with time is one definition. When the sample shows no adverse effects from ambient conditions is another. Most gels will show a readsorption of water when heated up to 150°C (45). When heated above 150°C, the surface area reaches a maximum at the so-called activated temperature (25). No matter which way drying is accomplished, the gel is not truly dry until after some stabilizing heat treatment (46).

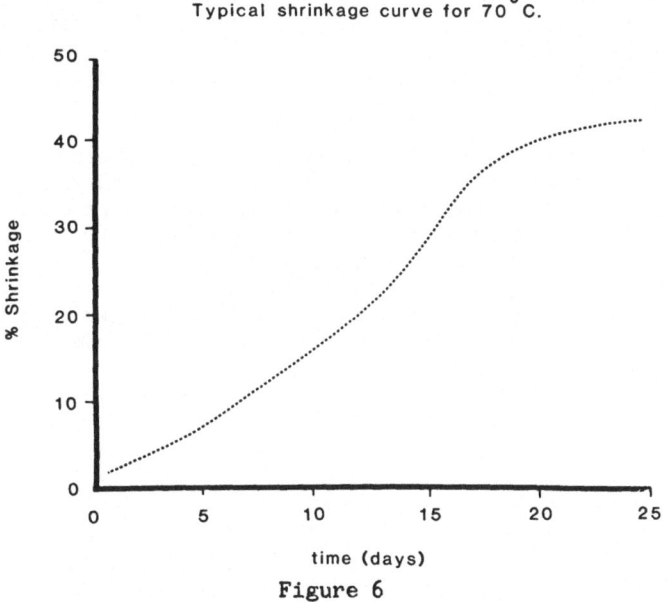

Typical shrinkage curve for 70°C.

Figure 6

Typical shrinkage curve for multicomponent gel showing transition from excess liquid to liquid-filled pores and transition from liquid-filled pores to pores with receding liquid.

Physical Characterization - Drying studies involve careful measurement of weight change, length change and volume change on a population of gel samples. There are many assumptions in these studies (47). This is because of little direct information on microstructure or physical properties.

In one such study, samples were placed in polypropylene tubes. With slow drying (tight caps), the dimensions of the gels stopped changing after 40 days. With rapid drying (loose caps), it takes about nine days to have the same final dimensions as the first group.

In both cases, there are two parts to the S-curve during drying which indicate two transitions. The first transition begins when the percent weight loss deviates from the diameter shrinkage. This is when there is no more liquid exuded from the gel and all evolving species are evaporating. The second transition is the critical stage of drying when cracking normally occurs. Here the solvent recedes into the pore and a meniscus is formed.

For slow and fast drying in the intermediate stage, the volume of water vs shrinkage data fall on the same curve. This indicates that the shrinkage is independent of drying speed. The contraction per gram of water removed varies with the amount of water remaining in the gel. In both cases, there is no further drying shrinkage after 85% weight loss.

For the critical-drying stage if the samples are dried too rapidly or the temperature is too high, the samples develop cracks.

The $Li_2O-Al_2O_3-SiO_2$ System   This system is used to illustrate drying behavior. The technological interest in this system is that lithium aluminosilicate glasses are suitable ionic conductors (48), low thermal expansion glass-ceramics (49) and a desirable matrix for ceramic fiber composites (50). Gels with lithium have been prepared from salts (51,52) and alkoxides (49,52).

Solutions with 15 mole % Li_2O-2 mole % Al_2O_3-83 mole % SiO_2 were prepared with 4 or 8 moles of water per mole TEOS. The TEOS is hydrolyzed before adding $LiNO_3$ and aluminum nitrate. Solutions were poured into polypropylene cylinders and dried fast (loose caps) or slow (tight caps). A drying study was performed as described above. Overall, the samples followed the typical S-shaped curve shown in Figure 6. The physical changes occurring during intermediate and final stage drying are shown in Figure 7. These data are being used to modify existing drying models.

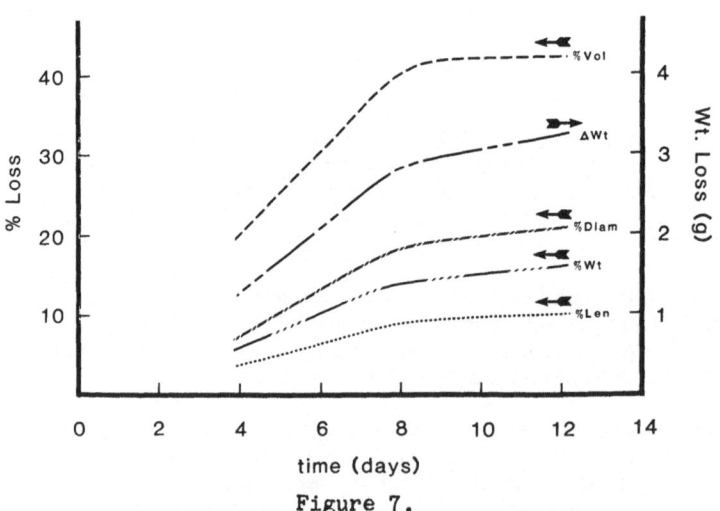

Figure 7.

Changes in dimensions and weight
of cylindrical samples

Model: Drying in Low Solids Systems   No drying model exists exclusively for gels. In fact, few models exist for analogous low solids systems (47). As shown in Figure 5, the approaches tried have been critical point drying which produces aerogels and natural evaporation which produces xerogels. Xerogels are the product of controlled humidity drying with or without so-called DCCA (drying control chemical additives). DCCA's are mixtures of solvents used to adjust the volume contraction during shrinkage.

According to the model (47), the linear shrinkage in a drying gel gives an S-shaped curve. Initially the shrinkage is slow. Then the shrinkage accelerates to a linear rate for a period of time. This portion of the shrinkage curve is modeled by diffusion. The evaporation rate at the surface is set equal to the diffusive flux to the surface. An assumption is made that the volume is proportional to the moisture content, and the moisture content varies with position. The result is a strain in the material. The surface stress can be lowered by increasing the diffusivity. When samples are heated to lower the solvent viscosity, the diffusivity of water increases by almost a factor of 10 between 0 and 20°C.

One assumption specific to gels is that particle-particle contact exists from the start of drying in gels. In the gels studied here, the pore size is small and the particle size is small. What is meant by a particle is the thickness of gel material between pores or the width of the wall separating two capillaries. In xerogels, the size of the pore is nearly the same as the size of the particle, as defined here. This makes the gel structure stronger than expected during the early stages of drying.

Microstructure of $Li_2O-Al_2O-SiO_2$ Gels   With this basic picture of drying behavior, it is becoming possible to model the behavior observed. The gel as it begins to dry is quite pliable. Most likely the water is squeezed toward the surface. Otherwise it follows a diffusion gradient. As the network shrinks, more hydroxyls are brought close together so that more condensation occurs. When the diffusivity of water is independent of concentration, a parabolic profile results. When the diffusivity decreases with decreasing water, the profile is flattened. Eventually, the concentration gradient is eliminated, thus eliminating any further driving force for transport. At that point, the question is whether or not strains in the material are great enough to finally cause fracture. The transition from the linear rate period to the final period is the critical time for cracks.

The ways to avoid cracks are said to be avoid small pores, keep pores all the same size and use low surface tension solvents. These warnings apply in xerogels, and the fact that crack-free xerogels are achieved at all is an indication of the uniformity of the pores despite small size. Aerogels, on the other hand, are formed crack-free because hypercritical evacuation eliminates capillary stresses. Since, xerogels have textures that are required for applications such as adsorbents and catalyst supports, the problems of drying bulk xerogels need to be overcome.

TEXTURE

Once drying is accomplished, the dried gels are classified by texture. Their texture is either fine or coarse. Acid-catalyzed gels have high bulk densities. Base-catalyzed gels are very friable. Acid-catalyzed gels have hydroxyl contents that scale with surface area. Base-catalyzed gels do not. In basic solutions, the solubility of silicic acid leads to internal condensation. In base-catalyzed solutions the polymer does not gel as a unit. Instead, there is a sediment of clusters at the bottom of the solution.

The texture of gels is difficult to resolve with the electron microscope. Aerogels of very low bulk density can be observed (53,54), but xerogels which are 50% dense present the experimental problem of looking through several layers at one time (55). Generally, the texture of gels is probed with nitrogen sorption (56). Nitrogen is used for microporosity (below 1.5 nm) and mesoporosity (below 50 nm). Mercury porosimetry is used for macroporosity (above 50 nm). The adsorption isotherm gives the surface area according to the BET equation. The desorption gives the pore size distribution from the Kelvin equation. Hysteresis in these curves is interpreted according to previously characterized materials to indicate texture. Acid-catalyzed gels show little hysteresis which is interpreted to mean uniform cross-section porosity. Base-catalyzed gels show hysteresis which is interpreted to mean narrow-neck or "ink bottle" pores.

Nitrogen Sorption Analysis - Samples for nitrogen adsorption treatments were ground in an alumina mortar and pestle. A sample size of 20 mg was used. Samples (-200 to +325 mesh) were weighed into the sample cell. The cell was then attached to the outgassing station of the Quantasorb[TM] Surface Area Analyser (Quantachrome Corp., Greenvale, NY). High purity dry nitrogen was flowed over the sample at a rate of 5 ml/min. Samples were outgassed for 12 hours at 200°C. Standard procedures recommended by the manufacturer for obtaining nitrogen adsorption-desorption isotherms were used. The volume of pores with radius less than 50 nm was determined by first adsorbing pure nitrogen then switching to a 98% N_2-2% He mixture at

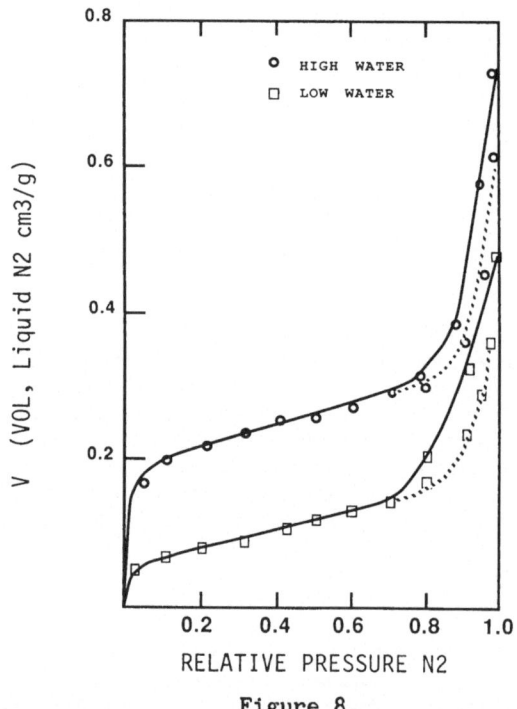

Figure 8.

Typical nitrogen sorption curves for
high water and low water sol-gel silica

77K. As the sample was heated to room temperature, the desorption signal was integrated and calibrated. Isotherms were obtained with ten prepared mixtures of helium and nitrogen. Surface area was estimated from the isotherms.

Typical nitrogen sorption curves for base-catalyzed samples are shown in Figure 8. These curves show slight hysteresis where curves for acid-catalyzed samples show none. The adsorption isotherm was used to estimate surface area and pore fraction.

In acid-catalyzed solutions, when the water to TEOS ratio was increased from 4 to 16 there was an increase in porosity from 10 to 40%. The trend in the surface area was an increase from 20 to 640 m^2/g with increasing water. A lack of hysteresis indicated that the pores were cylindrical. The cylinder radius was calculated from the Kelvin equation. A rough estimate for the average pore radius is 1.5 nm for 4 or more moles water. As with all measurements of pore size and distribution, these are sensitive to outgassing conditions and sample preparation.

For acid catalyzed solutions the volume of liquid nitrogen per gram gel (V_s) increased for all nitrogen partial pressures as the water level increased. The pore volume was read directly from the isotherm at the point corresponding to 98% N_2. The single point BET value was obtained using the V_s value corresponding to 10% N_2. Each isotherm had an inflection point at intermediate values of nitrogen partial pressure which became more pronounced with increased water. The plot of the isotherm approached the y-axis at right angles at high nitrogen partial pressures for acid-catalyzed solutions but not for base-catalyzed solutions.

The isotherms for base-catalyzed gels showed increasing hysteresis at high nitrogen partial pressures with increasing water level (Figure 8). This implies non-uniform cross-section pores. Using the desorption portion of the isotherm, the cumulative pore volume was calculated. The calculation was normalized to 50 nm. Normalizing the cumulative porosity to 50 nm is a fair assumption for acid-catalyzed gels. For base-catalyzed gels, the representation of cumulative porosity between 1 and 50 nm is not accurate. Acid-catalyzed gels are transparent, and base-catalyzed gels scatter light to produce opacity.

High Water-Low Water Silica System - This system is used to illustrate gel textures. The technological interest in its texture is that porous silicates are used for membranes (57,58), catalyst supports (59) and infiltrated composites (60). Silica gels have been prepared with over 800 m^2/g specific surface area.

To study texture, solutions were prepared where the water to TEOS ratio was varied. The ratio was 2:1, 4:1, 8:1, 16:1 and 32:1 for both HCl and NH_4OH with molar concentrations of 1×10^{-3}. The volume ratio of ethanol to TEOS for this series was 4:1 to permit solubility of the water.

Twenty ml Pyrex[TM] scintillation vials with airtight polyethylene lids were used to mix, gel and dry the samples. Each vial was filled with 8 ml of TEOS and the appropriate volume of ethanol. Then the aqueous electrolyte was added, and finally the water was added. The vials were shaken and placed in a drier at 80°C to react for two days. In this time all acid-catalyzed solutions gelled. No base-catalyzed solutions had gelled but most had sediment. After two days of reacting at 80°C, samples were uncapped and placed in a bell jar into which dry nitrogen was flowed. The bell jar was maintained at 80°C. After 24 hours the bell jar was heated to 200°C. After another 24 hours, samples were withdrawn, capped and allowed to cool. The high drying temperature eliminated most physically trapped water and ethanol.

The samples were characterized by appearance. A complete texture analysis was performed using nitrogen sorption techniques described above.

Model: Texture of Microporous Silica - There is a noticeable difference in appearance in newly gelled solutions which can be traced to the hydrolysis mechanisms (22). In acid-catalyzed solutions, the first hydrolysis of the TEOS monomer is easier than the second, so a growing polymer will have an even distribution of hydroxyls. The polymer can form an occasional crosslink by a condensation reaction, and any structural features which might scatter light remain too small during gelling or drying. In base-catalyzed solutions, the successive hydrolysis of the TEOS monomer becomes easier, so that condensed polymers co-exist with unreacted monomer. Structural features develop in the overall polymer network that scatter light, both in solution while gelling and in the drying sediment (30,61).

In general, base-catalyzed solutions are cloudy. All samples show a tendency to sediment. The thickness of the sediment layer increases with increasing base addition. The cloudiness decreases as the water level is increased. When base-catalyzed gels are dried, the result is weakly coalesced powder.

When comparing the effect of catalyst addition vs water level in acid-catalyzed solutions, the acid addition has little effect while the water level has a strong effect on surface area and porosity. The low water solution gels to a polymer which is so weakly reacted that it continues to shrink after 12 hours at 200°C. In base-catalyzed solutions, both the base and the water level have strong effects. High water solutions are more completely hydrolyzed and condensed, as are high base solutions. Also, dilution with ethanol allows more complete hydrolysis.

When comparing acid vs base, not only are the mechanisms different, the kinetics are different. For acid-catalysis, hydrolysis is complete and the number of unreacted hydroxyls per silicon decreases with decreasing acid concentration. For base-catalysis, the effect of polymerization is large.

Microstructure of Silica Gels - For the moment, direct evidence of the size of features from techniques such as electron microscopy are rare. The microstructure of dried gels has to be inferred from surface area and pore volume measurements. From this, it is possible to suggest that acid-catalyzed gels have uniform interconnected porosity. Base-catalyzed gels have pores within clusters and between clusters making up the sediment. Since silicic acid is more soluble in basic solutions, the surface area decreases over time with increasing base and increasing water. With this increased solubility, silica dissolves and reattaches far more easily in basic than acidic aqueous medium.

The adsorption-desorption isotherms can be used to plot a pore size distribution. The result for acid-catalyzed gels is a narrow distribution, with the width and mode of the peak decreasing in radius and the height increasing as the water level is increased. The absence of hysteresis in the isotherm indicates cylindrical pores. The result for base-catalyzed gels is a broad distribution at low water levels which becomes bimodal at intermediate water levels. The distribution peaks are located at 1.5 and 5 nm. The 5 nm peak increases and the 1.5 nm peak decreases as the water level is increased further. The distribution is shown in Figure 9.

The microstructure of dried gels has an effect on the microstructure of fired gels, especially when bulk samples are considered. Continuous open porosity allows oxidation of residual organics, where closed porosity usually gives bloating. Bloating is the foaming of glasses when gases become trapped.

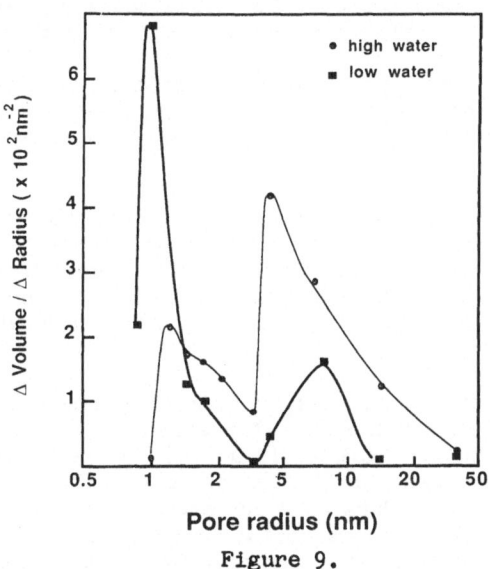

Figure 9.

Pore size distribution for high water and low water base catalyzed sol-gel silica.

The texture of gels is important for modeling the densification of gels. At the same time, the texture is important for the applications cited for microporous materials.

By far the dominant theme of sol-gel processing of pure silica has been duplication of the physical properties of fused silica. The advantages for this process over conventional ones are working in uncontaminated conditions and at lower temperatures than the melting point (8,45,62). The mechanism for densification is viscous sintering (63) and a model for initially porous bodies has been used successfully. Along with viscous flow, volume relaxation and condensation reactions were considered as well (64,65). Densification has been achieved with dehydration treatments (66,67). The result is in fact a chemical and physical duplicate of the melted glass composition.

In several studies, the densification of silica gels has been described in a qualitative way. The effects of heating rate and atmosphere have been noted. These effects, along with measurements of pore volume and surface area have been used to identify the temperature regions in which chemical and physical processes operate. Chemical processes dominate at low temperatures and physical processes dominate at high temperatures.

Sol-Gel Derived Fused Silica - This system is used to study densification. The technological interest in its densification is that fused silica is used universally for optics, its low thermal expansion and its chemical durability.

The behavior of silica gel during densification depends on its microstructure. As described previously, its microstructure depends on the solvent, amount of water reacted with TEOS, reaction conditions and drying conditions, along with the nature of acid or base catalyst. To focus on changes in microstructure during heating, all samples used in these studies (62,68,69) were prepared in the same way.

The solution contained an equal volume of ethanol and TEOS, 16 mol. of distilled water and 0.01 mol HNO_3. The solution was refluxed for three hours. All samples used were gelled in a bell jar heated to $70^\circ C$, after which the samples were dried for 40 days at $70^\circ C$. Their appearance was slightly cloudy monoliths. Bars were cut and dry polished from the dried gel monoliths. The physical characteristics of this gel after drying for 40 days at $70^\circ C$ were a volume porosity of 0.36, a bulk density of 1.52 g/cm^3, surface area of 640 m^2/g and an average pore radius of 1.5 nm. Linear shrinkage was measured in an Orton Model 1500 Automatic Recording Dilatometer with an external programmable temperature controller. Samples were 1.27 cm long and 0.35 cm in diameter. About 22% shrinkage is needed to reach full density.

To measure viscosities directly, a beam bending viscosimeter was used. The technique measures the deformation of a centrally loaded, end supported cylinder. The cylinders are 10 cm in length and 0.35 cm in diameter for viscosity measurements between 10^{12} and 10^{15} poise. The ends of the specimen are supported on a muffle in the central portion of a resistance furnace, equipped with a proportional temperature controller. The temperature variation in the specimen region over the time of experimental measurements is within $\pm 1C$.

Heat Treatments - Silica has been densified using constant rate heating (62), isothermal heating and step-heat treatments (68,69). In each case, as the temperature is increased the viscosity decreases making viscous flow easier. At the same time, the viscosity is tending to increase due to decomposition of silanols. This decomposition also increases the surface tension which, in turn increases the driving force for sintering.

Balancing these phenomena what results is an optimum heating rate.
During constant rate heating it was found that the heating rate 2.5°C/min
gave the largest linear shrinkage below 1000°C. Faster heating rates gave
rapid densification beginning at about 850°C, but in all cases the samples
bloated. Slower heating rates gave densification above 900° without bloating but the linear shrinkage at 1000°C was only 20%, not the full 22%. The
rate 2.5°C/min was fast enough to give shrinkage between 300 and 800°C
where the suppressed viscosity permits flow. It was also fast enough to
keep up with the decomposition of silanols which increase the viscosity.
At the same time, the rate was slow enough to allow the elimination of
gases before viscous flow traps the gases in isolated pores.

During isothermal heat treatments, linear shrinkage was measured in
order to estimate relative bulk density. This value is used in a sintering
model which calculates viscosities. The model is described below.

Samples were fired to temperatures 750, 800, and 850°C and held for
16 hours. All heat treatments were performed using a heating rate of
2.0°C/min to reach the soak temperature in oxygen at a flow rate of 30
cm³/min. The dilatometer pushrod pressure was not greater than 40 grams.
At temperatures below 800°C, the calculated viscosity is increasing even
after 1000 minutes. At 850°C, the viscosity increased for about the first
300 minutes, but then leveled off (Figure 10). This time corresponds to
the time after which there is no measurable water loss.

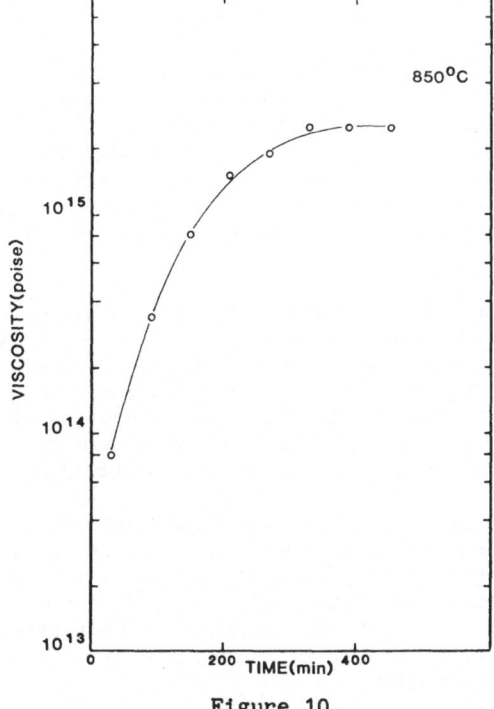

Figure 10.

Time dependence of viscosity during
densification of sol-gel silica.

It is clear that water has an effect on viscosity, but the results of isothermal heat treatments lead to the wrong conclusion that anhydrous silica has the same viscosity over the range 750 to 850°C. What this really means is that something besides water suppresses the viscosity to a large degree and this effect is not erased in a 16 hour heat treatment.

The viscosity is suppressed by what is called its excess free volume. The gel structure is inflated and the silica skeleton has a low specific density. Another way of describing this is an apparent fictive temperature. The gel has an apparent fictive temperature above 2000°C.

To remove the effect of the high apparent fictive temperature and isolate the effect of water, step heat treatments were tried to anneal out some of the excess free volume. Samples were fired to multiple temperatures and held for 4 hours at each temperature. The difference in viscosity for a given temperature between isothermally and step heat treated samples is a measure of the amount of free volume annealed out. The difference is small for the 750°C samples but increases for the 800 and 850°C samples. Samples which were heat treated in steps have a higher water content at a given temperature because they are less dense and have higher surface areas.

A comparison of isothermal shrinkage data and shrinkage data during heating in steps reveals contributions to the viscosity of gels from both dehydration and structural relaxation. Skeletal densities of gels are remarkably low even at 500°C. The low skeletal density can be assigned to a glass with a high fictive temperature. Taking the high fictive temperature into account, it is not surprising that the viscosity of gels is suppressed and increases for a long period of time before only the effects of hydroxyl are left. By removing the effects of relaxation by successive temperature steps, the effect of dehydration can be isolated.

In the same way that samples were treated in the dilatometer, samples were held isothermally for 4 hours in the viscosimeter. A flowing oxygen atmosphere was maintained. Also, samples were treated in steps such as an isothermal treatment at 700°C followed by an isothermal treatment at 800°C. Typically, a sample held at 700°C would have a measured surface area of 550 m^2/g and an average pore diameter of 2.7 nm. In the viscosimeter, the deflection rate of a loaded beam was measured and the viscosity was determined directly.

When a sample was held isothermally, the viscosity increased rapidly at first. When the sample was heated to a higher temperature, the initial increase in viscosity was not as great. This has been attributed to the fact that during an isothermal hold, a certain amount of excess free volume is annealed out. The continued increase in viscosity with time is largely due to dehydration. The time dependence of the measured viscosity was the same as the time dependence of the calculated viscosity.

Model: Viscous Sintering   The sintering model used calculates isothermal viscosities when both structural changes due to loss of free volume and compositional changes due to dehydration are occurring. In an attempt to separate these effects, isothermal shrinkage data were analyzed using this model. This model is based on a geometry consisting of a cubic array of intersecting cylinders. According to the model by Scherer (63), the quantity K is calculated using:

$$K = s/nl_o \ (p_s/p_o)^{1/3}$$

where s is the surface tension, n is the viscosity, l_o is the initial cylinder length, p_s is the skeletal density, and p_o is the initial bulk density. The reciprocal of K is proportional to viscosity, and for this gel, n is between 5×10^{12} and 5×10^{14} poise. As the temperature is increased the viscosity decreases, but then the viscosity tends to increase with time at temperature. A plot of K^{-1}, the sintering parameter, versus time shows that the viscosity of the gel is increasing with time. Even after 6 hours, in most cases, a stable value has not been reached. A plot of $\log(K^{-1})$ versus residual water gives a linear relationship for isothermal heat treatments. This is taken to mean that the increase in activation energy is largely due to dehydration. When water loss is no longer measurable, the viscosity still increases due to slight skeletal changes in the gel.

When calculated and measured viscosities are compared, the measured viscosities are consistently higher. The energy dissipated in viscous flow during sintering is a result of removing surface energy. There is also a contribution to the energy for flow from the structural relaxation and dehydration. For these reasons, the sintering model gives a calculated viscosity which predicts more rapid flow than the measured viscosity.

In either case, a useful construction is a time-temperature viscosity curve. On a time-temperature field, a locus of constant viscosity can be plotted so that the ease of sintering can be predicted. By knowing the temperature and time dependence of the viscosity, mechanical stability can be predicted for various stages of densification. On the one hand, a fully dense silica can be achieved by heating to a temperature where the viscosity is below 10^{13} poise. On the other hand, a fully dehydrated, microporous silica can be achieved by isothermally heat treating at a temperature where the viscosity is greater than 10^{14} poise. A viscosity value of 10^{13} poise is usually picked as a boundary. Lower viscosities permit sintering in a matter of hours. Higher viscosities do not result in significant sintering (70). According to these values for viscosity, sol-gel silica should sinter above 1000°C. At these temperatures, sol-gel silica and melted silica are indistinguishable in their physical properties (71).

<u>Microstructure of SiO$_2$</u>   During densification the pore volume and surface area were obtained from the nitrogen adsorption isotherms (62). The pore fraction increased from 0.36 to 0.60 at about 400°C and then decreased continuously. The surface area increased to a maximum at 400°C but leveled off before decreasing above 700°C. The increase in pore volume and surface area up to 400°C represented the removal of water and alcohol trapped in the dried gel. The decrease in pore volume and surface area above 400°C indicated an increase in the degree of polymerization.

When the microstructure is particulate, the surface area decreases continuously as the structure ripens. When the surface area levels off before decreasing at higher temperatures, the microstructure is not particulate. Exactly how to describe this microstructure is still not resolved. One explanation is the average pore size changes slightly from an oval to a circle. This slight change makes the surface area decrease sharply around 700°C. Since this temperature is still too low for significant viscous flow, the other factor is that the silica skeleton in the dried gel has a high free volume. If the skeleton loses this free volume by densifying to more nearly the density of fused silica, the remainder of the decrease in surface area above 700°C follows.

The density of conventionally melted silica is 2.202 g/cm^3. The skeletal density of the gel heated to 250°C is only about 1.85 g/cm^3. This corresponds to a fictive temperature of 2500°C. The skeletal density increases to about 1.99 g/cm^3 at 550°C and then levels off. This corresponds to a fictive temperature of 2200°C. At much higher temperatures the gel undergoes further structural relaxation. The skeletal density is known to increase again above 800°C. At the same time, the bulk density increases rapidly above 800°C until full density is reached at 1000°C. Fully dense silica from the sol-gel process is then identical to fused silica.

APPLICATIONS

Microporous silicates have many applications. These applications may require the overall geometry of a fiber, a thin film or a bulk shape. The sol-gel process as defined at the outset allows the fabrication of the preforms in the desired geometry. The sections on time-to-gel, drying, texture and densification were intended to show how these stages in the process relate to the microstructure of these preforms.

In each stage, factors such as catalyst, water level, and solvent have been shown to influence the microstructure. When these ideas are put together, a few general trends emerge. One of these trends is illustrated on Figure 11.

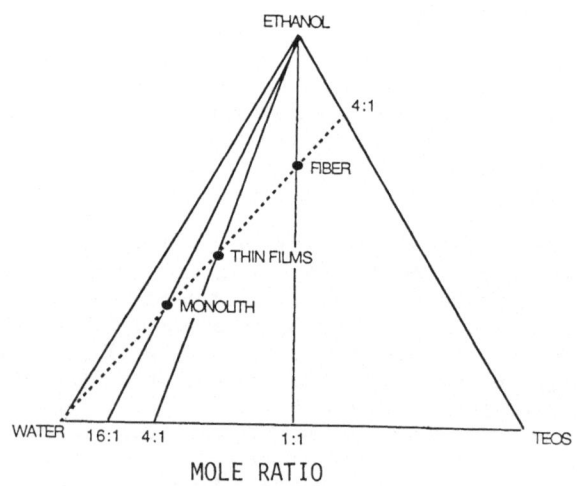

MOLE RATIO

Figure 11.

Triangle construction for
TEOS-ethanol-water system
indicating regions for fibers,
thin films and monoliths

Several of the compositions discussed above can be located on the triangular construction for the TEOS-ethanol-water system (Figure 11). Three lines are drawn from the ethanol apex to the TEOS-water binary. Along these lines, the water to TEOS ratios are constant 1:1, 4:1, and 16:1. Another line is drawn from the water apex to the TEOS-ethanol binary. Along this line, the ethanol to TEOS ratio is 4:1. The intersections of the constant water/TEOS lines with the ethanol/TEOS line equal to 4 represent compositions which form fibers (spinnable), thin films (coatable) and monoliths (castable). Excluding the region of immiscibility, solutions are spinnable with less than 40 mole pct water, are coatable with between 40 and 70 mole pct water and castable with greater than 70 mole pct water. To a large extent, the water to TEOS ratio controls the preform.

Fibers   Gels have been fabricated in the form of fibers for thermal insulation (28,35) and for optical transmission (72,73). For insulation materials the fibers are spun at the right time in the viscosity-time cycle and the diameter is 20 to 100 microns.

More recent applications suggest the use of the microporous fiber for light transmission. In one case, a gel microballoon (74,75) is filled with a gas which changes color with pressure. The microballoon is used as a pressure sensor, and the attached fiber is the transmission medium to allow remote sensing.

In another case, waveguides have been prepared by the gel process (73). In this application the gel is the porous preform of the fiber. In commercial chemically deposited waveguides, the preform is hundreds of times larger than the final fiber. Using a sol-gel process, the gel preform is perhaps ten times larger in diameter. While the gel preform is porous, dopants can be introduced. In this way the optical properties such as index of refraction can be modified using a low temperature process. Eventually the fiber is heated to collapse the porosity. The densification can be accomplished without contacting the fiber with any potential contaminants or sources of surface flaws. True containerless processing for these preforms is possible.

Thin Films   Gels have been applied as thin films on ceramic, metal and plastic substrates. The uses for these micron sized coatings are typically protective layers, optical filters and surface enhancers (6).

One application which incorporates the microporosity is antireflective coatings (76,77,78). There are several models for calculating the index of refraction for porous layers. By choosing the right heat treatment, a gel film can be designed with the right degree of porosity or right gradient in porosity.

In addition to the thin films which are applied to substrates, there are the free standing films or membranes (58,79,80). In these cases, the thickness is one tenth millimeter. The porosity is continuous to permit use of the membrane as a separation filter. In one case (58), the membrane is formed by pouring the solution onto a dense liquid and effecting the gelling in the atmosphere. This leads to anisotropic microstructures with macroporosity on the exposed surface.

There has been some work on using the sol-gel process to make contact lenses (79). The possibilities with these membranes are the higher oxygen permeability in the gel lens than acrylic lenses and the compatibility of gel lenses with fluids in the eye. The proper choice of the gel precursors allows the fabricated lens to be an organic-inorganic hybrid material.

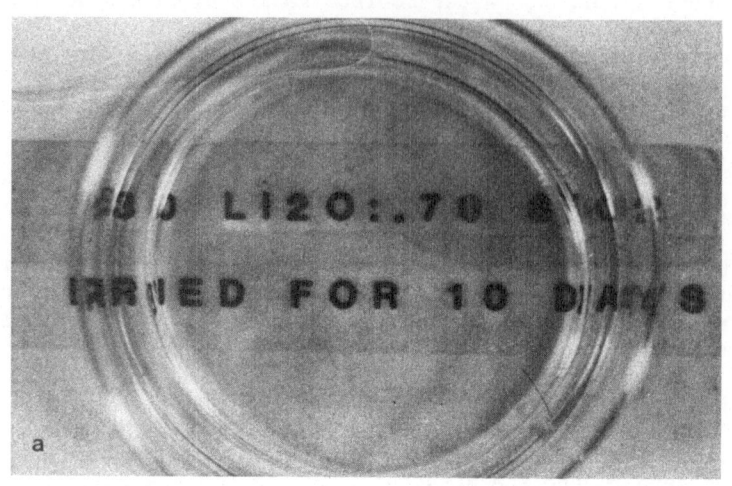

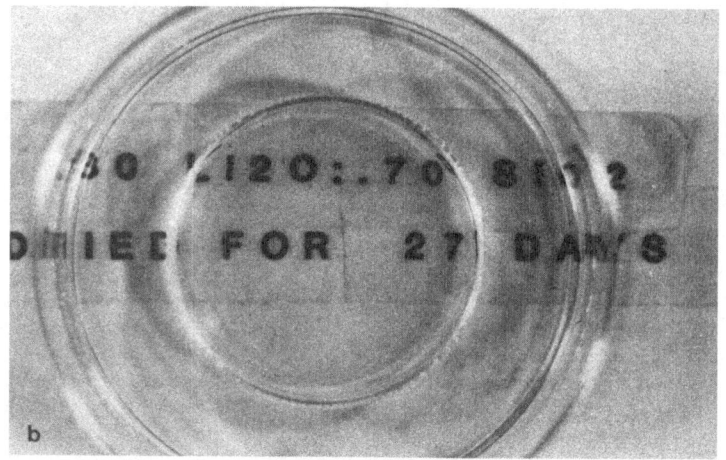

Figure 12.

Sol-gel miniature of mold.
(a) after 10 days and
(b) after 27 days.

Another organic-inorganic hybrid material has been proposed for fast-ion conductors. It is known that alkali silicates are reasonable ionic conductors (48) and that organics like polyethylene oxide (PEO) can enhance conductivity. These two concepts have been combined by using a sol-gel process to introduce lithium and PEO into a silica skeleton (81).

This is not the first time that the sol-gel process has been used for fast ion conductors. Nasicon compounds have been synthesized (82,83). These systems generally have 4 or more components. The ionic conductivity is sensitive to phase homogeneity and absence of grain boundary phase in polycrystalline samples. In this case, gel preparation ensures distribution of components and controlled stoichiometry.

Monoliths Gels have been prepared as monoliths, and the size of these castings gets larger with technological advances in drying. Monoliths can be prepared by natural evaporation (8,62) DCCA drying (21) and hypercritical evacuation (7,41).

The gel monolith can be used as a porous preform for a composite (84,85). The preform can be infiltrated with a second phase which becomes encapsulated by the gel matrix or a precursor can be introduced which is pyrolyzed in the pores.

The gel monolith can also serve as the host for organic dye lasers (86). The silica porosity is a rigid matrix into which the organic compound can be intruded. The silica does not interfere with transmission of the laser light output, nor does the handling of the silica degrade the organic compounds.

For some time, silica aerogel has been used for thermal insulation windows (53,54,87). Because the aerogel transmits high amounts of incident solar radiation, such windows are extremely good for passive solar heating. The silica skeleton in the aerogel remains intact when a vacuum is drawn on a layer sandwiched between two glass panels. These windows outperform standard so-called thermal pane double windows.

As a final application, xerogels which are the product of natural evaporation are capable of replicating all features of their mold and miniaturizing them (88). The xerogel monoliths often shrink 50% in a linear dimension, but the features are not distorted. This behavior was used to make a diffraction grating by machining the grating into the mold and then replicating and miniaturizing the pattern. Figures 12a and 12b show the magnitude of shrinkage experienced in xerogels.

SUMMARY

The applications which are mentioned here are intended to highlight aspects of the sol-gel process which influence the microstructure. First of all, it should be apparent that the chemistry of alkoxide solutions influences the microstructure. The sharpest contrast in microstructure is between acid and base catalyzed solutions. Second, the chemically generated microstructure exists in whatever preform is chosen. The preform may be a fiber, a thin film or a monolith. The overall geometry depends largely on the water content of the solution. Third, the handling of the preform during drying and densification establishes the final microstructure.

Many applications can be found for microporous silicates. Applications, such as catalyst supports, thermal insulation, or dispersed phase composites, are not new. Other processing schemes including chemical leaching and sintering have been used to obtain oxides with interconnected porosity. These processing schemes will continue to be used in the majority of cases.

In a few cases today and perhaps in an increasing number of cases in the future, a sol-gel process may be used. The reason for this is that both polycrystalline and amorphous oxides can be obtained with interconnected porosity by a sol-gel process. The porosity is finer than that obtained by chemical leaching and more uniform in cross-section than that generally found in sintered compacts. Compared to conventional powder processing, the processing equipment is simple for preparing and heat treating gels.

ACKNOWLEDGEMENTS

The author wishes to thank P. Anderson and H. de Lambilly for their drying studies, R. Donaldson for the solution viscosity measurements, T. Gallo for his beam-bending viscosimeter data, and T. Lombardi for help with texture studies, all of Rutgers University; and Visiting Scientists H. Wautier - (Solvay) and J.-Y. Chane-Ching (Rhone-Poulenc) for helpful discussions. A special thanks to Claudia Kuchinow for careful preparation of the manuscript.

Keeping the pore sizes and their uniformity in mind, there are many reasons to investigate gel textures further. An understanding of the solution chemistry and an understanding of the processing stages beyond the sol-gel transition need to be combined to be able to design microstructures in alkoxide gels.

REFERENCES

(1) Ulrich, D. R., Am. Ceram. Soc. Bull. (1985) 64, 1444-1448.
(2) Fegley, B.; Barringer, E., in "Better Ceramics through Chemistry"; Elsevier: New York, 1984, pp. 187-197.
(3) Danforth, S. C.; Velazquez, M., Advances in Ceramics (1984) 9, 105-114.
(4) Barringer, E. A.; Bowen, H. K., J. Am. Ceram. Soc. (1982) 65, C199-201.
(5) Sakka, S.; Kamiya K., J. Non-Cryst. Solids (1982) 48, 31-46.
(6) Dislich, H., Hinz, P., J. Non-Cryst. Solids (1982) 48, 11-16.
(7) Zarzycki, J.; Prassas, M.; Phalippou, J., J. Mater. Sci. (1982) 17, 3371-3379.
(8) Yamane, M.; et al., J. Mater. Sci. (1979) 14, 607-611.
(9) Matijevic, E., Prog. Coll. Poly. Sci. (1976) 61, 24-35.

(10) Rabinovich, E. M.; et al., J. Non-Cryst. Solids (1982) 47, 435-439.
(11) Scherer, G. W.; Luong, J. C., J. Non-Cryst. Solids (1984) 63, 163-172.
(12) Yoldas, B. F., J. Mater. Sci. (1979) 14, 1843-1849.
(13) Johnson, D. W., Am. Ceram. Soc. Bull. (1985) 64, 1597-1602.
(14) Sakka, S., Am. Ceram. Soc. Bull. (1985) 64, 1463-66.
(15) Zelinski, B. J.; Uhlmann, D. R. J. Phys. Chem. Solids (1984) 45, 1069-1090.
(16) Klein, L. C., Ann. Rev. Mater. Sci. (1985) 15, 227-48.
(17) Yoldas, B. E., This Volume
(18) Klein, L. C.; Garvey, G. J., in "Better Ceramics through Chemistry"; Elsevier: New York, 1984, pp. 33-39.
(19) Strawbridge, I.; Craievich, A. F.; James, P. F., J. Non-Cryst. Solids (1985) 72, 139-156.
(20) Schmidt, H.; Scholze, H.; Kaiser, A., J. Non-Cryst. Solids (1984) 63, 1-11.
(21) Wallace, S.; Hench, L. L., in "Better Ceramics through Chemistry," Elsevier: New York, 1984, pp. 47-52.
(22) Keefer, K. D., in "Better Ceramics through Chemistry," Elsevier: New York, 1984, pp. 15-24.
(23) Nogami, M.; Moriya, Y., J. Non-Cryst. Solids (1980) 37, 191-201.
(24) Brinker, C. J.; et al., J. Non-Cryst. Solids (1982) 48, 47-64.
(25) Iler, R. K., "The Chemistry of Silica," Wiley: New York, 1979.
(26) Brinker, C. J.; Scherer G. W., J. Non-Cryst. Solids (1985) 70, 301-322.
(27) Nicolaon, G. A.; Teichner, S. J., Bull. Soc. Chim. (France) (1968) 8, 3107-3113.
(28) Sakka, S.; et al., J. Non-Cryst. Solids (1984) 63, 223-235.
(29) Kamiya, K.; Sakka, S.; Mizutani, M., Yogyo-Kyokai-Shi (1978) 86, 553-559.
(30) Brinker, C. J., et al., J. Non-Cryst. Solids (1984) 63, 45-49.
(31) Schaefer, D. W.; Keefer, K. D., in "Better Ceramics through Chemistry," Elsevier: New York, 1984, pp. 1-14.
(32) LaCourse, W. C., in "Better Ceramics through Chemistry," Elsevier: New York, 1984, pp. 53-58.
(33) Mizuno, T.; Phalippou, J; Zarzycki, J., Glass Technology (1984) 26, 39-45.
(34) Kamiya, K.; Sakka, S.; Tatemichi, Y., J. Mater. Sci. (1980) 15, 1765-1771.
(35) Nogami, M., J. Non-Cryst. Solids (1985) 69, 415-423.
(36) Nogami, M., J. Am. Ceram. Soc. (1984) 67, C258-259.
(37) Nogami, M.; Tomozawa, M., J. Am. Ceram. Soc. (1986) 69, 99-102.
(38) Guglielmi, M.; Maddalena, A., J. Mater. Sci. Lett. (1985) 4, 123-124.
(39) Partlow, D. P.; Yoldas, B. E., J. Non-Cryst. Solids (1981) 46, 153-161.
(40) Garvey, G. J.; Klein, L. C., J. Physique (1982) 43C9, 271-274.
(41) Prassas, M.; Phalippou, J.; Zarzycki, J., J. Mater. Sci. (1984) 19, 1656-1665.
(42) Zarzycki, J., in "Ultrastructure Processing of Ceramics, Glasses and Composites," Wiley: New York, 1984, pp. 27-42.
(43) Rabinovich, E. M.; et al., J. Am. Ceram. Soc. (1983) 66, 683-688.
(44) Rabinovich, E. M.; et al., J. Non-Crystal. Solids (1984) 63 155-161.
(45) Krol, D. M.; van Lierop, J. G., J. Non-Crystal. Solids (1984) 63 131-144.
(46) Klein, L. C.; Nelson, C.; Higgins, K. L., in "Better Ceramics through Chemistry" Elsevier: New York, 1984, pp. 293-299.
(47) Cooper, A. R., in "Ceramic Processing Before Firing," Wiley: New York, 1960, pp.261-276.
(48) Ravaine, D., J. Non-Cryst. Solids (1985) 73, 287.

(49) Phalippou, J.; Prassas, M.; Zarzycki, J., J. Non-Cryst. Solids (1982) 48, 17-30.
(50) Brennan, J.; Prewo, K., J. Mater. Sci. (1982) 17, 2371-83.
(51) Schwartz, I.; Klein, L. C.; de Lambilly, H.; Anderson, P., to appear in J. Non-Cryst. Solids (1986).
(52) Wallace, S.; Hench, L. L., Cer. Eng. Sci. Proc. (1984) 5, 568-573.
(53) Rubin, M.; Lampert, C. M., Solar Energy Materials (1983) 7, 393-400.
(54) Caps, R.; Fricke, J., Int. J. Solar Energy (1984) 3, 13.
(55) Dumas, J; et al., J. Mater. Sci. Lett. (1984) 4, 1089-91.
(56) Lecloux, A. J., in "Catalysis-Science and Technology, Vol. 2," Springer-Verlag: New York, 1981, pp. 172-230.
(57) Kaiser, A.; Schmidt, H., J. Non-Cryst. Solids (1984) 63, 261-271.
(58) Gallagher, D.; Klein, L. C., J. Colloid Interface Sci. (1986) 101, 40-45.
(59) Carturan, G.; et al., J. Non-Cryst. Solids (1984) 63, 273-281.
(60) Hoffman, D.; Komarneni, S.; Roy, R., J. Mater. Sci. Lett. (1984) 3, 439-442.
(61) Brinker, C. J.; Scherer, G. W.; Roth, E. P., J. Non-Cryst. Solids (1985) 72, 345-368.
(62) Klein, L. C.; Gallo, T. A.; Garvey, G. J., J. Non-Cryst. Solids (1984) 63, 23-33.
(63) Scherer, G. W., J. Am. Ceram. Soc., (1977) 60, 236-239.
(64) Scherer, G. W.; Brinker, C. J.; Roth, E. P., J. Non-Cryst. Solids (1985) 72, 369-389.
(65) Brinker, C. J.; Roth, E. P.; Scherer, G. W.; Tallant, D. R., J. Non-Cryst. Solids (1985) 71, 171-185.
(66) Gallo, T. A.; Brinker, C. J.; Klein, L. C.; Scherer, G. W., in "Better Ceramics through Chemistry," (Elsevier: New York) 1984, pp. 85-90.
(67) Satoh, S.; Susa, K.; Matsuyama, I.; Suganuma, T., J. Am. Ceram. Soc., (1985) 68, 399-402.
(68) Gallo, T. A.; Klein, L. C., "Apparent viscosity of sol-gel processed silica," to appear in J. Non-Cryst. Solids (1986).
(69) Gallo, T. A.; Klein, L. C., "Calculated vs. measured viscosity of sol-gel processed silica," to appear in "Better Ceramics through Chemistry II," (Elsevier: New York) 1986.
(70) Rabinovich, E. M., J. Mater. Sci., (1985) 20, 4259-4297.
(71) Scherer, G. W.; Brinker, C. J.; Roth, E. P., "Structural relaxation in gel-derived glasses," to appear in J. Non-Cryst. Solids (1986).
(72) Benjamin, R. F.; Mayer, E. J.; Maynard, R. L., SPIE (1984) 506, 20.
(73) Harmer, A. C.; Puyane, R.; Gonzalez-Oliver, C., IFOC (1982) Nov./Dec., 40-44.
(74) Nogami, M.; Hayakawa, J.; Moriya, Y., J. Mater. Sci., (1982) 17, 2845-2849.
(75) Downs, R. L.; Ebner, M. A.; Homyk, B. D.; Nolen, R. L., J. Vac. Sci. Technol. (1981) 18 1272-1275.
(76) Brinker, C. J.; Harrington, M. S.; Solar Energy Materials (1981) 5, 159-172.
(77) Mukherjee, S. P.; Lowdermilk, W. H., J. Non-Cryst. Solids (1982) 48, 177-184.
(78) Yoldas, B., Applied Optics (1980) 19, 1425-9.
(79) Schmidt, H.; Philipp, G., J. Non-Cryst. Solids (1984) 63, 283-292.
(80) Kaiser, A.; Schmidt, H.; Bottner, H., J. Membrane Sci., (1985) 22, 257-268.
(81) Ravaine, D.; Seminel, A.; Charbouillot, Y.; Vincens, M., "A new family of organically modified silicates prepared from gels," to appear in J. Non-Cryst. Solids (1986).
(82) Boilot, J. P.; Colomban, Ph., J. Mater. Sci. Lett., (1985) 4, 22-24.

(83) Perthius, H.; Colomban, Ph., <u>Mater. Res. Bull.</u>, (1984) <u>19</u>, 621-631.

(84) Hoffman, D. W.; Roy, R.; Komarneni, S., <u>J. Am. Ceram. Soc.</u>, (1984) <u>67</u>, 468-471.

(85) Chi, F. K., <u>Cer. Eng. Sci. Proc.</u>, (1983) <u>4</u>, 704-717.

(86) Avnir, D.; Levy, D.; Reisfeld, R., <u>J. Phys. Chem.</u>, (1984) <u>88</u>, 5956-5959.

(87) Henning, S.; Svensson, L., <u>Physica Scripta</u> (1981) <u>23</u>, 697-702.

(88) Ohno, M.; Yamada, T.; Kurokawa, T., <u>J. Appl. Phys.</u> (1985) <u>57</u>, 2951-2952.

INORGANIC MACROMOLECULES AND THE SEARCH FOR NEW ELECTROACTIVE AND STRUCTURAL MATERIALS

Harry R. Allcock

Department of Chemistry
The Pennsylvania State University
University Park, Pennsylvania 16802

GENERAL PRINCIPLES

Modern science and technology form a continuum of disciplines that extends from theoretical chemistry and physics on the one hand to medicine and engineering on the other. An understanding of molecular structure and reactivity is essential for progress in all these disciplines. Within this continuum it is possible to discern three distinct areas of chemical science, areas that are based on common principles, but require the use of different experimental techniques and different conceptual approaches. The three areas are (1) the chemistry of small molecules, (2) the chemistry of macromolecules, and (3) the chemistry of solids and their surfaces.

Small-molecule chemistry deals with the structure and reactions of molecules that contain from two to perhaps 500 atoms. It is an area that has traditionally comprised the intellectual core of the discipline. Small molecules are relatively easy to synthesize, characterize, and examine by spectroscopic and diffraction techniques. The study of small molecule chemistry is an excellent approach to training chemists at the undergraduate and introductory graduate levels. Yet, increasingly, the focus of frontier research is shifting toward more complex molecular systems.

The second large area of chemistry is based on the science of macro-molecules (polymers), both in the form of petrochemical-based synthetic polymers and their biological counterparts (1). Synthetic polymer chemistry alone is a vast, still expanding field, which underlies major segments of the research and technology effort in the developed countries. However, the fundamental scientific understanding of macromolecules is at a much less advanced level than that of small molecules, mainly because absolute molecular structural information is exceedingly difficult to obtain for polymers. Thus, fundamental investigations in this field rely heavily on data obtained for small molecules.

The third area of chemical-based science is the chemistry of solids and their surfaces. This field includes such diverse but converging disciplines as semiconductor and superconductor research, metallurgy, ceramic science, heterogeneous catalysis, and the materials science of structural solids.

Traditionally, polymer chemistry has developed as the science of organic macromolecules. This is because of the ready availability of large quantities of inexpensive monomers from the petrochemicals industry, coupled with the instinctive involvement of organic chemists in polymer synthesis during the past 40 years. On the other hand, solid state science is based mainly on inorganic chemistry. Until recently, there has been very little exchange of ideas between polymer chemists and solid state scientists.

Our interest is to explore the territory that lies between organic polymer chemistry and inorganic solid state science (2). The immediate objective is the synthesis of long chain macromolecules that have an inorganic backbone structure, with organic or organometallic side groups attached to that backbone. It is anticipated that such hybrid inorganic-organic macromolecules will combine the properties of organic high polymers (flexibility, elasticity, toughness, light weight, and ease of fabrication) with those of inorganic solids (stability at high temperatures, resistance to burning, electrical conductivity, and biocompatibility).

The synthesis of inorganic backbone polymers generally requires a different approach from the traditional synthesis of petrochemical-based organic polymers. Most conventional polymers are prepared by the polymerization of different organic monomers--different vinyl compounds, diacids, diamines, amino acids, diols, or cyclic molecules such as caprolactam or ethylene oxide. Few inorganic polymers can be assembled in this way because unsaturated monomers are rare in inorganic chemistry, and condensation processes can frequently be reversed in the presence of moisture. Instead, in our program, we have relied on a non-traditional method of synthesis. In this, a reactive, high polymeric intermediate is first prepared (usually by a ring-opening polymerization). This intermediate is then used as a substrate for the replacement of the reactive (inorganic) side groups by unreactive organic groups. The method is illustrated in Scheme I.

Scheme I

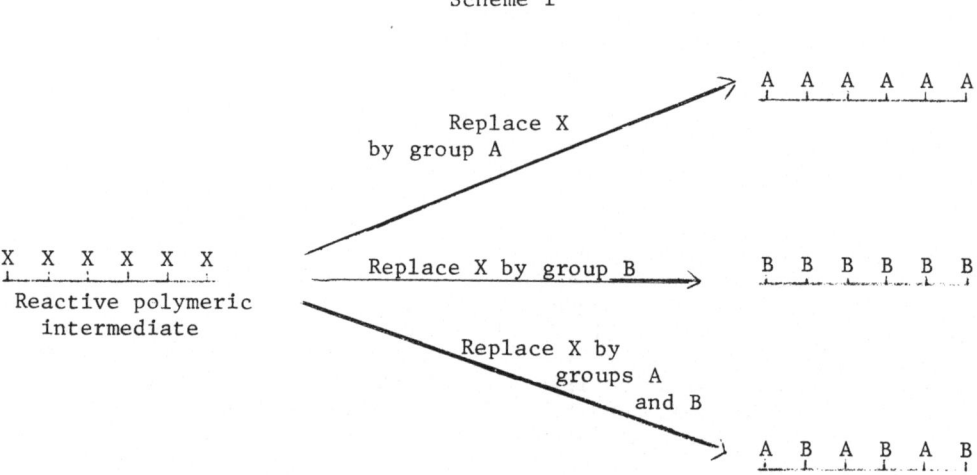

The advantage of this approach is that, once the polymer backbone has been assembled, many different types of side groups can be introduced to give either single-substituent or complex, mixed substituent polymers. Because some of the most important physical and chemical properties of a polymer depend on the types of side groups present, this method allows both broad and subtle tailoring of the properties to be achieved.

Because of this substitutive approach, it is essential to explore such reactions first at the small-molecule level. Small-molecule reactions are nearly always easier to monitor than their macromolecular counterparts. Moreover, molecular characterization, mechanistic data, and detailed structural information can be obtained much more easily at the small-molecule level. This information can then be used as an aid to the synthesis and structural characterization of the corresponding high polymers. Unless this procedure is followed, progress at the high polymer level is likely to be slow. A fundamental question that emerges from this approach is this: "What are the differences in reactivity, bond angles, bond lengths, and electronic behavior between small-molecules and their linear or macrocyclic, high polymeric counterparts?" Answering this question is one of the most interesting aspects of this field.

Molecular Structure and Macromolecular Properties

Polymers have unique properties because they possess long, filamentous molecular structures. Solids comprised of small-molecules are brittle because the structure is held together only by weak van der Waals or dipolar forces. The forces between polymer molecules are also weak, but the strength along the chain axis is high because covalent forces are involved. Thus, the solid state character of macromolecules is dominated by properties that result from the superimposition of one-dimensional order (in a structural rather than a conformational sense) on the essential fluidity or deformability of the intermolecular, van der Waals interactions. Chain entanglement, coupled with molecular chain flexibility, gives rise to material flexibility or elasticity. Crosslinks between chains stabilize the overall bulk geometry while still allowing thermal motion of the segments between crosslinks. Increasing crosslink density alters the properties of the polymer toward those of a ceramic.

Electrical conductivity in polymers can occur by two mechanisms that superficially resemble those in classical inorganic solids (Figure 1).

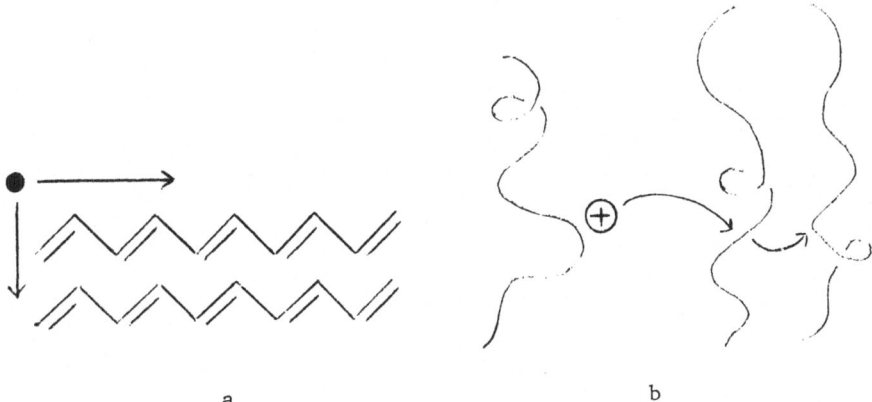

a   b

Fig. 1.   (a) Electronic conduction along polyconjugated polymer chains or between chains.   (b) Transport of cations, loosely bound to polymer molecules by transfer from chain to chain under the influence of thermally-induced chain motions.

Electronic unsaturation in a polymer chain backbone, within a highly crystalline matrix, can give rise to electronic conductivity (Figure 1a). Polyacetylene and poly(sulfur nitride) are well-known examples of such "covalent metals." Electron transfer between chains almost certainly contributes to this conductivity. However, electrical conductivity in a solid polymer can also take place by an electrolytic mechanism if the polymer possesses a highly mobile chain and can serve as a solid solvent for salts. Poly(ethylene oxide) is the classic example of a polymeric solid electrolyte. Both electronic and electrolytic conduction in solid polymers is of interest for the development of new semiconductors and rechargeable batteries.

How does the incorporation of inorganic elements into a macromolecular chain affect physical properties such as elasticity, flexibility, stability, or electrical conductivity? First, bonds between inorganic elements generally have different lengths, strengths, dipolar character, and torsional barriers from those in organic molecules. The bond angles may be different also, and a greater angular variability may be tolerated before a significant energy barrier is encountered. The electronic transmission properties of inorganic bonds may be different from those of C-C, C-O, or C-N bonds. Many inorganic elements are more resistant to oxidation than is carbon.

In principle, nearly every element in the Periodic Table, except for the lighter rare gases, could be incorporated into macromolecular structures. In practice, the Main Group elements of Group 3 (13), 4 (14), 5 (15), and 6 (16) offer the most promise for catenation or heteroatom polymer formation. Transition metal elements form coordination polymers in the presence of organic linkage groups.

Developing Areas of Inorganic Polymer Chemistry

Nine different areas of inorganic macromolecular chemistry have been studied in some detail. These are depicted as I-IX.

I (X = Na, K, C_2H_5, or H)

II

III

IV

V

VI

VII

VIII

IX

70

Of these, the polyphosphazenes (IX) are chemically the most diverse, and these will be considered separately in the following section. The others will be mentioned here briefly to provide perspective.

Mineralogical silicates, modified by the use of high temperatures, form the basis of classical ceramic science and technology. However, in recent years attempts have been made to prepare synthetic silicates in the laboratory from silicate esters by the so-called "sol-gel" process. In this approach, a monomer such as tetraethyl silicate (X) is hydrolyzed to generate various ethyl silicic acid species (XI). These condense to form a three-dimensional, expanded silicate gel structure (XII). Such gels can be shaped and then condensed further to yield an inflexible ceramic.

$$
\begin{array}{ccc}
\begin{array}{c}
OC_2H_5 \\
| \\
C_2H_5O - Si - OC_2H_5 \\
| \\
OC_2H_5
\end{array}
&
\xrightarrow[-C_2H_5OH]{H_2O}
&
\begin{array}{c}
OH \\
| \\
HO - Si - OH \\
| \\
OC_2H_5
\end{array}
\xrightarrow[-H_2O]{}
\begin{array}{c}
| \\
O \\
| \\
- O - Si - \\
| \\
O \\
| \\
- O - Si - \\
| \\
OC_2H_5
\end{array}
\end{array}
$$

$$\qquad X \qquad\qquad\qquad XI \qquad\qquad\qquad XII$$

In principle, other inorganic hydroxides can copolymerize in the same system so that, for example, aluminosilicates could be prepared in the same way. Inorganic polymers of this type form one extreme of the spectrum that extends from totally inorganic materials to the borderline of organic polymer chemistry.

Poly(organosiloxanes) (silicones) (II) represent the other extreme. These highly flexible polymers are synthesized by the ring opening polymerization of organocyclosiloxanes, such as $[OSi(CH_3)_2]_4$. The property differences between poly(dimethylsiloxane), for example, and sol-gel-derived silicates can be traced to the highly crosslinked structure of the silicate, which is a consequence of the tetra-functionality of silicon in $Si(OH)_4$, in contrast to the linear or macrocyclic structure of the poly-(organosiloxane). Moreover, the methyl side groups in II confer a host of "organic"-type properties to the molecules, superimposed on the stability and unique torsional flexibility of the linear siloxane chain.

Polysilazanes (III) are discussed elsewhere in this volume. Here, it is sufficient to point out that polysilazanes show many of the normal properties of short-chain polymers. When crosslinked, they develop ceramic-type character, ultimately yielding silicon nitride.

A similar situation exists with polysilanes (IV). Polymerization and pyrolysis of cyclic organosilanes, such as $[Si(CH_3)_2]_6$, yields first carbosilanes of approximate formula, $[Si(CH_3)(H)-CH_2]_n$, and these (after fabrication) can be pyrolyzed to silicon carbide ceramics. In silicon nitride and silicon carbide formation from polymers, the key requirements are (1) side groups that are reactive at high temperatures to form cross-links, and (2) a rate of crosslinking that is faster than the thermal breakdown of the inorganic chain to form, for example, cyclic oligomers.

Polymeric sulfur (V) is well-known, as is the behavior of poly(sulfur nitride) (VI) as a metallic, electronic conductor and superconductor. Transition metal coordination polymers (VII) have been known for a number of years, but compounds of this type tend to be insoluble and of low molecular weight. Doped polyphthalocyanines (VIII) are in the early stage of development as electrical conductors.

Polyphosphazenes

General Characteristics   Chemically and structurally, the poly-(organophosphazenes (IX) are the most advanced inorganic-organic macro-molecules. In spite of their recent origin, at least 300 different types of polyphosphazenes are known, some of which are already in use as technological materials (2-9).

Polyphosphazene chemistry epitomizes the principles discussed earlier-- an inorganic polymer backbone to which are connected organic side groups, and a synthesis route based on a ring-opening polymerization followed by a wide range of macromolecular substitution reactions. The macromolecular substitution process gives rise to polymers that range in properties from elastomers and thermoplastics to ceramics. Individual polymers are electrolytic conductors, semiconductors, biologically-active, biocompatible, or useful as membrane materials.

The fundamental synthesis route, discovered by us (10-12), is illustrated in Scheme II. Hexachlorocyclotriphosphazene (XIII), prepared from phosphorus pentachloride and ammonium chloride, is polymerized thermally to poly(dichlorophosphazene) (XIV). This rubbery high polymer dissolves in anhydrous organic solvents, and can then be treated with organic or organometallic nucleophiles to replace the chlorine atoms by organic or organometallic groups. The availability of a large number of suitable nucleophiles allows access to a broad range of different polymers. Different substituent groups generate different properties, and the range of different polymer characteristics rivals that of conventional organic polymers.

For example, methoxy, ethoxy, and mixed substituent CF_3CH_2O- and $HCF_2(CF_2)_2CH_2O-$ side groups give rise to elastomeric character, with low glass transition temperatures (-60°C to -90°C). Side groups, such as CF_3CH_2O or C_6H_5O, generate flexible, film- or fiber-forming polymers. Solubility in water occurs when the side groups are CH_3NH- or glucose residues (13,14). Biodegradation to non-toxic small molecules takes place when the side groups are amino acid ester units, such as $C_2H_5OOC-CH_2NH-$ (15,16). In addition, substitution chemistry can be carried out on the organic side groups themselves without affecting the inorganic chain, and this method has been employed to attach bioactive agents, such as antibacterial agents or enzymes (17-22) to the polymer, or to provide sites for coordination to transition metals (23,24).

A number of review articles are available that summarize the chemistry of poly(organophosphazenes) (2-9) and only two aspects will be mentioned here--the relationship between molecular structure and properties, and the use of these polymers to develop electroactive materials.

Polyphosphazene Molecular Structure and Properties   Why is one inorganic polymer a rubbery elastomer, whereas another is a glass at room temperature? What determines if a given polymer will depolymerize to small molecule cyclic oligomers at elevated temperatures or undergo side group condensation, crosslinking, and ceramicization? What factors determine if a certain polymer will form a highly crystalline matrix, while another will be totally amorphous?

Scheme II

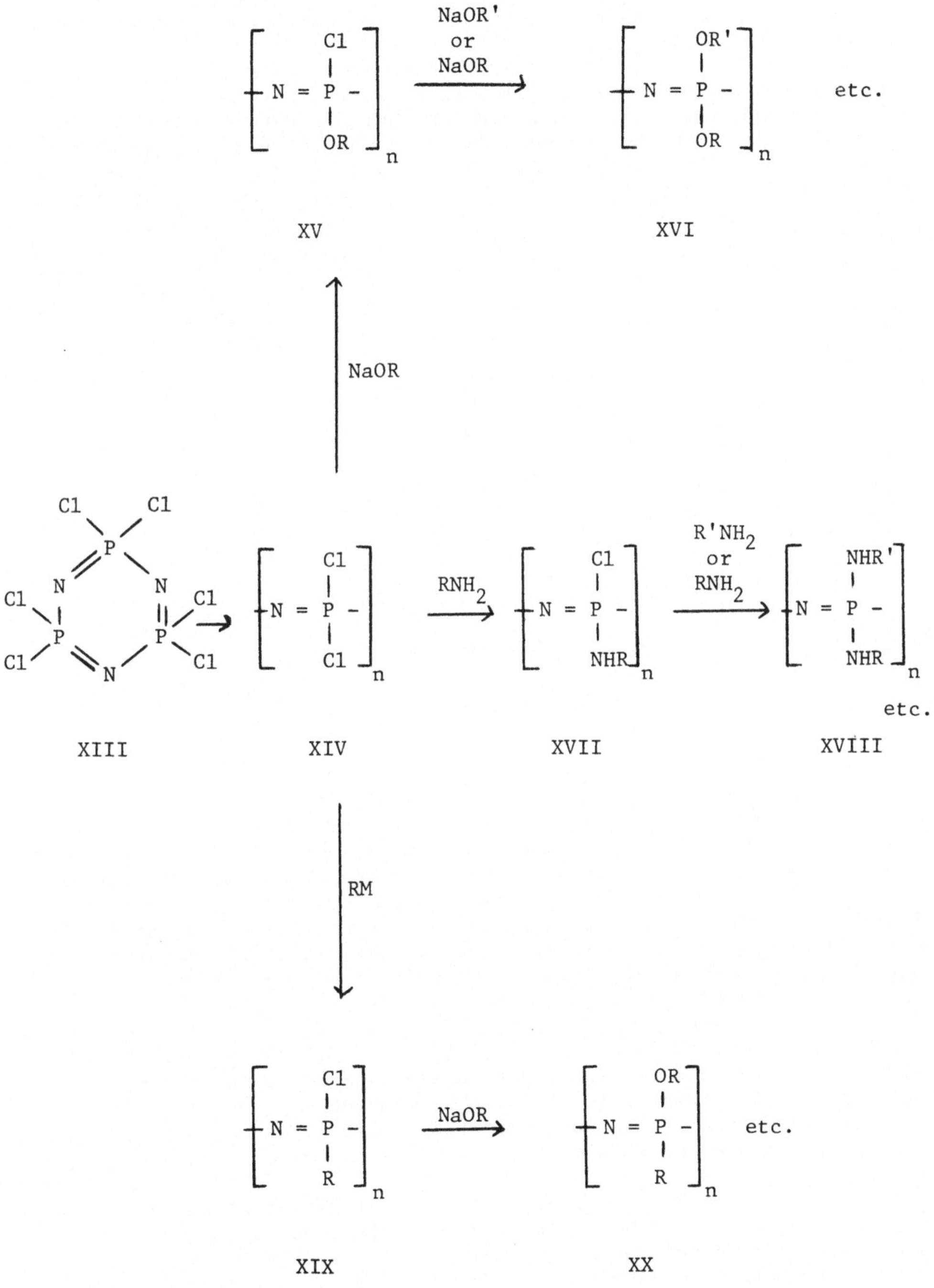

$$\left[\begin{array}{c} Cl \\ | \\ N = P - \\ | \\ OR \end{array}\right]_n \xrightarrow{\begin{array}{c} NaOR' \\ or \\ NaOR \end{array}} \left[\begin{array}{c} OR' \\ | \\ N = P - \\ | \\ OR \end{array}\right]_n \quad etc.$$

$$XV \qquad\qquad\qquad\qquad XVI$$

$$\uparrow\ NaOR$$

XIII: cyclotriphosphazene $N_3P_3Cl_6$

$$\xrightarrow{}\ \left[\begin{array}{c} Cl \\ | \\ N = P - \\ | \\ Cl \end{array}\right]_n \xrightarrow{RNH_2} \left[\begin{array}{c} Cl \\ | \\ N = P - \\ | \\ NHR \end{array}\right]_n \xrightarrow{\begin{array}{c} R'NH_2 \\ or \\ RNH_2 \end{array}} \left[\begin{array}{c} NHR' \\ | \\ N = P - \\ | \\ NHR \end{array}\right]_n$$

$$\qquad\qquad etc.$$

$$XIV \qquad\qquad XVII \qquad\qquad XVIII$$

$$\downarrow\ RM$$

$$\left[\begin{array}{c} Cl \\ | \\ N = P - \\ | \\ R \end{array}\right]_n \xrightarrow{NaOR} \left[\begin{array}{c} OR \\ | \\ N = P - \\ | \\ R \end{array}\right]_n \quad etc.$$

$$XIX \qquad\qquad\qquad\qquad XX$$

$n \simeq 15{,}000$ (molecular weights in the 1×10^6 to 1×10^7 range)

We have attempted to answer questions like these parallel with our synthetic work. Poly(organophosphazenes) are almost unique in the sense that many different types of polymers can be prepared by substitution from one backbone structure. Thus, the side groups can be varied over a wide range, while the average chain length and molecular weight distribution remains constant.

Molecular structural information is unavailable for most inorganic polymer systems, including polyphosphazenes. Remember that conventional single crystal X-ray diffraction analysis cannot be used to solve the structures of most synthetic polymers. Instead, fiber diffraction methods must be employed to determine polymer chain (c-axis) repeating distances, together with the number of monomer repeat units per turn of the helix. We have used fiber diffraction methods to show that, for polyphosphazenes such as $(NPF_2)_n$, $(NPCl_2)_n$, $[NP(OCH_2CF_3)_2]_n$, and $[NP(OC_6H_5)_2]_n$, the conformational repeat distance is close to 4.9 Å and that two monomer residues occupy one repeat of the helix (10,25-31). The chain conformation for all four polymers at room temperature appears to be that of a "cis-trans planar" arrangement (XXI).

IXX

The results of NMR and thermal analysis (Tg) measurements are compatible with the conclusions that (a) the phosphazene backbone has one of the highest torsional mobilities of any known polymer, and (b) that decreases in chain flexibility result from steric hindrance and polar interactions between the side groups. The cis-trans planar conformational structure minimizes intramolecular steric repulsions. Excessive steric hindrance between the side groups lowers the thermodynamic ceiling temperature and results in polymer depolymerization to cyclic trimer or tetramer at moderate temperatures (44,45).

Polyphosphazene chains can be crosslinked by reactions with diols or diamines during the polymer substitution step, or by coordination to transition metals. Pyrolysis of aminophosphazene high polymers results in the loss of amines and formation of the ceramic-like phospham or phosphorus-nitride precursors.

Electroactive Polyphosphazenes   Most polyphosphazenes do not conduct electricity, in spite of the presence of the unsaturated backbone structure. This behavior can be rationalized in terms of an "island"-type pi-bonding model (XXII) (34), coupled with the absence of free radicals, and the lack of extensive bulk crystallinity that might facilitate intramolecular electron transfer.

d_π-p_π bond

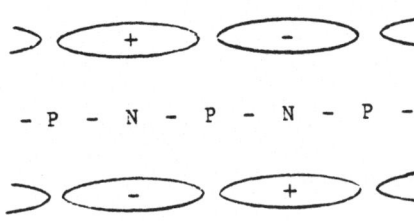

XXII

However, certain side groups attached to a phosphazene chain generate electrical conductivity.   For example, quaternized polyphosphazenes form salt-type adducts with tetracyanoquinodimethane (TCNQ), and such polymers are semiconductors (35).   A similar effect has been detected when copper phthalocyanine units are covalently bonded to a polyphosphazene chain and then doped with iodine (36).   Presumably, the chain flexibility allows the planar metallophthalocyanine units to form quasi-crystalline stacks that are responsible for the electroactivity.   Ferrocenylphosphazene high polymers, such as XXIII, show weak semiconductivity when doped with iodine (37).

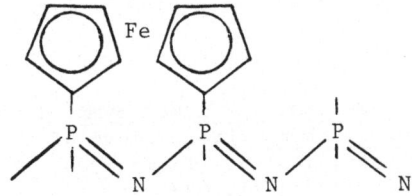

XXIII

We are currently exploring synthesis routes to polyphosphazenes in which transition metals are linked directly to the phosphazene chain.   The compounds shown in XXIV–XXVI are prototypes at the small-molecule level (38–44).

Finally, several alkoxyether-substituted polyphosphazenes, such as XXVII, function as solid electrolytes (45).   Such polymers are elastomers at room temperature.   They behave as solid solvents for salts, such as lithium or sodium triflate.   In a collaborative program with D. F. Shriver's group at Northwestern University, we have found that polymer XVII with

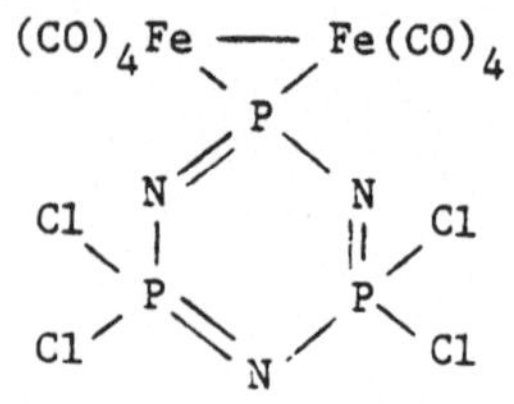

XXIV

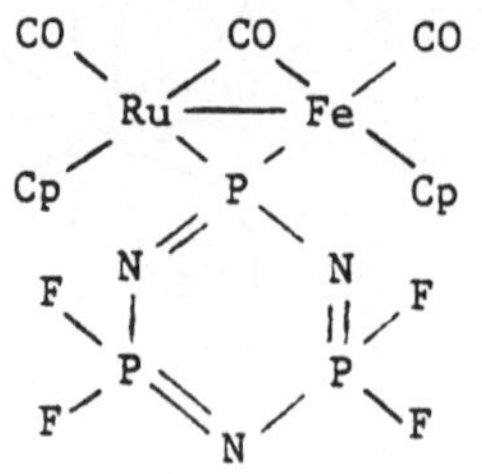

XXV

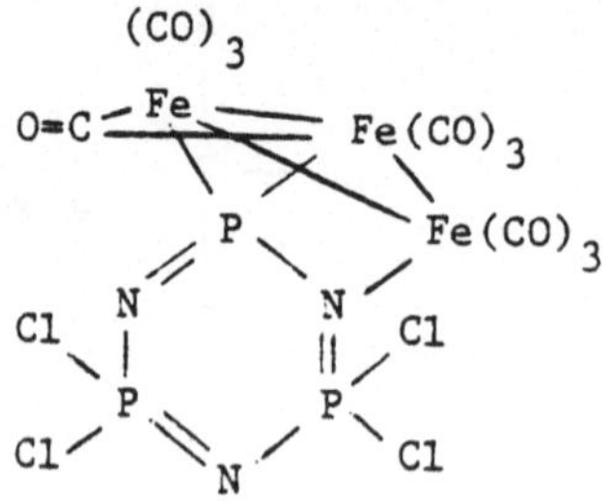

XXVI

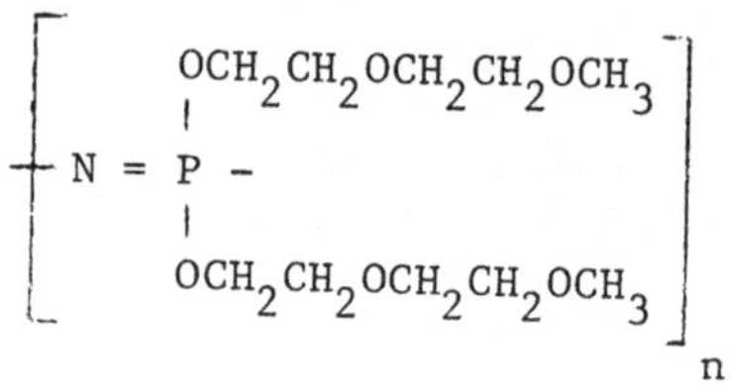

$$\left[-N = P - \begin{array}{c} OCH_2CH_2OCH_2CH_2OCH_3 \\ | \\ | \\ OCH_2CH_2OCH_2CH_2OCH_3 \end{array} \right]_n$$

XXVII

silver triflate has a higher room temperature conductivity (three orders of magnitude greater) than any previously studied polymeric electrolyte system. Thus, this polymer is a promising candidate for use in rechargeable lithium batteries.

References

(1)   Allcock, H. R.; Lampe, F. W. "Contemporary Polymer Chemistry," Prentice-Hall:Englewood Cliffs, New Jersey, 1981.
(2)   Allcock, H. R. "Inorganic Macromolecules," Chem. & Eng. News (1985) 63, 22.
(3)   Allcock, H. R. "Phosphorus-Nitrogen Compounds," Academic Press: New York, 1972.
(4)   Allcock, H. R. J. Polymer Sci., Polymer Symp. (1983) 70, 71.
(5)   Allcock, H. R. Makromol. Chem. (Proceedings of the IUPAC Conf. on Macromolecules, Florence, Italy), (1981) Suppl. 4, 3.
(6)   Allcock, H. R. Polymer (1980) 21, 673.
(7)   Allcock, H. R. Accounts of Chem. Res. (1979) 12, 351.
(8)   Singler, R. E.; Hagnauer, G. L.; Sicka, R. W. ACS Symp. Ser. (1984) 260 (J. C. Arthur, Ed.), 143.
(9)   Tate, D. P. J. Polymer Sci. Symp. (1974) 48, 33.
(10)   Allcock, H. R.; Kugel, R. L. J. Am Chem. Soc. (1965) 87, 4216.
(11)   Allcock, H. R.; Kugel, R. L.; Valan, K. J. Inorg. Chem. (1966), 5, 1709.
(12)   Allcock, H. R.; Kugel, R. L. Inorg. Chem. (1966), 5, 1716.
(13)   Allcock, H. R.; Cook, W. J.; Mack, D. P. Inorg. Chem. (1972) 11, 2584.
(14)   Allcock, H. R.; Scopelianos, A. G. Macromolecules (1983) 16, 715.
(15)   Allcock, H. R.; Fuller, T. J.; Mack, D. P.; Matsumura, K.; Smeltz, K. M. Macromolecules (1977) 10, 824.
(16)   Allcock, H. R.; Fuller, T. J.; Matsumura, K. Inorg. Chem. (1982) 21, 515.

(17) Allcock, H. R.; Austin, P. E. Macromolecules (1981) 14, 1616.
(18) Allcock, H. R.; Austin, P. E.; Rakowsky, T. F. Macromolecules (1981) 14, 1622.
(19) Allcock, H. R.; Austin, P. E.; Neenan, T. X. Macromolecules (1982) 15, 689.
(20) Allcock, H. R.; Neenan, T. X.; Kossa, W. C. Macromolecules (1982) 15, 693.
(21) Neenan, T. X.; Allcock, H. R. Biomaterials (1982), 3, 2, 78.
(22) Allcock, H. R.; Hymer, W. C.; Austin, P. E. Macromolecules (1983) 16, 1401.
(23) Allcock, H. R.; Lavin, K. D.; Tollefson, N. M.; Evans, T. L. Organometallics (1983) 2, 267.
(24) Dubois, R. A.; Garrou, P. E.; Lavin, K. D.; Allcock, H. R. Organometallics (1984) 3, 649.
(25) Allcock, H. R.; Konopski, G. F.; Kugel, R. L.; Stroh, E. G. J. Chem. Soc., Chem. Commun. (1970) 16, 985.
(26) Allcock, H. R.; Kugel, R. L.; Stroh, E. G. Inorg. Chem. (1972) 11, 1120.
(27) Allcock, H. R.; Allen, R. W.; Meister, J. J. Macromolecules (1976) 9, 950.
(28) Allen, R. W.; Allcock, H. R. Macromolecules (1976) 9, 956.
(29) Allcock, H. R.; Arcus, R. A. Macromolecules (1979) 12, 1130.
(30) Allcock, H. R.; Arcus, R. A.; Stroh, E. G. Macromolecules (1980) 13, 919.
(31) Allcock, H. R.; Tollefson, N. M.; Arcus, R. A.; Whittle, R. R. J. Am. Chem. Soc. (1985) 107, 5166.
(32) Allcock, H. R.; Cook, W. J. Macromolecules (1974) 7, 284.
(33) Allcock, H. R.; Moore, G. Y.; Cook, W. J. Macromolecules (1974) 7, 571.
(34) Dewar, M. J. S.; Lucken, E. A. C.; Whitehead, M. A. J. Chem. Soc., London, (1964) 5385.
(35) Allcock, H. R.; Levin, M. L.; Austin, P. E. Inorg. Chem. (in press).
(36) Allcock, H. R.; Neenan, T. X. Macromolecules (in press).
(37) Allcock, H. R.; Riding, G. H.; Lavin, K. D. Macromolecules (in press).
(38) Suszko, P. R.; Whittle, R. R.; Allcock, H. R. J. Chem. Soc. Chem. Commun. (1982) 960.
(39) Suszko, P. R.; Whittle, R. R.; Allcock, H. R. J. Chem. Soc., Chem. Commun. (1982) 649.
(40) Allcock, H. R.; Wagner, L. J.; Levin, M. L. J. Am. Chem. Soc. (1983) 105, 1321.
(41) Nissan, R. A.; Connolly, M. S.; Mirabelli, M. G. M.; Whittle, R. R.; Allcock, H. R. J. Chem. Soc., Chem. Commun. (1983) 822.
(42) Allcock H. R.; Riding, G. H.; Whittle, R. R. J. Am. Chem. Soc. (1984) 106, 5561.
(43) Allcock, H. R.; Suszko, P. R.; Wagner, L. J.; Whittle, R. R.; Boso, B. J. Am. Chem. Soc. (1984) 106, 4966.
(44) Allcock, H. R.; Suszko, P. R.; Wagner, L. J.; Whittle, R. R.; Boso, B. Organometallics (1985) 4, 446.
(45) Blonsky, P. M.; Shriver, D. F.; Austin, P. E.; Allcock, H. R. J. Am. Chem. Soc. (1984) 106, 6854.

THE PRECERAMIC POLYMER ROUTE TO SILICON-CONTAINING CERAMICS

Dietmar Seyferth and Yuan-Fu Yu

Department of Chemistry
Massachusetts Institute of Technology
Cambridge, Massachusetts 02139

INTRODUCTION AND GENERAL COMMENTS

Silicon-containing ceramics include the oxide materials, silica and the silicates; the binary compounds of silicon with non-metals, principally silicon carbide and silicon nitride, silicon oxynitride and the sialons; main group and transition metal silicides, and, finally, elemental silicon itself. While there is vigorous research activity throughout the world on the preparation of all of these classes of solid silicon compounds by the newer preparative techniques, in this report we will focus our attention on silicon carbide and silicon nitride.

Silicon carbide, SiC (1) and silicon nitride, Si_3N_4 (2), have been known for some time. Their properties (Tables 1 and 2), especially their high thermal and chemical stability, their hardness, their high strength, as well as other properties have led to useful applications for both of these materials. Silicon carbide has been an article of commerce since the development of the Acheson process for its manufacture just before the turn of the century, but silicon nitride is a relative newcomer as far as commercial utilization goes (3). Its modern applications include: gas turbine parts, rocket housings, electric insulators, thermocouple tubes, high temperature bearings, laser nozzles, and thin films in microelectronics.

The "conventional" methods for the preparation of SiC and Si_3N_4, the high temperature reaction of fine grade sand and coke (with additions of sawdust and NaCl) in an electric furnace (the Acheson process) for the former and usually the reaction of silicon tetrachloride with ammonia (in the gas phase or in solution) for the latter, do not involve soluble or fusible intermediates. For many applications of these materials this is not necessarily a disadvantage (e.g., for the application of SiC as an abrasive), but for some of the more recent desired applications soluble or fusible (i.e., processable) intermediates are required.

Table 1. Properties of Silicon Carbide

High thermal stability:
  decomposition temperature
        2985°C (cubic form)
        2825°C (α form)
High oxidation resistance
Great hardness (between 9 (Al_2O_3) and
    10 (diamond) on the Moh scale
High thermal conductivity
Low thermal expansion
High corrosion resistance
High erosion resistance
High temperature semi-conductor

Table 2. Properties of Silicon Nitride

High thermal stability (to 1800°C in an
    inert atmosphere)
High oxidative stability (to 1500°C in air)
Great hardness
Excellent thermal shock and creep resistance
High strength at elevated temperatures
Good corrosion resistance
Low coefficient of thermal expansion
Low electrical conductivity

The need for soluble or fusible precursors whose pyrolysis will give the desired ceramic material has led to a new area of macromolecular science, that of preceramic polymers (4). Such polymers are needed for a number of different applications. Ceramic powders by themselves are difficult to form into bulk bodies of complex shape. Although ceramists have addressed this problem using the more conventional ceramics techniques with some success, preceramic polymers could, in principle, serve in such applications, either as the sole material from which the shaped body is made or as a binder for the ceramic powder from which the shaped body is to be made. In either case, pyrolysis of the green body would then convert the polymer to a ceramic material, hopefully of the desired composition. In the latter alternative, shrinkage during pyrolysis should not be great, but when the green body is made entirely of preceramic polymer, shrinkage on pyrolysis could be considerable.

Ceramic fibers of diverse chemical compositions are sought for application in the production of metal-, ceramic-, glass- and polymer-matrix composites (5). The presence of such ceramic fibers in a matrix, provided they have the right length-to-diameter ratio and are distributed uniformly throughout the matrix, can result in very considerable increases in the strength (i.e., fracture toughness) of the resulting material. To prepare such ceramic fibers, a suitable polymeric precursor is needed which can be spun (by melt-spinning, dry-spinning, or wet-spinning techniques) to fibers which then can be pyrolyzed (with or without a prior cure step).

Some materials with otherwise very useful properties such as high thermal stability and great strength and toughness are unstable with respect to oxidation at high temperatures. A notable example of such a class of materials is that of the carbon-carbon composites. If these materials could be protected against oxidation by infiltration of their pores and the effective coating of their surface by a polymer whose pyrolysis gives an oxidation-resistant ceramic material, then one would have available new dimensions of applicability of such carbon-carbon composite materials.

These are three of the more important potential areas of applicability of preceramic polymers. There are others and it is clear that in this new approach to ceramics we have a major new research area which should be very fruitful. In order to have a _useful_ preceramic polymer, considerations of structure and reactivity are of paramount importance. Not every inorganic or organometallic polymer will be a useful preceramic polymer.

Some more general considerations merit discussion at this point. Although preceramic polymers are potentially "high value" products _if_ the desired properties result from their use, the more generally useful and practical systems will be those based on commercially available, relatively cheap starting monomers. Preferably, the polymer synthesis should involve simple, easily effected chemistry which proceeds in high yield. The preceramic polymer itself should be liquid or, if a solid, it should be fusible and/or soluble in at least some organic solvents, i.e., it should be processable. It would simplify matters if the polymer were stable on storage at room temperature and stable toward atmospheric oxygen and moisture. Its pyrolysis should provide a high yield of ceramic residue and the pyrolysis volatiles preferably should be non-hazardous and non-toxic. In the requirement of high ceramic yield, economic considerations are only secondary. If the weight loss on pyrolysis is low, shrinkage will be minimized as will be the destructive effects of the gases evolved during the pyrolysis.

There are important considerations as far as the chemistry is concerned. First, the design of the preceramic polymer is of crucial importance. Many linear organometallic and inorganic polymers, even if they are of high molecular weight, decompose thermally by formation and evolution of small cyclic molecules, and thus the ceramic yield is low or even zero. In such thermolyses, chain scission is followed by "back-biting" of the reactive terminus thus generated at a bond further along the chain. Thus high molecular weight, linear poly(dimethylsiloxanes) decompose thermally principally by extruding small cyclic oligomers, $(Me_2SiO)_n$, n = 3,4,5 .... When a polymer is characterized by this type of thermal decomposition, the ceramic yield will be low and it will be necessary to convert the linear polymer structure to a cross-linked one by suitable chemical reactions prior to its pyrolysis. In terms of the high ceramic yield requirement, the ideal preceramic polymer is one which has functional substituent groups which will give an efficient thermal cross-linking process so that on pyrolysis non-volatile, three-dimensional networks - which lead to maximum weight retention - are formed. Thus, preceramic polymer design requires the introduction of reactive or potentially reactive functionality.

In the design of preceramic polymers, achievement of the desired elemental composition in the ceramic obtained from them (SiC and Si_3N_4 in the present cases) is a major problem. For instance, in the case of polymers aimed at the production of SiC on pyrolysis, it is more usual than not to obtain solid residues after pyrolysis which, in addition to SiC, contain an excess either of free carbon or free silicon. In order to get close to the desired elemental composition, two approaches have been found useful in our research: (1) The use of two comonomers in the appropriate ratio in preparation of the polymer, and (2) the use of chemical or physical combinations of two different polymers in the appropriate ratio.

Preceramic polymers intended for melt-spinning require a compromise. If the thermal cross-linking process is too effective at relatively low temperatures (100-200°C), then melt-spinning will not be possible since heating will induce cross-linking and will produce an infusible material prior to the spinning. A less effective cross-linking process is required so that the polymer forms a stable melt which can be extruded through the holes of the spinneret. The resulting polymer fiber, however, must then be "cured", i.e., cross-linked, chemically or by irradiation, to render it infusible so that the fiber form is retained on pyrolysis. Finally, there still are chemical options in the pyrolysis step. Certainly, the rate of pyrolysis, i.e., the temperature profile of the pyrolysis, is extremely important. However, the gas stream used in the pyrolysis also is of great importance. One may carry out "inert" or "reactive" gas pyrolyses. An example of how one may in this way change the nature of the ceramic product is provided by one of our preceramic polymers which will be discussed in more detail later in this paper. This polymer, of composition $[(CH_3SiHNH)_a(CH_3SiN)_b]_m$, gives a <u>black</u> solid, a mixture of SiC, Si_3N_4, and some free carbon, on pyrolysis to 1000°C in an inert gas stream (nitrogen or argon). However, when the pyrolysis is carried out in a stream of ammonia, a <u>white</u> solid remains which usually contains less than 0.5% total carbon and is essentially pure silicon nitride. At higher temperatures (>400°C), the NH_3 molecules effect nucleophilic cleavage of the Si-C bonds present in the polymer and the methyl groups are lost as CH_4. Such chemistry at higher temperatures can be an important and sometimes useful part of the pyrolysis process.

The first useful organosilicon preceramic polymer, a silicon carbide precursor, was developed by S. Yajima and his coworkers at Tohoku University in Japan (6). In the preparation of their carbosilane polymer the initial step involves the sodium condensation of commercially available dimethyldichlorosilane to give a solid polysilane, $+(CH_3)_2Si+_n$ ($\tilde{n}$ ~30) in which the polymer backbone is made up of Si-Si bonds. Pyrolysis of this material leaves very little solid residue since the thermal degradation mechanism involves the formation of small cyclic polysilanes. This poly(dimethylsilylene) may, however, be converted to a useful preceramic polymer by heating at about 450°C under argon for some time. At this temperature a free radical rearrangement takes place as illustrated in equation 1 for such a $(CH_3)_2Si$ chain unit. It may be noted that high

$$
\begin{array}{cc}
CH_3 & CH_3 \\
| & | \\
\text{+Si---Si+} \\
| & | \\
CH_3 & CH_3
\end{array}
\longrightarrow
\begin{array}{cc}
CH_3 & CH_3 \\
| & | \\
\text{+Si-CH}_2\text{-Si+} \\
| & | \\
H & CH_3
\end{array}
\qquad (1)
$$

<u>1</u>

molecular weight poly(dimethylsilylmethylenes) of type $+(CH_3)_2SiCH_2+_n$ are not useful preceramic polymers. In structure <u>1</u> it is the Si-H bonds which are the reactive functional groups whose reactions on pyrolysis result in further cross-linking and thus in higher (~60%) ceramic yields. After removal of volatiles formed in this step a residue of relatively low molecular weight (~1500) remains behind. This material may be melt-spun at ca. 350°C to give polymer fibers. These must be subjected to a cure step (simplest is heating in air at 190°C) in order to provide the infusible fibers which then can be pyrolyzed to give ceramic fibers. As might be expected on the basis of the initial 2C/1Si ratio, these fibers do not contain only silicon carbide. A typical analysis (6) showed a composition of 1 SiC/0.78 carbon/0.22 SiO$_2$. (The latter is introduced in the cure step.) These fibers are sold under the trade-name Nicalon®. The product information provided by the vendor says that the fiber has excellent strength and modulus properties and that it retains its proper-ties at high temperatures; that it is highly resistant to oxidation and chemical attack; that it is readily wet by organic resins and metals. Some other modifications of the Yajima polycarbosilane have been developed.

The Yajima polycarbosilane, while it was the first, is not the only polymeric precursor to silicon carbide which has been developed. Another useful system which merits mention is the polycarbosilane which resulted from research carried out by C.L. Schilling and his coworkers in the Union Carbide Laboratories in Tarrytown, New York (7).

NEW PRECERAMIC POLYMER SYSTEMS: RECENT RESEARCH AT M.I.T.

Our own research aimed at the development of new polymeric precursors for silicon carbide began with an examination of a potential starting material in which the C:Si ratio was 1, the ratio desired in the derived ceramic product.

Available methylsilicon compounds with a 1 C/1 Si stoichiometry are CH_3SiCl_3 and CH_3SiHCl_2. The former gives highly cross-linked, insoluble products on treatment with an alkali metal in a suitable dilutent. The latter, in principle, could give $[CH_3SiH]_n$ cyclic oligomers and linear polymers on reaction with an alkali metal. In practice, the Si-H linkages also are reactive toward alkali metals. Thus, mixed organochlorosilane systems containing some CH_3SiHCl_2 have been treated with metallic potassium by Schilling and Williams (8). It was reported that the CH_3SiHCl_2-based contribution to the final product was $(CH_3SiH)_{0.2}(CH_3Si)_{0.8}$, i.e., about 80% of the available Si-H bonds had reacted. Such Si-H reactions lead to cross-linking in the product, or to formation of polycyclic species if cyclic products are preferred. Nevertheless, we have used this known reaction of CH_3SiHCl_2 with an alkali metal as an entry to new preceramic polymers.

When the reaction of CH_3SiHCl_2 with sodium pieces was carried out in tetrahydrofuran medium, a white solid was isolated in 48% yield. This solid was poorly soluble in hexane, somewhat soluble in benzene, and quite soluble in THF. Its molecular weight could not be determined by cryoscopy in benzene because of its limited solubility in that solvent. Its ^{1}H NMR spectrum (in CDCl$_3$) indicated that extensive reaction of Si-H bonds had occurred. The $\delta(SiH)/\delta(SiCH_3)$ integration led to a constitution $[(CH_3SiH)_{0.4}(CH_3Si)_{0.6}]_n$. Here the CH_3SiH units are ring and chain members which are not branching sites; the CH_3Si units are ring and chain members which are branching sites. For instance, a simple discrete compound of this type would be the one shown here:

$$(R = CH_3)$$

This would be $(CH_3SiH)_{10}(CH_3Si)_2$. In our reactions it is expected that mixtures of such polycyclic polysilanes will be formed. (Attempts to distill out pure compounds from our preparations were not successful. Less than 10% of the product was volatile at higher temperatures at 10^{-4} torr.) For the bicyclic species shown above the x/y ratio for the general formula $(CH_3SiH)_x(CH_3Si)_y$ is 5. For the material obtained in the CH_3SiHCl_2/Na reaction carried out in THF the x/y ratio is 0.67, which indicates much more extensive ring fusion.

The ceramic yield obtained when the $[(CH_3SiH)_{0.4}(CH_3Si)_{0.6}]_n$ polymer was pyrolyzed (TGA to 1000°C) was 60%; a gray-black solid was obtained whose analysis indicated a composition 1.0 SiC + 0.49 Si.

Slow evaporation of a toluene solution of $[(CH_3SiH)_{0.4}(CH_3Si)_{0.6}]_n$ gave a viscous, gummy residue from which fibers could be pulled. These could be converted to think, black ceramic fibers by first photolyzing in air (254 nm) for 1 h and pyrolyzing to 1000°C. When photolysis was carried out in a nitrogen atmosphere, pyrolysis of the fibers yielded a black powder. Obviously, formation of a thin SiO_2 coating on the polymer fiber surface is a crucial factor for obtaining a ceramic fiber. Since the unphotolyzed polymer melts at 130-140°C (in vacuo), the photolysis probably serves to render the material infusible, allowing it to retain its shape on pyrolysis.

The reaction of methyldichlorosilane with sodium in a solvent system composed of six parts of hexane and one of THF gave a higher yield of product which was soluble in organic solvents. Such reactions give a colorless, cloudy oil in 75 to over 80% yield which is soluble in many organic solvents. In various experiments the molecular weight (cryoscopic in benzene) averaged 520-740 and the constitution (by ^1H NMR) $[(CH_3SiH)_{0.76}(CH_3Si)_{0.24}]_n$ to $[(CH_3Si)_{0.9}(CH_3Si)_{0.1}]_n$. It was found that the slower the rate of addition of CH_3SiHCl_2 to the sodium suspension, the smaller the extent of reaction at Si-H. This less cross-linked material (compared to the product obtained in THF alone) gave much lower yields of ceramic product on pyrolysis to 1000°C (TGA yields ranging from 12-27% in various runs). Again, the product was (by analysis) a mixture of SiC and elemental silicon, 1.0 SiC + 0.42 Si being a typical composition.

The reaction of CH_3SiHCl_2 with sodium in refluxing xylene gave a liquid product, generally in yields of only around 40%, molecular weight 520, with a constitution $[(CH_3SiH)_{0.65}(CH_3Si)_{0.35}]_n$. On pyrolysis to 1000°C the ceramic yield was 25-30% to give a gray-black material analyzing for 1.0 SiC + 0.70 Si.

The results described above are not especially promising. The CH_3SiHCl_2/Na product which on pyrolysis gives a reasonable ceramic yield is of limited solubility in organic solvents and its conversion to ceramic fibers requires a photolysis/oxidation cure step. The CH_3SiHCl_2/Na product in which crosslinking is not as extensive and which is very soluble in organic solvents gives unacceptably low ceramic yields on pyrolysis. Furthermore, only in the case of the preparations carried out in 6-7/1 hexane/THF solvent medium were the yields of soluble product satisfactory. Finally, in all cases the ceramic product contained a considerable excess of "free" silicon over the ideal SiC composition. It was obvious that further chemical modification of the $[(CH_3SiH)_x(CH_3Si)_y]_n$ products obtained in the CH_3SiHCl_2/Na reactions was required. We report here two approaches which have converted such materials into much more useful products.

A number of approaches which we tried did not lead to success, but during the course of our studies we found that treatment of the $[(CH_3SiH)_x(CH_3Si)_y]_n$ products with alkali metal amides (catalytic quantities) serves to convert them to materials of higher molecular weight whose pyrolysis gives significantly higher ceramic yields. Thus, in one example, to 0.05 mol of liquid $[(CH_3SiH)_{0.85}(CH_3Si)_{0.15}]$ in THF was added, under nitrogen, a solution of about 1.25 mmol (2.5 mol%) of $[(CH_3)_3Si]_2NK$ in THF. The resulting red solution was treated with methyl iodide. Subsequent nonhydrolytic workup gave a soluble white powder in 68% yield, molecular weight 1000, whose pyrolysis to 1000°C gave a ceramic yield of 63%.

Similar experiments were carried out using catalytic quantities of other alkali metal silylamides: $[(CH_3)_2(CH_2=CH)Si]_2NK$, the reaction products of one molar equivalent of KH with one of cyclo-$[(CH_3)_2SiNH]_3$ and of one molar equivalent of KH with one of cyclo-$[(CH_3)_2(CH_2=CH)SiNH]_3$. In each case a white powder of higher molecular weight was produced in yields of 80-91%. Their molecular weights ranged from 630-910 g/mol. Of importance to the objectives of this work, the ceramic yields obtained on their pyrolysis to 1000°C ranged from 56-67%.

The proton NMR spectra of these products showed only broad resonances in the Si-H and Si-CH$_3$ regions. In the starting $[(CH_3SiH)_x(CH_3Si)_y]_n$ materials observed proton NMR integration ratios, SiCH$_3$/SiH, ranged from 3.27-3.74. This ratio was quite different in the case of the products of the silylamide-catalyzed processes, ranging from 8.8 to 14. Both Si-H and Si-Si bonds are reactive toward nucleophilic reagents. In the case of the alkali metal silylamides, the following processes can be envisioned:

$$(R_3Si)_2NK \; + \; -\overset{|}{\underset{|}{Si}}H \; \longrightarrow \; -\overset{|}{\underset{|}{Si}}N(SiR_3)_2 \; + \; KH \qquad (2)$$

$$(R_3Si)_2NK \; + \; -\overset{|}{\underset{|}{Si}}H \; \longrightarrow \; (R_3Si)_2NH \; + \; -\overset{|}{\underset{|}{Si}}K \qquad (3)$$

$$(R_3Si)_2NK \; + \; -\overset{|}{\underset{|}{Si}}-\overset{|}{\underset{|}{Si}}- \; \longrightarrow \; (R_3Si)_2N-\overset{|}{\underset{|}{Si}}- \; + \; -\overset{|}{\underset{|}{Si}}K \qquad (4)$$

In each process, a new reactive nucleophile is generated: KH in equation 2, a silyl alkali metal function in reactions 3 and 4. These also could undergo nucleophilic attack on the $[(CH_3SiH)_x(CH_3Si)_y]_n$ system and during these reactions some of the oligomeric species which comprise the starting material would be linked together, giving products of higher molecular weight. Other processes are possible as well, e.g., a silylene process as shown in equation 5. Thus not only anionic species but also neutral silylenes could be involved as intermediates. In any case, extensive loss of Si-H takes place during this catalyzed process: it is more than a

$$(R_3Si)_2NK \quad + \quad \underset{\underset{H}{|}}{\overset{|}{-}}Si\underset{|}{\overset{|}{-}}Si\underset{}{\overset{|}{-}} \quad \longrightarrow \quad (R_3Si)_2NSi\underset{|}{\overset{|}{-}} \quad + \quad \underset{\underset{H}{|}}{\overset{|}{-}}Si\underset{}{\overset{|}{-}}K \qquad (5)$$

$$\downarrow$$

$$\diagdown Si: \; + \; KH$$

simple redistribution reaction. Further studies relating to the mechanism of this process must be carried out.

While these silylamide-catalyzed reactions have provided a good way to solve the problem of the low ceramic yield in the pyrolysis of $[(CH_3SiH)_x(CH_3Si)_y]_n$, the problem of the elemental composition of the ceramic product remained (i.e., the problem of Si/C ratios greater than one) since only catalytic quantities of the silylamide were used.

In earlier work (9) we had reported the formation of novel preceramic polysilazanes by the treatment of the (mostly) cyclic ammonolysis product of CH_3SiHCl_2, $[CH_3SiHNH]_n$, with catalytic quantities of KH in THF. This, after a CH_3I quench, produced useful preceramic polymers of type $[(CH_3SiHNH)_a(CH_3Si)_b(CH_3SiHNCH_3)_c]_n$. However, prior to addition of CH_3I, the reaction solution contained a reactive, polymeric alkali metal silylamide, $[(CH_3SiHNH)_a(CH_3SiN)_b(CH_3SiHNK)_c]_n$. This species reacts with electrophiles other than methyl iodide, e.g., with diverse chlorosilanes, and it has been isolated and analyzed. Since it is a silylamide, we expected that it also would react with $[(CH_3SiH)_x(CH_3Si)_y]_n$ polysilane-type materials. Not only would it be expected to convert the latter into material of higher molecular weight, but it also would be expected to improve the Si/C ratio (i.e., bring it closer to 1). Our studies had shown that pyrolysis of $[(CH_3SiHNH)_a(CH_3SiN)_b(CH_3SiHNCH_3)_c]_n$ gives a ceramic product in 80-85% yield containing Si_3N_4, SiC and excess carbon. A typical composition is 1.27 SiC; 0.88 Si_3N_4 and 0.75 C. Thus, combination of the two species in the appropriate stoichiometry, i.e., of $[(CH_3SiH)_x(CH_3Si)_y]_n$ and $[(CH_3SiHNH)_a(CH_3SiN)_b(CH_3SiHNK)_c]_n$, and pyrolysis of the product (which we will call a "graft" polymer) after CH_3I quench could, in principle, lead to a ceramic product in which the excess Si obtained in pyrolysis of the former and the excess C obtained in the pyrolysis of the latter combine to give SiC. A further benefit might be expected from such a combination in the formation of ceramic fibers: The $[(CH_3SiHNH)_a(CH_3SiN)_b(CH_3SiHNCH_3)_c]_2$ system is self-curing as the temperature is raised on the way to the production of a ceramic material; the $[(CH_3SiH)_x(CH_3Si)_y]_n$ system, as described above, is not. It might be expected that a combination of the two would give a self-curing system.

With these ideas in mind, experiments were carried out in which the two polymer systems, $[(CH_3SiH)_x(CH_3Si)_y]_n$ and the "living" polymer-silyl amide, $[(CH_3SiHNH)_a(CH_3SiN)_b(CH_3SiHNK)_c]_n$, were mixed in THF solution in varying proportions (2.41:1 to 1:2 mole ratio) and allowed to react at room temperature for 1 h and at reflux for 1 h. (Such experiments were carried out with the $[(CH_3SiH)_x(CH_3Si)_y]_n$ materials prepared in hexane/THF as well as with those prepared in THF alone.) After quenching with methyl iodide, nonhydrolytic workup gave a new polymer in nearly quantitative yield (based on weight of material charged). The molecular weight of these products was in the 1800-2500 range. Their pyrolysis under nitrogen gave ceramic products in 74-83% yield. Thus, the reaction of the two polymer systems gives a new polymer in close to quantitative yield which seems to be an excellent new preceramic polymer in terms of ceramic yield.

In an alternative method of synthesis of $[(CH_3SiH)_x(CH_3Si)_y]/$ $[(CH_3SiHNH)_a(CH_3SiN)_b]$ "combined" polymers, the polysilyl amide was generated in situ in the presence of $[(CH_3SiH)_x(CH_3Si)_y]_n$. This, however, gave materials that were somewhat different. In one such experiment, a mixture of $(CH_3SiHNH)_n$ cyclics (as obtained in the ammonolysis of CH_3SiHCl_2 in THF) and the $[(CH_3SiH)_x(CH_3Si)_y]_n$ material (x = 0.76; y = 0.26) in THF was treated with a catalytic amount of KH. After the reaction mixture had been treated with methyl iodide, the usual workup gave an 89% yield of hexane-soluble white powder, molecular weight ~2750. On pyrolysis, this material gave a 73% yield of a black ceramic.

The "combined" polymer prepared in this way ("in situ polymer") was in some ways different from the "combined" polymer prepared by the first method ("graft" polymer). The principal differences were observed in their proton NMR spectra and in the form of their TGA curves. Figure 1 shows the proton NMR spectra of a "graft" polymer and of an "in situ" polymer. Both have very similar starting molar ratios of $[(CH_3SiH)_x(CH_3Si)_y]$ to (CH_3SiHNH), 1.5 and 1.45, respectively, in terms of initial reactants used. The region in which the Si-H proton resonances occur are of interest. The signals around $\delta 5.1$ and 4.7 are due to CH_3SiHN protons, those around $\delta 4.0$ are due to CH_3SiH protons which are attached to Si atoms not bound to nitrogen. In the "in situ" polymer the intensity ratio of the $\delta 5.1$, 4.7 to the $\delta 4.0$ protons is 12; in the "graft" polymer it is 1. This suggests that the two differently prepared polymers have different structures. It is likely that in the "in situ" preparation intermediates are formed by the action of KH on the $(CH_3SiHNH)_n$ cyclics are intercepted by reaction with the $[(CH_3SiH)_x(CH_3Si)_y]_n$ also present before the $[(CH_3SiHNH)_a(CH_3SiH)_b(CH_3SiHNK)_c]_n$ polymer (which is the starting reactant used in the "graft" procedure) has a chance to be formed to the extent of its usual molecular weight. Thus, less of the original CH_3SiHNH protons are lost and/or more of those of the $[(CH_3SiH)_x(CH_3Si)_y]_n$ system are reacted.

The TGA curves of the "graft" polymer and the "in situ" polymer are different as well. Noteworthy in the former is a small weight loss between 100°C and 200°C, which begins at around 100°C. This initial small weight loss occurs only at higher temperature (beginning at ~175°C) in the case of the "in situ" polymer. This difference in initial thermal stability could well have chemical consequences of importance with respect to ceramics and both kinds of polymers may be useful as preceramic materials.

Physical blends of $[(CH_3SiH)_x(CH_3Si)_y]_n$ (solid polymer, THF preparation) and $[(CH_3SiHNH)_a(CH_3SiN)_b(CH_3SiHNCH_3)_c]_n$ also were examined. When about equimolar quantities of each were mixed and finely ground together, pyrolysis to 1000°C gave a 70% ceramic yield (TGA). It appears that a reaction between the two polymers already occurs at lower temperatures. When such mixtures were heated, either in the absence of a solvent at 100°C under nitrogen or in toluene solution at reflux, white powders were obtained which were insoluble in hexane, benzene, and THF. The ceramic yields (by TGA) were 67% and 75%, respectively.

Further experiments showed that the "combined" polymers may be converted to black ceramic fibers. Pyrolysis of pressed bars of the "combined" polymer to 1000°C gave a black, foam product of irregular shape (74-76% ceramic yield). In other experiments, SiC powder was dispersed in toluene containing 20% by weight of the "combined" polymer. The solution was evaporated and the residue, a fine powder of SiC with the "combined" polymer binder, was pressed into bars and pyrolyzed at 1000°C. A ceramic bar (6% weight loss, slightly shrunk in size) was obtained.

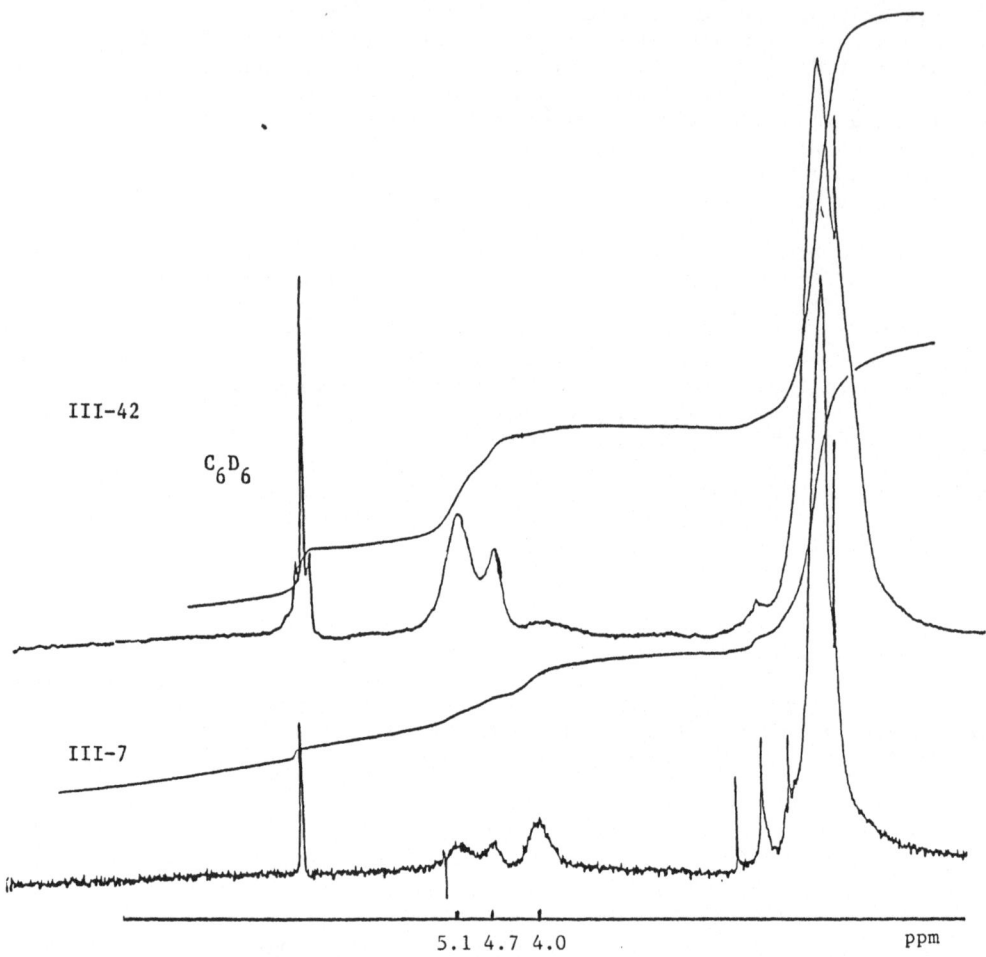

Figure 1. ^{1}H NMR spectra of "graft" and "in situ" polymers.
III-42: "in situ" polymer.
III-7: "graft" polymer.

This conversion of $[(CH_3SiH)_x(CH_3Si)_y]_n$ to a material with superior properties is not restricted to alkali metal silylamide catalysts. Simple organic alkali metal amides may be used as well. Such experiments were carried using potassium piperidide, potassium ethylamide, and with potassium propylamide. In each case, white solids of higher molecular weight than the starting material were obtained. The ceramic yields obtained in their pyrolysis to 1000°C were below 60%.

The ceramic products obtained in the pyrolysis of the "combined" polymers have not been studied in detail, but some of them have been analyzed for C, N, and Si. The compositions of the ceramic materials obtained cover the range 1 Si_3N_4 + 3.3 to 6.6 SiC + 0.74 to 0.85 C. Thus, as expected, they are rich in silicon carbide and the excess Si which is obtained in the pyrolysis of the $[(CH_3SiH)_x(CH_3Si)_y]_n$ materials alone is not present, so that objective has been achieved. By proper adjustment of starting material ratios, we find that the excess carbon content can be minimized.

The Yajima polycarbosilane discussed earlier, as obtained by thermal rearrangement of the poly(dimethylsilylene) is a polymeric silicon hydride, with $[(CH_3)(H)SiCH_2]$ as the main repeating unit. An idealized structure, 2, has been written for this polymer. As such, it also might be expected

to react with our $[(CH_3SiHNH)_a(CH_3SiN)_b(CH_3SiHNK)_c]$ poly(silylamide). This was found to be the case. The commercially available Yajima poly-carbosilane (sold in the U.S. by Dow Corning Corporation; our sample, a white solid, had a molecular weight of 1210 and a ceramic yield of 58% was obtained on pyrolysis) was found to react with our poly(silylamide). A reaction carried out in THF solution, initially at room temperature, then at reflux, followed by treatment of the reaction mixture with methyl iodide, gave after appropriate workup a nearly quantitative yield of a white solid which was very soluble in common organic solvents including hexane, benzene, and THF. When the polycarbosilane-to-polysilylamide ratio was approximately one, pyrolysis of the product polymer gave a black ceramic solid in 84% yield which analysis showed to have a composition $(1\ SiC + 0.22\ Si_3N_4 + 0.7\ C)$. When the polycarbosilane/polysilylamide ratio was ~5, the ceramic yield was lower (67%). In these experiments the cyclo-$(CH_3SiHNH)_n$ starting materials used to synthesize the polysilylamide had been prepared by CH_3SiHCl_2 ammonolysis in diethyl ether. When this preparation was carried out in THF, the final ceramic yields obtained by pyrolysis of the polycarbosilane/polysilylamide hybrid polymer were 88% (1:1 reactant ratio) and 64% (5:1 reactant ratio).

The "in situ" procedure in which the CH_3SiHCl_2 ammonolysis product, cyclo-$(CH_3SiHNH)_m$, was treated with a catalytic quantity of KH in the presence of the polycarbosilane, followed by a CH_3I quench and the usual work-up, gave equally good results in terms of high final ceramic yields, whether the starting cyclo-$(CH_3SiHNH)_m$ was prepared in Et_2O or THF.

It is clear from these results that a new polymer is formed when the polycarbosilane and the polymeric silylamide are heated together in soluton and then quenched with methyl iodide. Proton NMR spectroscopy brings further evidence of such a reaction. In Figure 2 is shown (at the bottom) the ^1H NMR spectrum of a 1:1 by weight physical mixture of the polycarbo-silane and the polysilazane obtained by methyl iodide quench of the poly-meric silylamide. The NMR spectrum above it shows the polymer produced in the 1:1 polycarbosilane/polymeric silylamide (followed by CH_3I) reaction. First of all, the CH_3Si/HSi integrated ratio is different: 8.7 in the former, 9.4 in the latter. Secondly, the CH_3SiHNH proton (at $\delta5.06$) to CH_3SiH (at $\delta4.50$) proton ratio has changed from about 2 in the 1:1 physical mixture to 1 in the reaction product polymer.

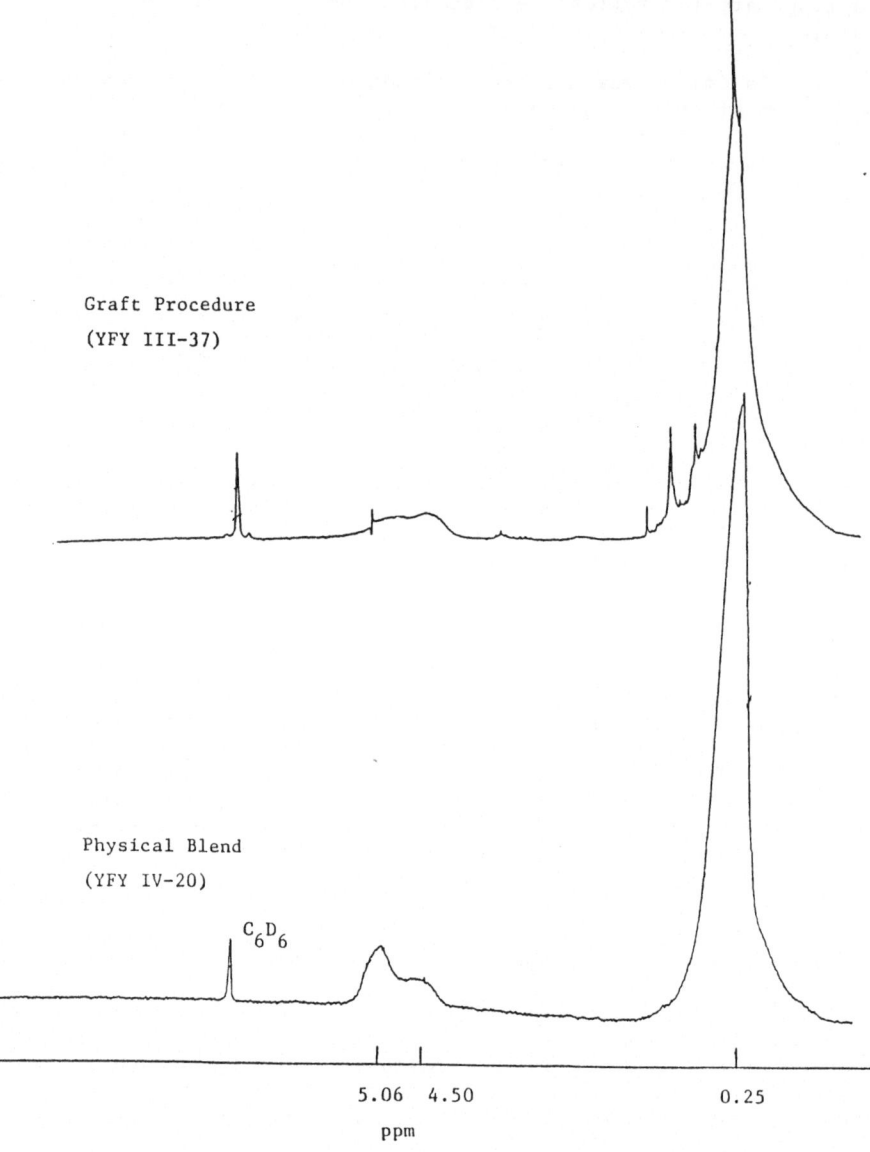

Graft Procedure
(YFY III-37)

Physical Blend
(YFY IV-20)

C_6D_6

5.06 4.50   0.25

ppm

Figure 2. ^{1}H NMR spectra of a 1:1 by weight physical mixture of the
polycarbosilane and the polysilazane (YFY-IV-20) and of the
hybrid polymer obtained by reaction of the polycarbosilane
and the poly(silylamide) (1:1 by weight, "graft" procedure,
CH_3I quench) (YFY-III-37).

Poly(silylamides) derived from other polysilazanes also may be used.
In one such example, the polymer obtained by heating $[(CH_3)_2SiNH]_m$ cyclic
oligomers with NH_4Br at 160°C, followed by treatment of the product with
NH_3 in diethyl ether and removal of volatiles (10) was partially
deprotonated with KH in THF and the resulting solution was added to a

solution of the polycarbosilane in THF. After a 4 h reflux period, methyl iodide was added. The usual work-up gave a soluble, solid polymer in 93% yield, molecular weight 1570. Pyrolysis of this product to 1000°C gave a black ceramic material.

A physical mixture of the polycarbosilane and the polysilazane $[(CH_3SiHNH)_a(CH_3SiN)_b(CH_3SiHNCH_3)_c]_m$, also was found to react when heated to 1000°C, giving good yields of ceramic product. That appreciable reaction had occurred by 200°C was shown in an experiment in which a 1:1 by weight mixture of the initially soluble polymers was converted to a white, foamy solid which no longer was soluble in organic solvents.

The combined polymers obtained by the "graft", "in situ", and physical blend methods can be converted to black ceramic fibers. Pyrolysis of pressed bars of the combined polymers to 1000°C provides a black solid product. In other experiments, SiC powder was dispersed in a toluene solution containing 25% by weight of the combined powders. The solvent was evaporated and the residue, a fine powder of silicon carbide with combined polymer binder, was pressed into bars and pyrolyzed at 1000°C. A ceramic bar was obtained showing a low weight loss and slightly shrunken size. Thus, the usual ceramic applications seem indicated. While this approach in which chemical combination of the Yajima polycarbosilane and our poly-(silylamide) leads to new hybrid polymers successfully addresses the problem of ceramic yield, it does not deal with the problem of chemical composition. Pyrolysis of the polycarbosilane and of the polysilazane separately gives ceramic materials which in each case contain an excess of free carbon. As expected, the hybrid polymers produced by reaction of the polycarbosilane with the poly(silylamide) by either the "graft" or "in situ" procedures gives ceramic products which contain an excess of free carbon.

Returning now to the $[(CH_3SiH)_x(CH_3Si)_y]_n$-type polysilanes, there are other options which may be considered for their conversion to more useful preceramic polymers. These polysilanes are very reactive and their Si-Si as well as their Si-H bonds are potential sites of reactivity toward nucleophilic, electrophilic, radical and low valent, coordinatively unsaturated transition metal reagents. We have investigated an approach based on the well-known addition of Si-H bonds to C=C bonds, olefin hydrosilylation (11). This is a well-known reaction (equation 6) which

$$-\overset{|}{\underset{|}{Si}}-H \quad + \quad \overset{\diagdown}{\diagup}C=C\overset{\diagup}{\diagdown} \quad \xrightarrow{\text{catalyst}} \quad -\overset{|}{\underset{|}{Si}}-\overset{|}{\underset{|}{C}}-\overset{|}{\underset{|}{C}}-H \qquad (6)$$

proceeds best in the presence of a suitable catalyst. There are two general mechanistic possibilities: (1) free radical initiated additions; (2) transition metal-catalyzed additions. The latter may involve homogeneous or heterogeneous catalysis. A very relevant problem with transition metal catalysis in our intended reactions is the fact that they also activate Si-Si bond reactivity. In any case, for the purposes of preceramic polymer preparation, one would prefer a catalyzed hydrosilylation which does not introduce metal species as impurities into the product. This makes transition metal catalysts less attractive since their removal from polymeric products is always a problem. In the present case, a hydrolytic work-up which might serve to remove metal species may not be possible and should, in any case, be avoided because of possible Si-H hydrolysis. Since the final product still will contain Si-H and Si-Si bonds, both of which generally are activated by the same transition metal catalytst, leaving such metal species in the preceramic polymer may have undesirable consequences when the polymer is heated.

Free radical catalysts which are used in organic solvents in general
are peroxides and azo compounds. Peroxides, either directly or through
secondary reactions, would introduce oxygen into the polymer via formation of Si-O bonds. As a result, oxygen would be present in the final
ceramic material. Thus, the preferred catalyst is an azo compound, for
instance, the commercially available azobisisobutyronitrile, AIBN,
$(CH_3)_2(CN)CN=NC(CN)(CH_3)_2$. Its thermolysis releases N_2 and forms
organic radicals, $C(CN)(CH_3)_2$

The type of polymer which is a very effective precursor for a ceramic material seems to be one in which cyclic species are connected to
one another by a multiplicity of links - a sheet-like network polymer
with reactive functionality on the periphery which can be activated
thermally to extend the network (cf. our successful preceramic polymers of ref. 9). Thus, it was felt that the best "olefins" for reaction with $[(CH_3SiH)_x(CH_3Si)_y]_n$ would be cyclic compounds containing
(preferably) at least three vinyl substituents. In strictly organic
structures such poly(vinyl) compounds are not easy to prepare, but in
the realm of organosilicon compounds, they are easily accessible.

The poly(vinyl) compounds of silicon chosen as reactants for the
purposes of this investigation could be prepared easily from the
commercially available methylvinyldichlorosilane by ammonolysis, hydrolysis
and thiolysis (Scheme 1). All three of these compounds reacted readily

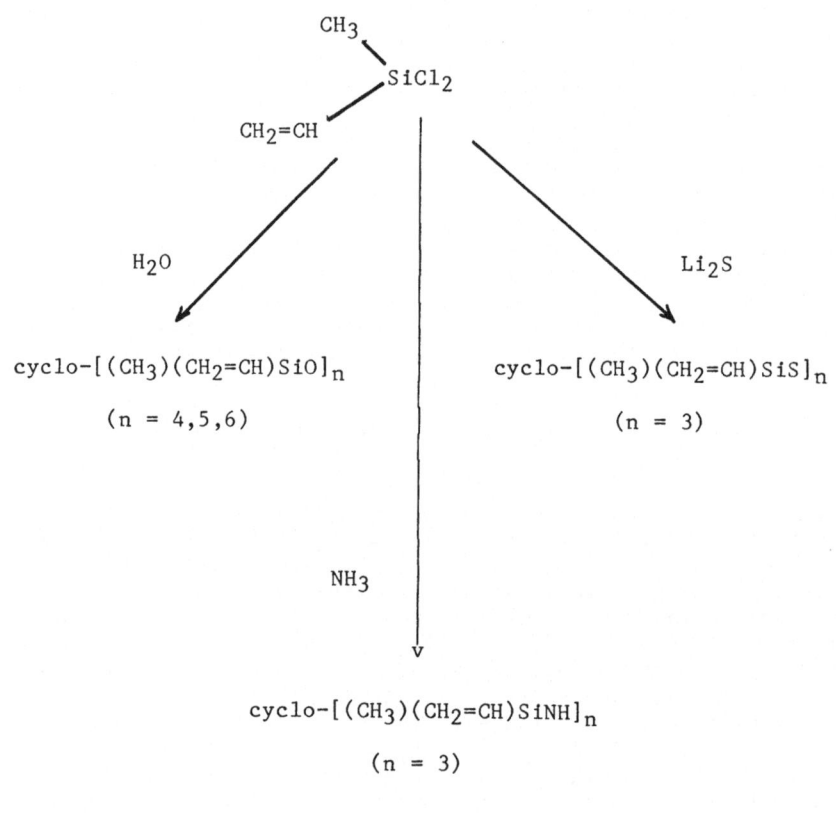

Scheme 1.

with $[(CH_3SiH)_x(CH_3Si)_y]_n$ in benzene solution at reflux in the presence of a catalytic amount of AIBN. From the results obtained when $[(CH_3(CH_2=CH)SiNH]_3$ was the vinyl compound used, it was clear that low (2 or 3) $SiH/SiCH=CH_2$ ratios are not practical: too many vinyl groups lead to insoluble products. When less of the $[(CH_3)(CH_2=CH)SiNH]_2$ is used (relative to $[(CH_3SiH)_x(CH_3Si)_y]_n$) $(SiH/SiCH=CH_2) = 6$ or larger), the products were white solids which all were soluble in organic solvents such as hexane, benzene, and THF. The yields were quantitative; no material was lost. The products are of fairly low molecular weight and, in some cases, quite satisfactory (68%m 77%) ceramic yields (by TGA) were obtained when these materials were pyrolyzed to $1000°C$.

Pyrolysis of one of the products whose proton NMR spectrum and elemental analysis indicated a constitution $[(CH_3SiH)_{0.73}(CH_3Si)_{0.1}(CH_3SiCH_2CH_2Si(CH_3)NH)_{1.17}]$, gave a black solid whose elemental analysis indicated a composition 1 SiC/0.033 Si_3N_4/0.04 C. Thus the stoichiometry and reaction conditions in this experiment gave a high yield of a ceramic product which was silicon carbide contaminated with only minor amounts of silicon nitride and free carbon. When such a pyrolysis was effected to $1500°C$, the ceramic product was at least partly crystalline. X-ray diffraction showed only lines due to β-SiC.

Similar experiments were carried out with $[(CH_3SiH)_x(CH_3Si)_y]_n$ and cyclo-$[(CH_3)(CH_2=CH)SiO]_n$ using AIBN as initiator. The products were organic-soluble white powders. Here also, a $SiH/SiCH=CH_2$ ratio of 6:1 gave best results. One such product, a white powder, molecular weight ~3500, had a constitution (NMR) of $[(CH_3SiH)_{0.67}(CH_3Si)_{0.16}(CH_3SiCH_2CH_2Si(CH_3)O)_{0.17}]$. On pyrolysis it gave a black ceramic product in 73% yield (1 SiC + 0.1 SiO_2 + 0.23 C). AIBN-initiated addition of $[(CH_3SiH)_x(CH_3Si)_y]_n$ to $[(CH_3)(CH_2=CH)SiS]_3$ was carried out under various conditions. The material prepared using a 6:1 $SiH/SiCH=CH_2$ ratio (10 min reflux in benzene) had the composition $[(CH_3SiH)_{0.67}(CH_3Si)_{0.16}(CH_3SiCH_2CH_2Si(CH_3)S)_{0.17}]$ (by NMR). The ceramic yields of this polymer and of the others prepared were only 50-60% and thus this system based on the vinylsilthiane is less satisfactory than the others.

CONCLUSIONS

We have shown in this paper that an organosilicon polymer which seemed quite unpromising as far as application as a preceramic polymer is concerned could, through further chemistry, be incorporated into new polymers whose properties in terms of ceramic yield and elemental composition were quite acceptable for use as precursors for ceramic materials. It is obvious that the chemist can make a significant impact on this area of ceramics. However, it should be stressed that the useful applications of this chemistry can only be developed by close collaboration between the chemist and the ceramist.

ACKNOWLEDGEMENTS

The work reported in this paper was carried out with generous support from the Air Force Office of Scientific Research. The poly(silylamide) and the polysilazane used in this research had been developed earlier with support from the Office of Naval Research.

REFERENCES

1. Gmelin Handbook of Inorganic Chemistry, 8th Edition, Springer-Verlag: Berlin, Silicon, Supplement Volumes B2, 1984, and B3, 1986.

2. Messier, D.R.; Croft, W.J. in "Preparation and Properties of Soild-State Materials", Vol. 7, Wilcox, W.R., ed.; Dekker: New York, 1982, Chapter 2.

3. As recently as 1975, the statement "Silicon nitride has no established use." was applicable: Rochow, E.G., "The Chemistry of Silicon", Pergamon Texts in Inorganic Chemistry, Volume 9, Pergamon: Oxford, 1975, p. 1417.

4. (a) Wynne, K.J.; Rice, R.W. Ann. Rev. Mater. Sci. (1984) 14, 297.
   (b) Rice, R.W. Am. Ceram. Soc. Bull. (1983) 62, 889.

5. Rice, R.W. Chem. Tech. (1983) 230.

6. Yajima, S. Am. Ceram. Soc. Bull. (1983) 62, 893.

7. Schilling, C.L., Jr.; Wesson, J.P.; Williams, T.C. Am. Ceram. Soc. Bull. (1983) 62, 912.

8. (a) Schilling, C.L., Jr.; Williams, T.C., Report 1983, TR-83-1, Order No. AD-A141546; Chem. Abstr. 101, 196820p;
   (b) U.S. Patent 4,472,591 (18 Sept. 1984).

9. (a) Seyferth, D.; Wiseman, G.H. J. Am. Ceram. Soc. (1984) 67, C-132.
   (b) U.S. Patent 4,482,669 (13 Nov. 1984);
   (c) Seyferth, D.; Wiseman, G.H. in "Ultrastructure Processing of Ceramics, Glasses and Composites", 2, edited by L.L. Hench and D.R. Ulrich, Wiley: New York, 1986, Chapter 38.

10. (a) Krüger, C.; Rochow, E.G. J. Polymer Sci., Part A (1964) 2, 3179.
    (b) Rochow, E.G. Monatsh. (1964) 95, 750

11. (a) Eaborn, C.; Bott, R.W., in "Organometallic Compounds of the Group IV Elements", Vol. 1, "The Bond to Carbon", A.G. MacDiarmid, editor, Dekker: New York, 1968, pp. 213-278.
    (b) Lukevics, E.; Belyakova, Z.V.; Pomerantseva, M.G.; Voronkov, M.G. J. Organometal. Chem. Library (1977) 5, 1.

SYNTHESIS OF CERAMIC POWDERS AND THIN FILMS FROM LASER HEATED GASES

John S. Haggerty

Energy Laboratory and
Materials Science and Engineering
Massachusetts Institute of Technology
Cambridge, MA 02139

ABSTRACT

Two laser initiated gas phase synthesis processes have been developed and the characteristics of resulting materials related to process variables; highly perfect ceramic powders are made in one and thin films in the other. Both were developed because they permit unusual reaction conditions to be achieved with great precision and thereby produce superior reaction products.

Silicon, SiC and Si_3N_4 powders principally have been made from appropriate combinations of SiH_4, CH_3SiH_3, NH_3, CH_4 and C_2H_6 laser heated reactants. Particle size, agglomeration, crystallinity and stoichiometry can be controlled with the process which has been modelled with respect to nucleation and growth kinetics. Manufacturing costs of these powders appear acceptable for high quality structural ceramics.

Thin-films of amorphous hydrogenated silicon (a-Si:H) and Si_3N_4 have been made by operating under conditions where heterogeneous rather than homogeneous nucleation occurs. The virtually unique combination of high gas temperatures and low substrate temperatures permits rapid deposition rates and controlled film properties. Resulting films have demonstrated superior electrical, optical, structural and mechanical properties.

INTRODUCTION

Throughout the world, lasers are being applied to a rapidly expanding list of materials processing topics. Most are innovative, many are technically interesting and a very few have some promise of commercial importance.

The more important synthesis applications are listed in Table 1. One finds several fundamental differences between the ways lasers are applied; these are grouped according to their use as a primary energy source for the process, as a diagnostic for the process and as a means of enhancing the process with catalytic "like" effects. The latter is probably a subset of the first in most instances because usually energy is transferred to the region in which the reaction is taking place.

The interactions between the reactants and the photons change as wavelengths and intensities are varied. Low to medium intensity IR lasers cause photothermal reactions. Although vibrationally excited, the reactant molecules remain in their electronic ground states. The absorbed energy is thermallized through intermolecular collisions. Ultraviolet and visible lasers are capable of inducing true photochemical reactions because the higher energy photons can cause reactants to be electronically excited. At sufficiently high intensities, all three wavelengths can cause formation of ionic species via multiphoton absorption or dielectric breakdown.

TABLE 1. Laser Applications in Synthesis Processing

Type of Application	How Applied
Energy Source	Bulk Chemical Synthesis
	Chain Reactions
Photothermal (IR)	Purification
	• Isotope Separation
	• Eliminate Minor Contaminants
Photochemical (UV, visible)	Specific Product Characteristics
	• Powders
	• Films
Plasma/Dielectric Breakdown	2 Dimensional Sheets
(high I)	1 Dimensional Lines
Diagnostics	Study Reaction Kinetics,
	Identify Species, and Monitor
	Process Characteristics
	• Reaction Intermediates
	and Reaction Pathways
Analysis	• Concentrations of Major Species
	• Temperature
	• Velocity
Control	• Number Density
	• Diameter
	• Phases
Catalysis/Surface Activation	Enhanced Kinetics
	• Bulk Reactions
	• Film Deposition
	• Surface Etching

Lasers have been used as primary energy sources for the applications shown in the right-hand column of Table 1. (No horizontal correlations are intended.) Bulk chemicals have been synthesized by several researchers[1], generally by pyrolysis type reactions. No new materials are made; but, metastable forms may result from rapid quenching. The synthesis of vitamin D is an example.[2] Chain reactions can be produced when spontaneous polymerization follows formation of photon induced free radicals. The formation of 10^4 vinylchloride molecules have been demonstrated[3] per photon absorbed by a dichloroethane molecule. Lasers can cause purification by selective excitation and/or rapid quenching.[1] Uranium isotope separation and model studies with SF_6 are the most studied examples[4-6] of this application. More recently, purifications of $AsCl_3$, BCl_3 (phosgene) and SiH_4 (diborane) have proven effective[7-9] because they permit selective interaction with minor constituents without entropy of mixing effects. Finally, lasers are used to produce reaction products having highly specific properties through a combination of process attributes rather than unusual chemical reaction paths. Synthesis of powders[10] and films[11-16] are two examples of this type of application. Films are made in two-dimensional sheets or in one-dimensional lines by transferring energy to the gas, the substrate, or some combination of the two.

Lasers of various types are used to diagnose both exotic and conventional processes.[1] This application is used for fundamental process analysis and for process control by identifying chemical species, providing kinetic data and providing monitorable process characteristics. Certainly one of the most important results is the identification of reaction intermediates permitting definition of reaction pathways.[17] Characterization of reactant and product concentrations, temperatures, velocities, number densities, diameters and phases are other parameters that are measured with laser techniques.

Photo generated catalytic species and surfaces can be produced with lasers. These effects have been used to enhance overall reaction kinetics for bulk synthesis[1], deposition on surfaces[18] and etching of surfaces.[19] These effects are caused by coupling to intermediates and/or causing photochemistry in adsorbed species; the latter with lasers directed toward the surfaces.

With all of this activity, what are the real penetrations of lasers into commercial synthesis processes? Diagnostics are being used extensively because they are cost effective and they provide unique data. As primary energy sources there are very few cases where lasers offer promise. The reason is cost, not technical feasibility or product attributes. This arises because usually only low mass-throughputs can be achieved through expensive, low-to-moderate power equipment that generally has lower than normal conversion efficiencies. Capital, energy and labor costs all become high when attributed to a low mass-throughput. Either quantum yield or material values must be high before lasers can be used effectively as primary energy sources.

We are investigating the synthesis of two high-value materials from gas phase reactants using lasers as the primary source of process energy. "High tech" ceramic powders[10],[20] and semiconductor thin films[11],[21] are being made. Both are CVD processes; one causes homogeneous nucleation the other heterogeneous nucleation. Referring to Table 1, we employ IR (CO_2) lasers to induce photothermal reactions in manners that highly specific product characteristics are achieved. In both cases, the desired material characteristics match technical process attributes uniquely achievable with laser heat sources.

For both powder and thin film processes, important process attributes include:

1. highly controlled atmosphere conditions,
2. cold wall reaction vessels,
3. unusually rapid heating and cooling rates,
4. precise control of process variables,
5. accessibility of the reaction to process diagnostics.

For the laser induced CVD process (LICVD), the ability to establish the substrate temperature at a level that is different from the gas reaction temperature is an important feature.

Each of these process configurations will be described separately.

Ceramic Powders From Laser Driven Reactions

Although ceramic powders are usually marketed as minimum cost commodity products, substantially higher costs are justified if in finished parts superior yields and better properties are achieved through their use. When typical cost factors for parts made of ceramics are compared[22] with those of both advanced metals and polymers, it is evident that raw material costs are disproportionally low and costs associated with rejects are disproportionally high for structural ceramics. Substantially increased raw material costs can be accepted if their use results in lower rejection rates. More importantly, ceramics are not used in many high value applications that are appropriate based on their intrinsic properties because the same factors which cause high rejection rates cause properties to be excessively unreliable. Many applications of ceramic materials; e.g., multilayer substrates, engine parts, cutting tools, and transducers, permit extremely high materials costs (10-100 $/kg).

MIT hypothesized[23] and has now demonstrated[24],[25] that the microstructural characteristics of polycrystalline ceramic bodies can be dramatically improved by using powders having uniform particle sizes. Also, the time-temperature cycles needed for complete densification are lower in temperature and shorter in time than are needed for conventional powders.[26] The superior microstructural and processing characteristics result directly from the defect-free, high coordination number green bodies that can be produced with uniform diameter, agglomerate free powders. An example is shown in Figure 1.

The powders required for ceramic processes based on ordered dispersions must exhibit several important characteristics. The required powder properties and gas phase process attributes are summarized in Table 2. These include small particle size (< 1 μm), narrow size distribution, absence of agglomerates, equiaxed shapes and high purity. Small size and narrow size distributions require high initial nucleation rates that terminate before the growth process has proceeded to an appreciable extent. Absence of agglomerates requires a minimum exposure to elevated temperatures where strong necks can form between colliding particles. Equiaxed shapes result from high growth rates. High purities are achievable with cold wall reaction vessels and high purity feed stock. Acceptable costs depend on achieving high mass flow rates. The laser heated gas phase powder synthesis process is capable of achieving all of these process attributes; consequently, the powders can have the desired characteristics.

Process and Powder Characteristics

Synthesis Apparatus Powders have been synthesized with crossflow and counterflow gas stream – laser beam configurations and with both static and flowing gases.[27,28] Most of the process research has been conducted with the reaction cell shown schematically in Figure 2.

In the crossflow configuration, the laser beam having a Gaussian shaped intensity profile orthogonally intersects the reactant gas stream possessing a parabolic velocity profile. The laser beam enters and exits

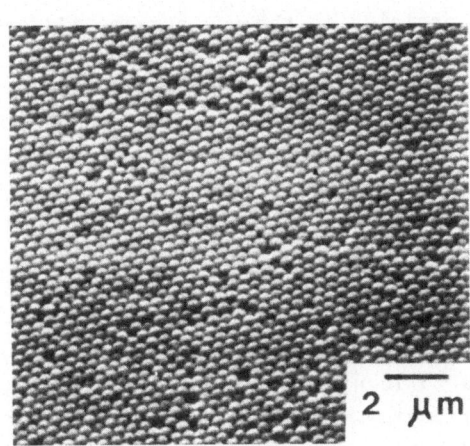

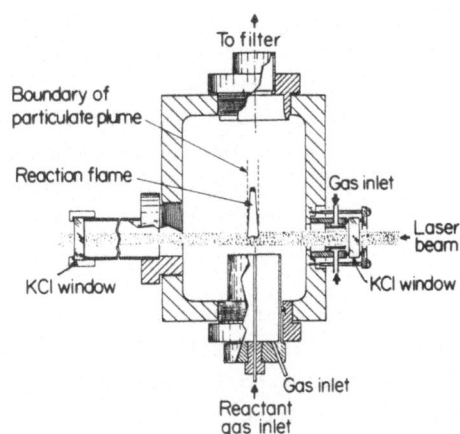

Fig. 1. Ordered packing of monodispersed 0.2 μm diameter TiO_2 spheres achieved by E. A. Barringer at MIT.

Fig. 2. Powder synthesis cell showing laser beam, reactant stream and location of reaction zone.

TABLE 2. Ideal Powder Characteristics and Required Process Atributes

Powder Characteristics	Process Requirements
Small Diameter	High Nucleation Rate
	High Supersaturation
	High Heating Rate
Spherical Shape	High Growth Rate
	Moderate Supersaturation
Uniform Diameter	Termination of Nucleation
	Premixed Reactants
	Uniform Time-Temperature
Non-Agglomeration	Minimum Time at High Temp
	Rapid Cooling Rate
Purity	
Stoichiometry	Controlled Reactant Flow
Crystalline State	Temperature Control
Contamination	Cold Vessel Surfaces
Cost	High Mass Flow Rate
	Exothermic Reaction
	High Value Product
	Acceptable Reactant Cost

the cell through KCl windows. The premixed reactant gases, under some conditions diluted with an inert gas, enter through a 0.75-10.0 mm ID stainless steel nozzle located 2-10 mm below the laser beam. A coaxial stream of argon is used to entrain the particles in the gas stream. Argon is sometimes passed across the inlet KCl windows to prevent powder build-up and possible window breakage. Cell pressures are maintained between 0.08 to 2.0 atm with a mechanical pump and throttling valve. At present, the powder is captured in a microfiber filter located between the reaction cell and vacuum pump. In the future, the particles will be collected by electrostatic precipitation or by separation with the fluid used for ultimate dispersion.

The counterflow geometry, with laser beam and reactant gas streams impinging on each other from opposite directions, has the important advantages of exposing all gas molecules to identical time-temperature histories and absorbing all of the laser energy. However, the achievement of a stable reaction requires that the reaction and gas stream velocities must be equal and opposite to one another. While this is readily accomplished once process conditions are defined, it is an impractical geometry for experimental work in which process conditions are varied over wide ranges.

Initial Process development has been based on a single laser beam appratus. In some cases the transmitted beam is reflected back to the reaction zone both increasing the local intensity and making it more uniform along the beam path. Most recently, we have constructed a pilot scale reaction cell which employs a 1200W laser and focusses four beams into the reaction zone. The inceased power and the more uniform heating should permit synthesis rates up to approximately 1-5 kg/hr/tip.

Chemistries. Principally, Si, Si_3N_4 and SiC powders have been made from appropriate combinations of SiH_4, NH_3, CH_3, and C_2H_4 gases. Alternative reactant gases such as $CH_3 \cdot SiH_3$ and Cl_2SiH_2 have been investigated for SiC.[29] B_2H_6 was used to add boron to the Si and SiC powders as a sintering aid.[30] Recently, we have explored the feasibility of synthesizing other powder materials from laser heatd gaseous reactants.[31,32] TiB_2 was successfully synthesized from $TiCl_4 + B_2H_6$ mixtures but not from $TiCl_4 + BCl_3$; the latter produced $TiCl_3$. B powder was made from both BCl_3 and B_2H_6. TiO_2 and Al_2O_3 have been made from alkoxides and reactants like $Al(CH_3)_3$. While most of this process research has focused on a limited set of compounds and reactants, it is apparent that laser induced reactions are applicable to a broad range of materials.

Process Modelling. A typical reaction zone, located where the laser beam and gas stream intersect, is shown[27] schematically in Figure 3. Average heating rates, nucleation rates, growth rates and temperatures can be estimated based on measured zone and particle dimensions, calculated gas velocities and uncorrected pyrometrically determined temperatures. For typical reaction conditions employing a total reactant gas flow rate of approximately 100 cm^3/min and a pressure of 0.2 atm, the velocity of the gas decreases from approximately 500 cm/sec at the nozzle to 350 cm/sec at the center of the laser beam. With a NH_3/SiH_4 flame, the reaction commences approximately 3 to 5 mm into the laser beam. Thus, the exposure time needed to initiate the reaction is nominally 10^{-3} sec. A reaction temperature of approximately 1000°C,

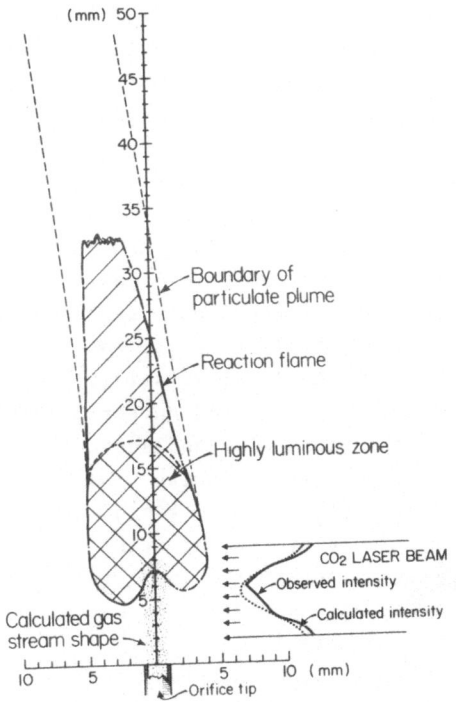

Fig. 3. Typical laser induced reaction zone showing inlet nozzle, position and intensity profile of CO_2 laser, regions in reaction zone and plume trajectory.

indicates a gas heating rate of approximately 10^6 °C/sec. With growth occurring primarily in the hot zone, growth rates are approximately 10^6 Å/sec and particles are exposed to the maximum reaction temperatures for less than 10^{-3} seconds.

The gas stream velocities and temperatures have been calculated as functions of axial and radial positions for varying process conditions and chemistries[27],[33]. Initially, all absorbed laser energy was assumed to be converted to sensible heat; more recent calculations include radiant losses. An example of calculated temperature and velocity distributions is given in Figure 4. These calculations have permitted the stable reaction domain to be defined and have provided essential information about the positions where reactions occur as well as radial mixing by convection and diffusion mechanisms. These analyses are being extended to include turbulant flow.

Diagnostics based on light scattering and transmittance measurements permit nucleation and growth processes to be analyzed in terms of temperature and reactant partial pressures with $\leqslant 1$ mm spacial resolution throughout the reaction zone and particulate plume. They also permit important observations about the crystalline state of the reaction product and the formation of agglomerates. Using an experimental technique developed by Flint[34],[35] and Marra[36],[37] and modifications of computational techniques reported by Sarofim[38], it is possible to calculate the local emissivity, particle diameter and number density from the extinction and scatter-extinction ratio results. Mass balance calculations, based on local number density and particle diameter, permit the local reactant concentrations to be calculated. These, combined with local temperatures, permit reaction kinetics to be interpreted in terms of rate controlling models. They also provide a basis for real time process control. Typical results[34-37] are given in Figures 5,

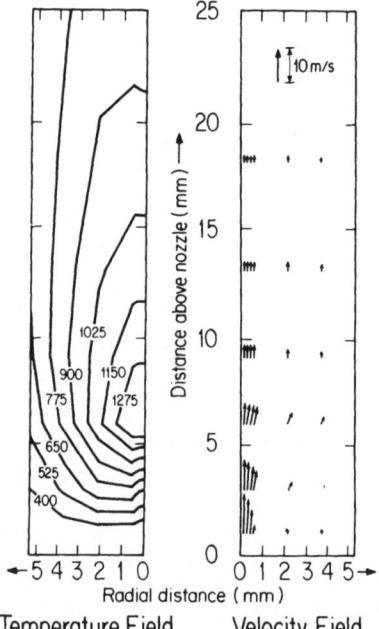

Temperature Field   Velocity Field

Fig. 4. Calculated axial and radial temperature, and flow distributions for laser heated SiH_4.[33] Maximum temperatures are located at the CO_2 laser beam; a cusp at the base of reaction zone is shown in the isotherms. The local velocity vectors illustrate the changes from a parabolic distribution, through radial expansion with heating to plug flow at the exit.

6 and 7 for Si powders made from SiH_4. The temperatures shown in Figure 5 are calculated from the pyrometrically determined brightness temperatures and the calculated emissivities.

These calculations employ the complex index of the particulate materials which are strongly dependent on crystallinity, temperature, stoichiometry and wavelength. This sensitivity can be used to gain insights about the reaction process because the actual number densities and diameters are measured independently by characterizing the powders. For either amorphous or polycrystalline indices, Figure 7, the calculated number densities decrease rapidly over a distance of approximately 1 mm ($\sim 3 \times 10^{-4}$ sec) and then remain constant. Supported by direct STEM observations and annealing studies[36],[39], these results show that the Si particles form with an amorphous structure then crystallize when the gases become hotter as they penetrate into the laser beam.[34] The actual number density follows the dotted line from the amorphous curve (~ 3 mm) to the polycrystalline curve (~ 4.5 mm) remaining essentially constant ($N \approx 4 \times 10^{12}$ cm^{-3}) both throughout the growth process where the diameters increase (Figure 6) and after growth terminates in the constant diameter region. The cessation of nucleation is an essential requirement for achieving the desired uniform particle size.

In close agreement with the simplified analyses, these direct observations show that the process is characterized by rapid heating rates ($\sim 10^6$ °C/sec), short reaction times ($\sim 5 \times 10^{-4}$ sec) and rapid cooling rates ($\sim 10^5$ °C/sec). The high heating rates and high supersaturation levels cause high nucleation rates and consequently small particle sizes. Growth occurs at a constant, high rate until it ceases usually at approximately the point where the temperature begins to decrease. The short time at high temperatures minimizes problems with uncontrollably forming hard agglomerates when particles collide through Brownian motion. The cooling rates are fast enough to quench in nonequilibrium crystalline or amorphous structures. The maximum exposure temperature can also be precisely controlled largely independent of heating and cooling rates. This unusual

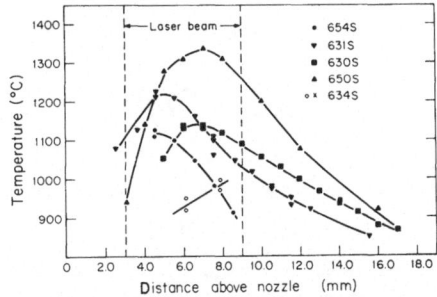

Fig. 5. Corrected Si particle temperature distributions for several synthesis conditions. The boundaries of the CO_2 laser beam are shown.

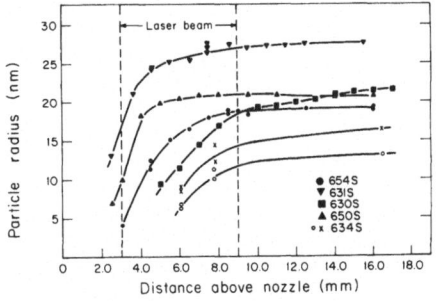

Fig. 6. Calculated Si particle sizes for several synthesis conditions. The boundaries of the CO_2 laser beam are shown.

TABLE 3. Range of Process Conditions Used for Si, Si_3N_4 and SiC Powders.

Process Variable	Si	Si_3N_4	SiC
Cell Pressure (atm)	0.2-2.0	0.2-0.9	0.2-2.0
SiH_4 Flow Rate (cc/min)	5.4-1280	5.4-40	11-700
NH_3 Flow Rate (cc/min)	0	44-110	0
C_2H_4 or CH_4 Flow Rate (cc/min)	0	0	9.0-800
Ar Flow Rate to Annulus plus Window (cc/min)	560-10,000	1000-1100	1000-8000
Laser Intensity (W/cm^2)	176-5.4x10^3	530-1x10^5	530-5.2x10^3
Reaction Zone Temp. (°C)	750-1650	675-1390	865-1930

feature permits the powder's crystalline state to be manipulated. In combination, the attributes of this process constitute a very precise means of producing ceramic powders.

These kinetic studies provide the basis for defining the physical nature of the particle formation and growth processes. Our previously reported models for Si powder synthesis were based on classical nucleation and growth theory.[34-37,40] These descriptions proved reasonably accurate representations for small particle diameter, low pressure Si synthesis conditions. We found increasingly clear discrepancies between experiment and theory as we attempted to increase the particle size, raise the pressure and extended the models[29] to SiC and Si_3N_4 synthesis. Within the past year, we have successfully described the synthesis processes by collision and coalescence models and have extended maximum mean particle diameters up to 3200Å. With this improved understanding, process conditions have been quickly established by which optimum diameter (~ 800-3000Å), non-agglomerated powders of Si and SiC can be produced (Figure 8). We anticipate very rapid progress when Si_3N_4 synthesis experiments are reinitiated.

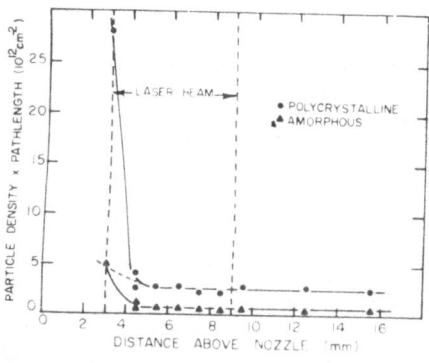

Fig. 7. Si particle number density for assumed polycrystalline and amorphous indexes of refraction.

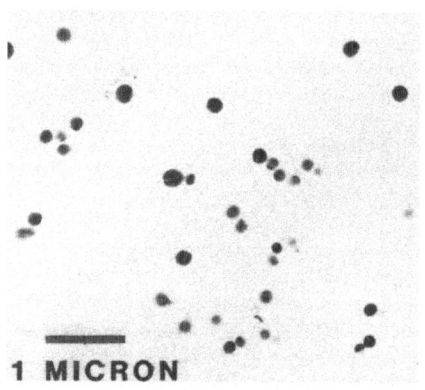

Fig. 8. TEM photomicrograph of dispersed Si powder type F-055.

TABLE 4. Summary of Powder Characteristics.

Powder Characteristics	Si	Si_3N_4	SiC
Mean Diameters (Å)	190–3050	75–500	200–1100
Standard Deviation of Diameters (% of Mean)	~50	~25	~25
Impurities O_2 (wt.%)	0.06–0.7	~ 0.3	0.3–1.3
Total Others (ppm)	<200	<100	<100
Major Elements	Ca,Cu,Fe	Al,Ca	Al,Cu,V
Stoichiometry	---	0–60% excess Si	0–10% excess Si or C
Crystallinity	crystalline –amorphous	amorphous –crystalline	crystalline (Si and SiC) –amorphous
Grain Size:Mean Diameter	1:5–1:3	~ 1:2	1:2–1:1

Synthesis Conditions. The specific process conditions employed to make the powders have been manipulated over an extensive range. The range of investigated conditions are summarized in Table 3. More recent synthesis runs have employed higher pressures, higher mass flow rates and near zero argon flow rates for the window stream. Fortunately, these latter conditions combine more desirable powder characteristics with economically more attractive process parameters.

Powder Characteristics. The Si, Si_3N_4 and SiC powders all exhibit the same general features and appear to match the idealized characteristics we sought. Table 4 summarizes the range of properties that have been achieved with these particulate materials.[25,29,40,41]

The particles are spherical and uniform in size. The Si_3N_4 and SiC powders are usually smaller and have a narrower size distribution than the Si powders. After being captured in a filter, the powders were observed in chainlike agglomerates. Neck formation has been observed by TEM between Si particles, but not usually with the Si_3N_4 and SiC powders. Dispersion results[25,42] have shown primary bonding does not always exist at necks between Si and SiC particles because light scattering and photon correlation spectrometer characterizations of the sols showed that the dispersed particle sizes can be equal to those of the individual particles.[25,41-43] It is presumed that the Si_3N_4 powders will be dispersable because they do not exhibit interparticle necks.

The BET equivalent spherical diameter, and the average TEM diameter corrected for the size distribution measured from micrographs, have always been nearly equal.[36,44] This indicates that the particles have smooth surfaces, no porosity accessible to the surface, nearly spherical shapes, and relatively narrow size distributions. Powder densities, measured by He pycnometry, indicated the particles had no internal porosity.[28]

Chemical analyses[40],[41] indicate that the oxygen content is generally less than 1.0% by weight and some powders are as low as 0.05 wt%. The total cation impurities are typically less than 200 ppm. For Si_3N_4 and SiC powders, the stoichiometry can be made to vary substantially depending on the process conditions. For Si_3N_4 powder, the stoichiometry ranged from nearly pure Si_3N_4 (< 1.0 wt% excess Si) to Si_3N_4 + 60 wt % excess Si. Near stoichiometric values are caused by increased laser intensity, increased pressure, and lower gas stream velocities. SiC powders have been produced with up to 10 wt% excess C or Si.[29],[40] The more stoichiometric powders were produced with increased laser intensity and close to stoichiometric reactant gas ratios. It appears the excess Si and C are distributed uniformly throughout the individual particles in both the Si_3N_4 and SiC powders.

Except when synthesized under very low laser intensities, the Si powders are crystalline; in most cases, the crystallite size was a fraction (1/5 - 1/3) of the BET equivalent diameter, indicating the individual particles are polycrystalline. Marra found[36] that the particle size to grain size ratio reflected the nucleation and growth of crystals in the amorphous particles. The synthesis temperatures used to produce large powders exceed the melting point of Si. These particles are usually single crystals or are twinned. Virtually all process conditions produced Si_3N_4 powder which was amorphous. High laser intensities and high reactant pressures resulted in stoichiometric, crystalline Si_3N_4 powders. Nearly all SiC powders were crystalline. Most process parameters produced polycrystalline SiC particles with the BET equivalent spherical diameter approximately twice as large as the crystallite size measured by X-ray line broadening.

Manufacturing Cost Analyses. The costs of manufacturing powders by the laser heated processes have been estimated (Aries and Newton[45] procedure) and compared to conventional powders. The results are very encouraging.

The operating conditions and yields summarized in Table 5 are the same as used in the laboratory scale process. The assumed 2.0 cm diameter gas tip is approximately the maximum laser beam penetration depth that can produce uniform local intensities with multiple, radially opposed laser beams. Preheated gases were presumed to minimize laser heating; but, no heat recovery was assumed. The plant was scheduled for full time operation (8760 hrs/yr maximum) and assumed to actually produce 91% of available time (8000 hrs/yr). Four full shifts having one person for three tips was assumed. Current bulk costs for NH_3 (0.52 $/kg) and C_2H_4 (0.5 $/kg) were used for these calculations. Silane costs were estimated informally from several sources.[46-48] The cost analysis was done based on presumed silane costs of 2.0 and 20.0 $/kg. A figure of 10-15 $/kg

TABLE 5. Manufacturing Rates and Costs.

Assumptions

Gas tip diameter	= 2.0 cm
Gas velocity	= 500 cm/sec
Gas pressure	= 1 atm
Reaction efficiency	= 100 %
Capture efficiency	= 100%
Operational efficiency	= 91% (8000 hr/yr)

Heating –
      0 – 700°C by resistance
      700 – 1000°C by laser
      no heat recovery

Results	Si	Si_3N_4	SiC
Production rate: kg/hr	7.2	5.1	6.8
ton/yr	57.6	40.8	54.4
Laser power requirement (kW)	1.5	1.3	1.5
Total process energy (kWhr/kg)	1.8	2.1	1.9
Costs ($/kg)			
Labor	0.77	1.09	0.82
Plant	1.66	2.34	1.76
Building	<0.01	<0.01	<0.01
Utilities	0.19	0.25	0.21
Raw materials	2.28–22.80	1.62–13.95	1.77–16.17
Total	4.29–25.44	5.32–17.65	4.57–18.97

appears reasonable for silane with somewhat higher impurity levels than present semiconductor grades.

The results of the manufacturing cost analysis are given in Table 5. They show that the assumed process configuration is well matched to available CO_2 laser equipment. The presumed gas tip diameter and gas flow rate require lasers in the 1.5 kW range. The output per tip is 40–57 tons per year, a level that is well into an acceptable manufacturing scale. Finally, calculated manufacturing costs range from 4.30 to 25.00 $/kg depending mostly on the silane cost. Generally we feel that this projected manufacturing cost is conservative and may well be much lower.

Submicron, uniform diameter powders are not commercially available at any price. However for comparison, Figure 9 summarizes 1982–83 sales prices of Acheson SiC in 5000 kg lots as a function of particle size (solid small circles).[49] The price is essentially constant for powder sizes greater than 100 μm. Below 100 μm, the cost increases rapidly with decreasing particle size. A 1/d extrapolation was used to project prices into the submicron range needed for sintering. This is the most conservative projection if prices are dominated by comminution energy per unit mass[50] as suggested by the prices in Figure 9. These powders will have the typical broad particle size distribution and contaminations that result from comminution processes.

Recently a number of silicon carbide manufacturers have announced their intention to market submicron powders for around $15-30/kg. The circular cross-hatched region in Figure 9 represents the range of these particle sizes and prices. These prices are significantly lower than the corresponding extrapolated prices in Figure 9. The difference probably results from the developing demand for submicron powders as heat-engine applications become established. It is unlikely that these prices will decrease much further, and it remains to be seen if high quality parts can be made from any of these ground powders.

The range of our SiC manufacturing costs is also shown in Figure 9 for comparison with the commercial powders. These costs are independent of particle diameter because control of the nucleation process has no obvious consequence to equipment costs or mass flow rates. Even with probably higher than actual cost figures, the laser heated gas phase synthesis process has a lower manufacturing cost than the sales price of present submicron SiC powders. The laser synthesized powders also have a narrow size distribution and are free of unwanted contaminates.

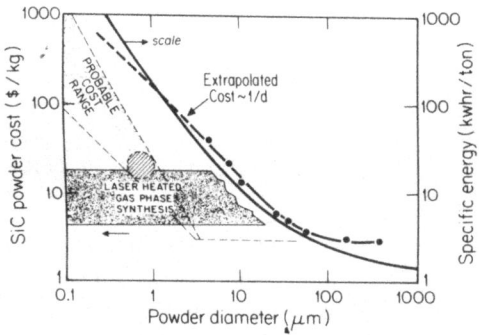

Fig. 9. 1982-83 sales price of Acheson SiC powder[49] in 5000 kg lots, a generalized specific comminution energy[50] and manufacturing cost for SiC powders made from laser heated gases as functions of particle size. The cross-hatched circle represents recent submicron powder prices and particle sizes. The extrapolations of probable cost boundaries assume $1/d$ and $1/d^2$ dependencies.

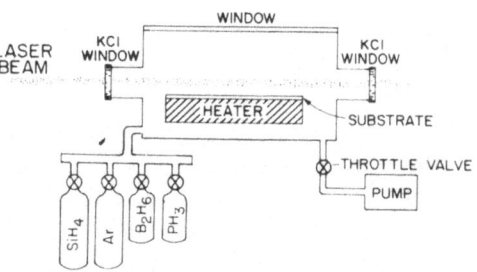

Fig. 10. Schematic of the LICVD reaction cell.

Laser Induced Chemical Vapor Deposition (LICVD) of Thin Films

Thin film materials generally have the highest known values per unit volume or mass. In electronic or optical applications they permit unique characteristics to be realized. In hard-face or corrosion applications they afford superior properties to structures which can be made of lesser quality and lower cost materials.

Subject to some exceptions, CVD processes generally deposit better quality films than plasma or sputtering processes. This superiority cannot always be realized in practice because substrate and reactant temperatures are usually made equal to one another and are usually elevated to achieve useful reaction kinetics. The necessarily high substrate temperatures can preclude the deposition of amorphous or low temperature polymorph films. This can result in excessive stresses with mismatches in thermal expansion coefficients, degrade junctions or other features in electronic devices, and exceed the upper temperature limitations for otherwise superior substrate materials. By permitting the separation of reactant gas and substrate temperatures, the LICVD process appeared capable of combining the best features of conventional CVD processes with those of plasma and sputtering processes which permit low substrate temperatures. CVD deposition can be extended into previously inaccessable process conditions while permitting film properties to be manipulated independent of the growth process.

We have used hydrogenated amorphous silicon as a model material for process research[11,21,51] because all of the LICVD process attributes are required to produce satisfactory films. Low substrate temperatures are needed to insure an amorphous structure, the retained hydrogen content (defined by substrate temperature) must be controlled to achieve correct optical and electronic properties, growth rates should be high, and purity levels are critical. Many other applications are apparent.

In the process we have developed, the laser beam passes parallel to the substrate through an optically absorbing gas which is heated by the laser (Figure 10). It differs fundamentally from those LICVD processes in which the laser beam intersects the substrate.[13-16] In these, locally elevated temperatures occur at the point of intersection even though the reactions may be proceeding largely by photochemical mechanisms. Film properties are as defined by the elevated substrate temperatures in these processes as in conventional CVD processes; moreover, they are subject to the deleterious effects of point heat sources in large planes.

Thin Film Apparatus. The apparatus used for the laser induced CVD process (LICVD) is shown[52] in Figure 10. Substrates are mounted on a temperature controlled stage in a hermetic reaction cell; substrate temperature is monitored by means of a thermocouple in contact with the stage. Although substrate heating is not necessary for deposition to occur, the substrate temperature has a significant effect on the properties of the resultant films and is an important variable in process optimization. Generally, substrate temperatures have been in the 200 to 400°C range for these experiments.[51,53] Untuned CO_2 and grating-tuned CO_2 lasers have been used in the cw mode. The axis of the laser beam is made to pass through the cell, parallel to the substrate plane at a distance of approximately one beam diameter. The unfocused beam diameters are 6 mm. A lens with a focal length of 12.7 cm was used in the experiments involving a focussed beam. Incident laser intensities have been in the 100 to 1000 watts/cm^2 range.

Pure silane (4 to 10 Torr), has been used in most of the experiments. SiH_4/He and SiH_4/H_2 mixtures have been used in a few experiments. Films have been deposited with both vertical and horizontal substrate orientations, and under both constant volume (static cell) and constant pressure (flowing gas) conditions. In the flowing gas experiments, silane is admitted through a nozzle below the substrate stage. The gases pass through the laser beam and out the top of the cell through a filter and throttling valve assembly that maintains a constant pressure. A variety of substrate materials including borosilicate glass, aluminum, aluminum oxide, single crystal silicon and borosilicate glass coated with platinum, molybdenum or indium-tin oxide has been used. Deposition rates were measured by an interferrometric technique employing reflected He-Ne laser light.

The LICVD technique has similarities to the HOMOCVD technique reported by Scott et al[54] in that the reactant gas and substrate temperatures can be different from one another and also it proceeds as a homogeneous thermal reaction. However, LICVD is not subject to parasitic reactions on the hot walls nor to contamination from heated surfaces. Bilenchi et al[55] have recently reported the deposition of a-Si:H films by a technique very similar to the one described here.

Process Characterization. The characteristics of the LICVD process have been modelled empirically and analytically.[11,21,51,56,57] Empirical characterization involved systematic manipulation of process variables and observation of their effects on process and film characteristics. The variables studied include gas composition, gas pressure, gas flow rate, laser power, laser intensity, beam-substrate spacing and substrate temperature. It is evident that the growth rate is directly related to the peak gas temperature now that relationships between peak gas temperature and process variables have been developed. Our process modelling consists of verifying the thermalization of the absorbed laser energy, calculating peak gas temperatures, stating probable molecular reactions, and measuring growth rates in terms of appropriate process variables.

Thermalization of Absorbed Photon Energy. When an infrared photon is absorbed, the SiH_4 molecule makes a transition to an excited vibrational level. Collisions redistribute the absorbed energy returning the absorbing molecule to its ground state where it is again able to absorb another photon. For LICVD conditions (~ 1 watt absorbed/cm, P_{SiH_4} ~ 10 torr and T_g ~ 800K), a molecule absorbs a photon every 700 μs

110

and the average time between collisions is 0.015 µs. Vibrational-vibrational (V→V), vibrational-rotational (V→R), and rotational-translational (R→T) relaxation times are 0.5 µs, 100 µs, and 0.1 µs respectively.[58],[59] Therefore, most of the molecules experience sufficient collisions to relax to a state where the absorbed vibrational energy is distributed equally among the various degrees of freedom of the molecule. Thus, under the conditions employed in LICVD, the overall effect of absorbing IR photons is simply to increase the gas temperature.

Peak Gas Temperature. The peak gas temperature, T_g, can be calculated from a steady-state energy balance between the energy absorbed by the gas and the energy lost by thermal conduction. For a highly simplified system geometry, the steady-state energy balance equation is:

$$W \alpha = \frac{2\pi}{\ln(2D/r)} \kappa (T_g - T_s) \qquad \text{(watts/cm)} \qquad (1)$$

where T_g is the gas temperature, T_s is the substrate temperature, κ is the thermal conductivity of the gas, α is the absorptivity of the gas, and W is the laser power. $2\pi/\ln(2D/r)$ is the thermal conduction shape factor for an isothermal cylinder (the laser beam) of radius r at a distance D from an isothermal surface (the substrate).[60]

The absorptivity, α, depends on the number of absorbing molecules ($P_{SiH_4} \equiv P$), the frequency difference between the laser line and the SiH_4 absorption line, and the respective line widths.[61] With LICVD processing conditions, the principal source of line overlap results from pressure broadening, which is approximately proportional to the SiH_4 pressure, P. Therefore, α is proportional to the square of the SiH_4 pressure:

$$\alpha = \alpha_o P^2 \qquad (2)$$

Equation (1) can be rewritten in a form that illustrates how the gas temperature depends on the various experimental parameters:

$$T_g = T_s + \frac{\ln(2D/r)}{2\pi\kappa} W\alpha_o P^2 \quad . \qquad (3)$$

Equation (3) indicates that T_g is dependent on the laser power (W), the square of the silane pressure, and inversely dependent on the gas thermal conductivity. To calculate T_g accurately we have solved a more general version of Equation (3), using temperature dependent values for α and κ, and our complete system geometry.[53] For the range of conditions used to deposit Si films, calculated gas temperatures range from approximately 450°C to 575°C.

Reaction Mechanism. Analogous to the HOMOCVD growth model proposed by Scott et al.,[54] we propose a model for LICVD in which the rate-limiting step is homogeneous thermal decomposition of the SiH_4:

$$SiH_4(g) \rightarrow SiH_2(g) + H_2(g) \quad . \qquad (4)$$

If SiH_4 decomposition is the rate-limiting step, the film growth rate G can be expressed as:

$$G \propto P_{SiH_4} e^{-E/kT_g} \qquad (5)$$

where E is the activation energy of the decomposition reaction. According to kinetic studies of homogeneous SiH$_4$ decomposition, the activation energy is 52–56 kcal/mole.[62]

Growth Rate Kinetics. Figure 11 summarizes the results of 80 a-Si:H film growth runs made under different process conditions; variables include laser power, gas pressure, gas composition, and substrate temperature. The growth rate per torr SiH$_4$ follows the Arrhenius dependence on T$_g$ predicted by Equation (5).

The value of the observed activation energy, 46 ± 5 kcal/mole, is slightly lower than the 52–56 kcal/mole reported for the pyrolysis of SiH$_4$. However, it is much higher than the ~ 35 kcal/mole reported for conventional CVD growth of Si from SiH$_4$ for which the rate-limiting step is reported to be surface diffusion.[63] Acknowledging the potential errors in calculated LICVD peak gas temperatures, we believe that the observed activation energy is in sufficient agreement with the higher values to conclude the process is indeed rate-controlled by SiH$_4$ dissociation.

Film Properties

Physical and Microstructural Properties. Adherent films have been deposited on fused silica, borosilicate glass, aluminum, and single-crystal silicon. All films were determined to be amorphous by electron diffraction, as expected from the low substrate temperatures used (T$_s$ < 400°C). Films 1 µm thick at the center of 2.5 cm x 5.0 cm substrates are typically 0.3 µm thick at the substrate edges which parallel the laser beam. Along the direction parallel to the beam axis, the films are very uniform.

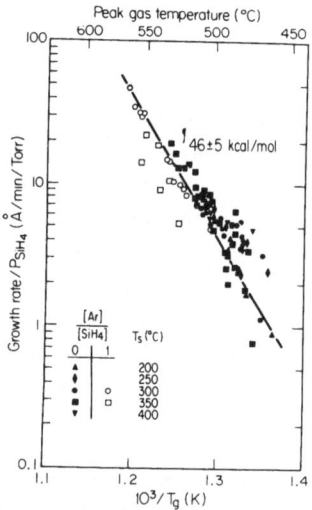

Fig. 11. Film growth rate per torr of SiH$_4$ versus reciprocal calculated gas temperature for various LICVD processing conditions.

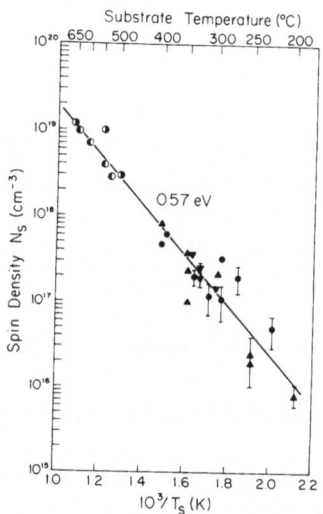

Fig. 12. Spin density versus reciprocal substrate temperature for LICVD films (●, ▼, ▲) and for conventional CVD films (◐ and ◑).

Electron Spins. The concentration of unpaired spins, N_s, evaluated by electron spin resonance (ESR), provides an important measure of the quality of the a-Si:H films.[64] The incorporation of hydrogen in a-Si films ordinarily decreases the defect concentration sufficiently to allow variation of the Fermi level by substitutional doping. Figure 12 shows an Arrhenius plot of N_s versus the substrate temperature for a-Si:H films deposited by the LICVD process. Values for conventional CVD films are also included.[65,66] The lowest N_s obtained thus far is $\approx 8 \times 10^{15} cm^{-3}$, which is nearly the same as the best films produced by glow-discharge[67] and HOMOCVD techniques.[68] The fact that CVD and LICVD values are on the same line emphasizes the similarity of the two processes. Moreover, the Arrhenius plot is quite linear which indicates that the residual dangling-bond concentration is in thermal equilibrium at the substrate temperature.

Electrical Conductivity in Doped Films. Recently, Bilenchi et al.[69] have reported sharp increases in the room-temperature conductivity of LICVD films after admixture of either PH_3 or B_2H_6 to the SiH_4. We have doped LICVD a-Si:H from light to very heavy levels and found the conductivity activation energy can be reduced to less than 0.2 eV for both n- and p-type material.[70,71] This should make the attainment of high open-circuit voltage in p-i-n LICVD solar cells possible.

Electrical conductivities were measured on films deposited over Al electrodes on fused silica. A four-contact geometry minimized contact resistance effects and the applied field was about 10^3 V/cm. Measurements were made in the dark at pressures between 5×10^{-7} Torr and 2×10^{-6} Torr. The films were first annealed at about 50°C below the deposition temperature for two hours to obtain an adsorbate-free surface[72] and to eliminate any changes due to the Staebler-Wronski effect.[73] The conductivity activation energy (E_a) and the room temperature conductivity (σ_{RT}) were computed by a least-squares fit to all points. Thermopower measurements at room temperature verified that B-doped and P-doped films have positive and negative majority carriers, respectively.

Figure 13 shows that by P-doping $T_s = 400°C$ films at PH_3/SiH_4 ratios from 0 to 8.8×10^{-3}, E_a and σ_{RT} are continuously controllable. An E_a as low as 0.19 eV indicates that P incorporation introduces a donor energy level less than 0.2 eV below the conduction band. The P-doping modulated σ_{RT} by a factor of up to 10^7. Both E_a and σ_{RT} show a P-doping efficiency in LICVD almost an order of magnitude lower than glow-discharge films;[74] i.e. considerably more PH_3 is needed in the reactant gas of LICVD to obtain a given E_a or σ_{RT} than in glow discharge films. Bilenchi et al[69] report an LICVD P-incorporation ratio of 0.2, compared to about 1.0 for glow discharge doping.[75] This lowered incorporation could explain most of the decrease in P-doping efficiency in LICVD a-Si:H.

Figure 14 shows comparable data for B doping of $T_s = 300°C$ films at B_2H_6/SiH_4 ratios from 0 to 1×10^{-4}. An E_a as low as 0.19 eV indicates that B incorporation introduces an acceptor energy level less than 0.2 eV above the valence band. The B-doping modulated σ_{RT} by a factor of up to 10^{13}.

Hydrogen. The hydrogen concentration, [H], was determined by the effusion technique.[67] Kampas and Griffith[76] have proposed a model for the growth of a-Si:H films which relates the ratio of the rates of $-SiH_3$ deactivition, r_1, and hydrogen elimination, r_2, to the function $[H]^{-1}-3$. An Arrhenius plot of this function versus T_s is shown in Figure 15 for both LICVD films and conventional CVD films.[76,77] The plot is reasonably linear with an activation energy of 0.59 eV. This value, essentially the same as for dangling bond creation, represents the difference in activation energies of the hydrogen elimination and $-SiH_3$ deactivation processes.

Infra-red spectra of the films on crystal Si wafers were taken with an IBM IR/85 FTIR using a crystal Si reference. Undoped LICVD films grown at $T_s = 400°C$ show little bonded H, while films grown at $T_s = 200°C$ exhibit peaks at about 2095, 890 and 845 cm^{-1} usually attributed to $(SiH_2)_n$. These spectra were unchanged by P-doping up to PH_3/SiH_4 ratios of 8.8×10^{-3}. As shown in Figure 16, the undoped and lightly B-doped 300°C LICVD films ($B_2H_6/SiH_4 < 5\times10^{-6}$) also exhibit $(SiH_2)_n$ IR peaks. However, the heavily doped film ($B_2H_6/SiH_4 = 1 \times 10^{-4}$) shows little bonded H, with small peaks at 2000 cm^{-1} and 2090 cm^{-1} suggesting a more equal distribution of H between SiH and SiH_2 configurations.

Stress. Stresses in the films were measured at the deposition temperature by measuring the curvature of the film-substrate system with a He-Ne laser beam reflected from the sample.[78] The LICVD a-Si:H films are under tensile stress for all process conditions studied. This contrasts

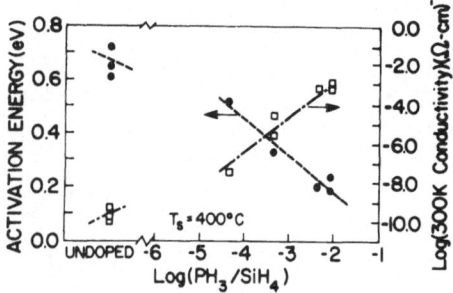

Fig. 13. Dark conductivity parameters E_a (●) and σ_{RT} (□) as functions of the PH_3/SiH_4 ratio in the reactant gas for $T_s = 400°C$ films.

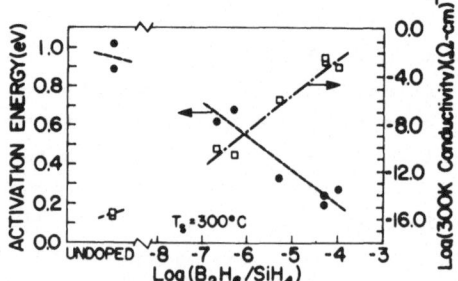

Fig. 14. Dark conductivity parameters E_a (●) and σ_{RT} (□) as functions of the B_2H_6/SiH_4 ratio in the reactant gas for $T_s = 300°$ films.

with glow discharge films, which are always under compressive stress.[7,9]
For undoped films grown at typical deposition rates ($\sim$ 40Å/min), the
stress appears to be a function of T_s only and exhibits a maximum of about
5×10^9 dynes/cm^2 at $T_s \approx 300°C$. The stress decreases to 1.5×10^9 and 5×10^8
dynes/cm^2 at deposition temperatures of 400°C and 200°C respectively.
Light to moderate B_2H_6-doping ($2 \times 10^{-7} < B_2H_6/SiH_4 < 5 \times 10^{-6}$) reduces the
stress levels to $\sim 8 \times 10^8$ dynes/cm^2 at $T_s = 300°C$. However, the stress
increases again to $4-7 \times 10^9$ dynes/cm^2 as the doping level is increased to a
B_2H_6/SiH_4 ratio of 1×10^{-4}. PH_3 doping does not affect the stress levels
appreciably for the range of conditions studied. Thermal expansion
coefficients of undoped a-Si:H films increase with decreasing substrate
temperature[78] and B_2H_6 and PH_3 dopants have no observed effects on these
values. This should ensure that LICVD solar cells will not crack during
cooling after growth.

CONCLUSIONS

The two developed laser synthesis processes offer considerable
promise for large scale production of superior ceramic powders and thin
films. Both processes have addressed applications that require high
quality materials. Achieving viable processes with laser energy sources
requires fortuitous combinations of emission/absorption lines, laser
efficiences and reaction enthalpies; these, in-addition to the superior
material properties, are realized for Si, SiC and Si_3N_4 powders as well as
a-Si:H films using a CO_2 laser.

By modelling the powder nucleation and growth processes, it has been
possible to manipulate powder characteristics such as size, agglomeration
crystallinity and stoichiometry. Most importantly, relatively large
($\sim$ 0.1–0.3 μm), spherical nonagglomerated powders of all three materials can

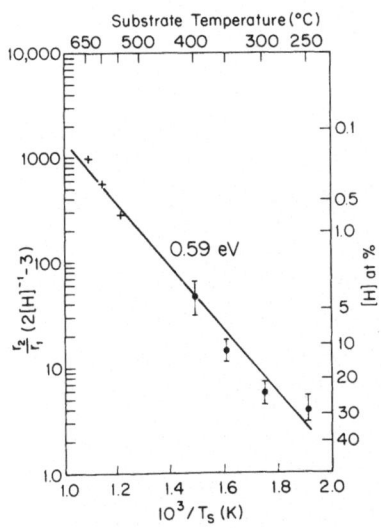

Fig. 15. $2[H]^{-1}-3$ versus $1/T_s$ for
LICVD and conventional CVD films.

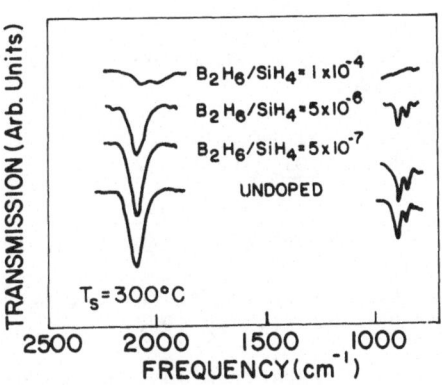

Fig. 16. Infra-red spectra of B-
doped and undoped LICVD films grown
at $T_s = 300°C$.

115

be produced under nearly maximum mass flow rate conditions. Using anerobic, anhydrous processes, the resulting powders can be dispersed and formed into high density ceramic parts without significant contamination. These nearly ideal ceramic powders make improved properties possible as well as critically needed improved reliability of the resulting ceramic parts. Manufacturing costs appear acceptable for structural ceramic parts.

The virtually unique separation of reactant gas and substrate temperature levels in a thermal CVD process has made possible high deposition rates of a-Si:H films whose properties can be manipulated by dopants and substrate temperature. Appropriate optical, electrical, structrual, microstructural and mechanical properties have been studied; in all cases the properties are superior. The process appears capable of covering large areas at acceptable deposition rates with device quality films.

For both the powder and thin film processes, better product charcteristics result directly from the unusual process conditions that are achieved. In each case the process attributes match conditions that yield the desired powders or films. Because of this match, the processes are stable, and reproducibly give materials having narrow ranges of nearly optimum properties.

ACKNOWLEDGEMENTS

Numerous staff, students and sponsors have made this work possible through their valued contributions. Hopefully, the references will properly reflect both individual research efforts and sponsorship of our research. All contributions are gratefully acknowledged.

REFERENCES

1. A. Kaldor and R. L. Woodin, Proceedings of the IEEE, 70, 565-578 (1982).

2. V. Malatesta, C. Willis, and P. A. Hackett, J. Am. Chem. Soc., 103, 6781 (1981).

3. J. Wolfrum, M. Kneba, and P. N. Clough, W. German Patent 293,853 (1981).

4. S. S. Miller, D. D. DeFord, T. J. Marks, and E. Weitz, J. Am. Chem. Soc., 101, 1036 (1979).

5. A. Kaldor, R. B. Hall, D. M. Cox, J. A. Horsley, P. Rabinowitz, and G. M. Kramer, J. Am. Chem. Soc., 101, 4465 (1979).

6. P. Rabinowitz, A. Kaldor, A. Gnauck, R. L. Woodin, and J. S. Gethner, Opt. Lett., May 1982.

7. R. V. Ambartzumian, Yu. A. Gorokhov, S. L. Grigorovich, V. S. Letokhov, G. G. Makarov, Yu. A. Malinin, A. A. Puretskii, E. P. Filippov, and N. P. Furzikov, Sov. J. Quantum, 7, 96 (1977).

8. J. A. Merrit and L. C. Robertson, J. Chem. Phys., $\underline{67}$, 3545 (1977).

9. S. M. Freund and W. C. Danen, Inorg. Nucl. Chem., $\underline{15}$, 45 (1979).

10. J. H. Flint and J. S. Haggerty, Applications of Lasers to Industrial Chemistry, Proceedings of SPIE, Vol. 458, The International Society for Optical Engineering, Bellingham, Washington, p.108, 1984.

11. J. H. Flint, M. Meunier, D. Adler, J. S. Haggerty, Laser Assisted Deposition, Etching, and Doping, Proceedings of SPIE, Vol. 459, The International Society for Optical Engineering, Bellingham, Washington, p. 66, 1984.

12. R. Bilenchi, I. Gianinoni and M. Musci, J. Appl. Phys., $\underline{53}$, 6479-81 (1982).

13. C. P. Christansen and K. M. Lakin, Appl. Phys. Lett., $\underline{32}$, 254 (1978).

14. M. Hanabusa, A. Namiki, K. Yoshihara, Appl. Phys. Lett., $\underline{35}$, 626 (1979).

15. S. D. Allen, J. Appl. Phys., $\underline{52}$, 6501 (1981).

16. C. F. Chen and R. M. Osgood, Laser Diagnostics and Photochemical Processing for Semiconductor Devices, Materials Research Society Symposia Proceeding's, Vol. 17, Ed. by R. M. Osgood, S. R. F. Brueck and H. R. Schlossbert, North-Holland, Amsterdam, p. 169-175, 1983.

17. I. Nadler, H. Reisler, and C. Wittig, Applications of Lasers to Industrial Chemistry, Proceedings of SPIE, Vol. 458, The International Society for Optical Engineering, Bellingham, Washington, p. 17, 1984.

18. J. Y. Tsao and D. J. Ehrilich, Laser Assisted Deposition, Etching, and Doping, Proceedings of SPIE, Vol. 459, The International Society for Optical Engineering, Bellingham, Washington, p. 1, 1984.

19. F. A. Houle, Laser Assisted Deposition, Etching, and Doping, Proceedings of SPIE, Vol. 459, The International Society for Optical Engineering, Bellingham, Washington, p. 110, 1984.

20. J. S. Haggerty and W. R. Cannon, Sinterable Powders from Laser Driven Reactions, MIT-EL 78-037, Annual Report N00014-77-C-0581, October 1978.

21. T. R. Gattuso, M. Meunier, and J. S. Haggerty, "Laser Induced Deposition of Thin Films," MIT-EL 82-022, Annual Report, May 1982.

22. E. A. Barringer and H. K. Bowen, Presented at 12th Automotive Materials Conference, March 1984. To be published in Ceramic Engineering and Science Proceedings.

23. H. K. Bowen, Mat. Sci. Eng., $\underline{44}$, 1-56 (1980).

24. E. A. Barringer, N. Jubb, B. Fegley, R. L. Pober and H. K. Bowen, Ultra-structure Porcessing of Ceramics, Glasses and Composites, Ed. L. L. Hench and D. R. Ulrich, John Wiley & Sons, New York, p. 315, 1984.

25. J. S. Haggerty, G. Garvey, J-M Lihrmann and J. E. Ritter, "Processing and Properties of Reaction Bonded Silicon Nitride made from Laser Synthesized Silicon Powders", Proceedings of MRS Symposium, Boston, MA, December 1985.

26. E. A. Barringer, "The Synthesis, Interfacial Electrochemistry, Ordering, and Sintering of Monodispersed TiO_2 Powders", Ph.D. Thesis, MIT, Cambridge, MA, September 1983.

27. W. R. Cannon, S. C. Danforth, J. H. Flint, J. S. Haggerty, and R. A. Marra, J. Am. Ceram. Soc., 65, 324-30 (1982).

28. W. R. Cannon, S. C. Danforth, J. S. Haggerty, and R. A. Marra, J. Am. Ceram. Soc., 65, 330-5 (1982).

29. K. Sawano, "Formation of Silicon Carbide Powder from Laser Induced Vapor Phase Reactions", Ph.D. Thesis, M.I.T., June 1985.

30. G. Greskovich, and J. H. Rosolowski, J. Am. Ceram. Soc., 59, 285-8 (1976).

31. D. Casey and J. S. Haggerty, "Laser-Induced Vapor-Phase Syntheses of Boron and Titanium Diboride Powders, submitted for publication in the Journal of Materials Science, 1986.

32. J. D. Casey and J. S. Haggerty, "Laser-Induced Vapor Phase Synthesis of Titanium Diaoxide", to be submitted for publication in the Journal of Materials Science, 1986.

33. S. Akmandor, "Theoretical and Computational Model of Reacting Silane Gas Flows: Laser Driven Pyrolysis of Subsonic and Supersonic Jets", Ph.D. Thesis, M.I.T., June 1985.

34. J. H. Flint, "Powder Temperature in Laser Driven Reactions", M. S. Thesis, M.I.T., February 1982.

35. J. H. Flint, R. A. Marra and J. S. Haggerty, "Powder Temperature, Size, and Number Density in Laser-Driven Reactions", Aerosol Science and Technology, Vol. 5, 2, 249-260 (1986).

36. R. A. Marra, "Homogeneous Nucleation and Growth of Silicon Powder from Laser Heated Gas Phase Reactants", Ph.D. Thesis, M.I.T., February 1983.

37. R. A. Marra and J. S. Haggerty, "Homogeneous Nucleation and Growth of Silicon Powder from S Laser Heated SiH_4", submitted for publication in the J. Am. Ceram. Soc., (1984).

38. A. D'Slessio, A. Dilorenzo, A. F. Sarofim, F. Beretta, S. Masi, and C. Venitozzi, "Soot Formation in Methane-Oxygen Flames", Fifteenth Symposium (International) on Combusion 1427, (1975), The Combustion Institute, Pittsburg, PA.

39. R. A. Marra, "The Crystal Structure of Silicon Powders Produced from Laser Heated Silane", submitted for publication in the J. Am. Ceram. Soc., (1986).

40. Y. Suyama, R. A. Marra, J. S. Haggerty and H. K. Bowen, "Synthesis of Ultrafine SiC Powders by Laser Driven Gas Phase Reactions", submitted for publication to the J. Am. Ceram. Soc., 64, 10, 1356-59 (1985).

41. J. S. Haggerty, Sinterable Powders from Laser Reactions, MIT-EL 82-002, Final Report N00014-77-C-0581, September 1981.

42. S. Mizuta, W. R. Cannon, A. Bleier, and J. S. Haggerty, "Wetting and Dispersion of Silicon Powder Without Deflocculants", Am. Ceram. Soc. Bull., 61, 872-5 (1982).

43. S. C. Danforth, (Rutgers University, NJ), and M. Dahlen, (Volvo, Sweden), unpublished results.

44. G. Garvey, M.I.T., unpublished results.

45. R. S. Aries, and R. D. Newton, Chemical Engineering Cost Estimation, Chemonomics, Inc., New York, April (1951).

46. R. H. Baney, Dow Corning, private communication.

47. F. Chambers, Standard Oil Co. (Indiana), private communication.

48. P. Orinsnshky, Union Carbide Co., private communication.

49. R. Cannon, M.I.T., (University of California, Berkeley CA), private communication.

50. "Comminution and Energy Consumption", National Materials Advisory Board (NAS-NAE), PB81-225708, May 1981

51. M. Meunier, T. R. Gattuso, D. Adler, and J. S. Haggerty, Appl. Phys. Lett., 43, 273-5 (1983).

52. T. R. Gattuso, M. Meunier, D. Adler, and J. S. Haggerty, "IR Laser-Induced Deposition of Thin Films", in Laser Diagnostics and Photochemical Processing for Semiconductor Devices, R. M. Osgood, S. R. J. Bruek, H. R. Schlossberg, Eds., North-Holland, 215-221,(1983).

53. M. Meunier, J. H. Flint, D. Adler, and J. S. Haggerty, Materials Research Symposia Proceedings, Laser-Controlled Chemical Processing of Surfaces, Ed. A. N. Johnson and D. J. Ehrlich, (in press)

54. B. A. Scott, R. M. Placenik and E. E. Simonyi, Appl. Phys. Lett., 39, 73 (1981).

55. R. Bilenchi, I. Gianinoni, M. Musci and R. Murri, "Laser Induced Chemical Vapor Deposition of Hydrogenated Amorphous Silicon", Laser Diagnostics and Photochemical Processing for Semiconductor Devices, R. M. Osgood, S. R. J. Brueck, H. R. Schlossberg, Eds., North-Holland, 199-205 (1983).

56. M. Meunier, "Amorphous Silicon Produced by Laser Induced Chemical Vapor Deposition", Ph.D. Thesis, M.I.T., September 1984.

57. J. H. Flint, M. Meunier, D. Adler and J. S. Haggerty, "a-Si:H Films Produced from Laser Heated Gases: Process Characteristics and Film Porperties", Laser Assisted Deposition, Etching, and Doping, SPIE, Vol. 459, 1984.

58. P. A. Longeway and F. W. Lampe, J. Am. Chem. Soc., 103, 6813-8 (1981).

59. R. D. Levine and R. B. Bernstein, Molecular Radiation Dynamics, (Oxford University Press, NY, Chap. 5, 1974.

60. J. P. Holman, Heat Transfer, McGraw-Hill, NY, NY 1976.

61. A. C. G. Mitchell and M. W. Zemansky, <u>Resonance Radiation and Excited Atoms</u>, (Cambridge University Press, Cambridge, U. K. 1961).

62. C. G. Newman, H. E. O'Neal, M. A. Ring, F. Leska and N. Shipley, Int. J. Chem. Kinetics, <u>11</u>, 1167 (1979).

63. A. M. Beers and J. Bloem, Appl. Phys. Lett., <u>41</u>, 153 (1982).

64. D. Adler, J. de Physique, <u>42</u>, C4, 3-14 (1981).

65. S. Hasegawa, T. Kasajima and T. Shimizu, Phil. Mag., B, <u>43</u>, 149-56 (1981).

66. P. Hey and B. O. Seraphin, Sol. En. Mat., <u>8</u>, 215-30 (1982).

67. H. Fritzsche, Sol. En. Mat., <u>3</u>, 447-501 (1980).

68. B. A. Scott, J. A. Reimer, R. M. Plecenik, E. E. Simonyi and W. Reuter, Appl. Phy. Lett., <u>40</u>, 973 (1982).

69. R. Bilenchi, I. Gianinoni, M. Musci, R. Murri and S. Tacchetti, Appl. Phys. Lett. <u>47</u>, 279 (1985).

70. H. M. Branz, S. Fan, J. H. Flint, B. T. Fiske, D. Adler and J. S. Haggerty, Appl. Phys. Lett. <u>48</u> (2), 171-3. (1986).

71. H. M. Branz, M. Meunier, S. Fan, J. h. Flint, J. S. Haggerty and D. Adler, Proceedings of the Eighteenth IEEE Photovoltaic Specialist Conference, Las Vegas Nevada, October 1985.

72. M. Tanielian, Philos. Mag. <u>B45</u>, 435 (1982).

73. D.L. Staebler and C.R. Wronski, Appl. Phys. Lett. <u>31</u>, 292 (1977).

74. W.E. Spear and P.G. LeComber, Philos. Mag. <u>33</u>, 935 (1976).

75. W. Beyer and H. Overhof, in <u>Semiconductors and Semimetals</u>, Vol. 21C, edited by J. I. Pankove, Academic, Orlando, 1984 p. 257.

76. F. J. Kampas and R.W. Griffith, Appl. Phys. Lett. <u>39</u>, 407 (1981).

77. P. C. Booth, D. D. Alfred and B. D. Seraphin, J. Non-Cryst. Solids, <u>35-36</u>, 213 (1980).

78. L.V. Atkins, "Mechanical Properties of Hydrogenated Amorphous Silicon Films," S.B. Thesis, June 1984, MIT, Cambridge, MA.

79. J.P. Harbison, A.J. Williams, and D.V. Lang, J. Appl. Phys. <u>55</u>, 946 (1984).

NEW APPROACHES TO THE DESIGN OF MATERIALS VIA PREPARATIVE INORGANIC CHEMISTRY

Abraham Clearfield

Department of Chemistry
Texas A&M University
College Station, Texas 77843

"An important goal of materials research is the ability to design and synthesize, in high yield, new materials whose structure and properties can be predicted, varied and controlled." This statement taken from "Trends and Opportunities in Materials Research," a report (1) to the National Science Foundation by its Advisory Committee, is the challenge held out to the synthetic chemist by the demands of materials technology. The report goes on to state that the goal can only be achieved when the structure and stability are well enough understood on an atomic and molecular scale to predict macroscopic properties. Frequently the processing following the synthesis determines the ultimate structure and properties of the end product. With this in mind, we have structured the symposium to primarily focus on innovative synthetic procedures, materials modifications and processing, and structure-property relationships.

Polymer chemistry and synthetic organic chemistry have benefited greatly from an understanding and control of molecular architecture. However, the inorganic chemist is not so fortunate. We do not yet have a sufficient knowledge of the effect of different synthetic procedures to predict the final outcome. Nevertheless the application of inorganic synthetic techniques to the broad area of materials preparation is yielding a cornucopia of new materials. We have already seen the power and possibilities of sol-gel techniques and the use of organometallic and inorganic polymer routes to ceramic materials in earlier chapters. In this brief essay, I wish to focus on some more conventional processes or combination of processes to arrive at new products.

Some of the synthetic procedures available to the inorganic chemist for the preparation of bulk solids are listed below:

1. coprecipitation
2. hydrothermal methods
3. ion exchange
4. electrochemical methods
5. intercalation
6. vapor phase transport
7. high pressure synthesis
8. sol-gel processes
9. pyrolysis
10. photolysis

Application of two or more of these procedures, simultaneously or in tandem, often produces results unobtainable by one procedure alone. Layered compounds in particular are susceptible to controlled manipulation and will form the major class of compounds dealt with. In this essay we shall give a brief description of the preparation and reactions of α-zirconium phosphate, sodium titanates and manganese dioxide.

Our odyssey begins with the preparation of α-zirconium phosphate, the prefix implying that other forms exist. Addition of a soluble zirconium salt to phosphoric acid results in the precipitation of an amorphous gel. However, refluxing the gel in excess phosphoric acid leads to the formation of a crystalline layered product of composition $Zr(HPO_4)_2 \cdot H_2O$ (2). The layers are clay-like in that they consist of ZrO_6 octahedra and phosphate tetrahedra with this difference; three oxygens of the tetrahedron bridge to three metal atoms. This leaves the fourth oxygen pointing away from the mean plane of the layer into the interlamellar space (Fig. 1). This oxygen bonds to the proton which is available for ion exchange reactions. In a clay, the silicate tetrahedra are in a reversed position to that of the phosphate groups in zirconium phosphate.

The process of crystallization by refluxing is a slow one requiring in excess of 500 hr in 12M H_3PO_4 for completion (3,4). As a consequence this system is ideal for demonstrating clearly the effect of crystal perfection on chemical and physical properties. It has been shown that the surface area can vary from about 150 m^2/g in the gels to less than 1 m^2/g in the crystals (5). Unit cell dimensions decrease with crystal perfection to a minimum at about 100 hr reflux time and then increase slightly (4). Proton conductivity is found to decrease with increased crystal perfection, the amorphous gels having values approaching 10^{-2} ohm^{-1}cm^{-1} at room temperature (6). This is

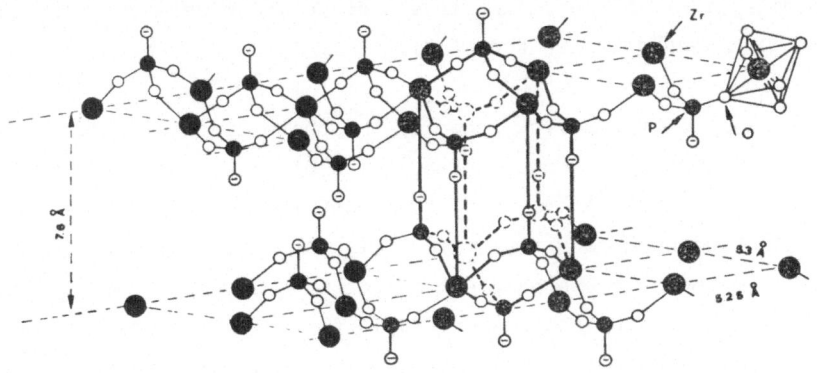

Fig. 1. A portion of the α-zirconium phosphate structure showing the relationship between adjacent layers. A cavity is shown in heavy outline and the negatively marked circles are oxygens bonded to ex-
changeable protons.

generally attributed to the freer movement of water in the gels. It is also
responsible for the conductivity on the surface of the crystals exceeding
that in the interior by a thousandfold (7) as the water between the layers is
thought to have a restricted motion.

The most striking illustration of the effect of crystallinity is evi-
denced in the ion exchange properties (8). Figure 2 shows the titration
curves for sodium ion exchange with an amorphous gel and a highly crystalline
sample. The strikingly different curves reflect structural differences in
the exchangers. The gel swells in water so that the exchanged sodium ion can
uniformly distribute itself by diffusion throughout the gel forming a solid
solution, $Na_xH_{2-x}Zr(PO_4)_2 \cdot yH_2O$. Each addition of sodium ion to the solid
changes the activity of the remaining protons in the gel allowing the
external pH to increase. In the case of the crystals the initial uptake of
sodium ions is on the surface but as diffusion into the interior takes place
a second phase forms as shown in eq. (1) (9).

$$Zr(HPO_4)_2 \cdot H_2O + Na^+ \longrightarrow ZrNaH(PO_4)_2 \cdot 5H_2O + H^+ \qquad (1)$$

The phase rule requires that the system be invariant [3 phases (two solids
and one liquid) and 3 components (i.e., Na^+ and H^+ in solution and Na^+ mole
fraction in the solid) at constant temperature and pressure]. As a result
the pH and sodium ion concentration in the external solution remains constant
until all the zirconium phosphate is converted to the half exchanged sodium
phase. Then the pH rises since one phase has disappeared. Further exchange
results in the conversion of the half sodium phase to $Zr(NaPO_4)_2 \cdot 3H_2O$ with

the system again invariant. All levels of behavior between these two extremes have been demonstrated (3,4).

Dehydration of the solid zirconium phosphate containing different levels of sodium ion leads to a complicated series of phase changes (Fig. 3) (10). For our purposes here we only wish to illustrate how the layered zirconium phosphate may be used as a synthetic tool, so the phases will not be further described. On heating the half exchanged phase to approximately 500°C, it is converted to the triphosphate $NaZr_2(PO_4)_3$ (9). It has recently been shown that triphosphates, and particularly those of the alkaline earth metals, exhibit a low coefficient of expansion (11-13). This ion exchange route represents an easy path to their preparation. However, with divalent cation exchanged zirconium phosphate a rather complicated series of phase changes is observed on heating. In the case of Ca^{2+}, Mn^{2+} and Zn^{2+} a triphosphate-like phase is obtained at about 700°C even though the composition is $ZrM(II)(PO_4)_2$ (14). These compounds require further characterization as they may lead to entire new families of low expansion solids.

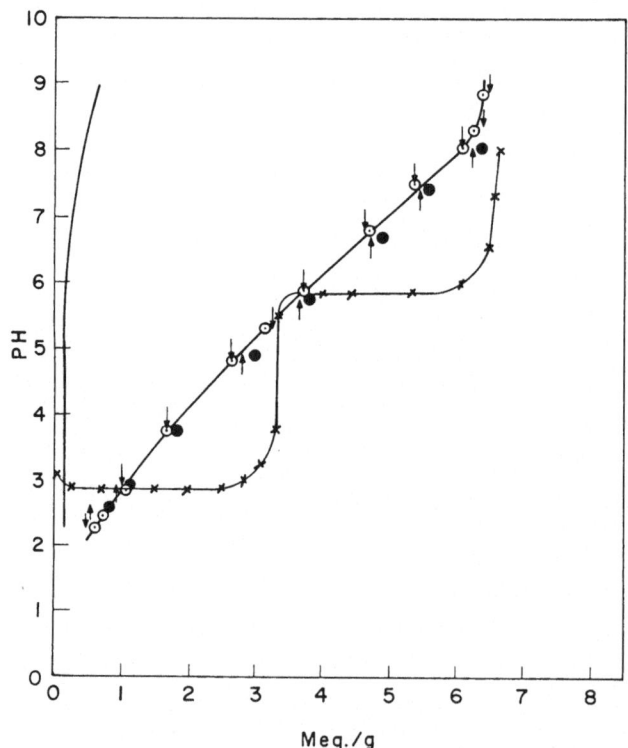

Fig. 2. Potentiometric titration curves for α-zirconium phosphate. Crystalline sample - x; amorphous sample - O, forward direction;
● - reverse direction. Titrant: 0.1M (NaCl + NaOH).

Another procedure which we have utilized as a synthetic tool is to precede the ion exchange reaction by intercalation. For example, the crystalline group(IV) phosphates have small interlayer distances of the order of 7.6Å (15) and do not swell in water. Thus it is difficult to incorporate large species between the layers. However, amines are easily intercalated into these compounds (16) forming a bilayer. In the case of the butylamine intercalate of zirconium phosphate the interlayer distance is 18.6Å. With the layers in this condition it is now possible to exchange large cations and complexes for alkyl ammonium ions (17). This procedure may be quite general for use with other layered compounds.

Ion exchange or intercalation techniques may be applied to a wide variety of compounds as a means of effecting novel syntheses. An interesting example is furnished by Tournoux, et al. (18). They synthesized $K_2Ti_4O_9$ from KNO_3 and TiO_2 at 1000°. On treatment with acid the corresponding acid $H_2Ti_4O_9 \cdot H_2O$ was obtained. Upon dehydration of this acid they obtained a new polymorph of TiO_2. At least three homologous series of alkali metal titanates are known with general formulae $Na_2Ti_nO_{2n+1}$, $Na_{2n+2}Ti_nO_{3n+1}$ and $Na_4Ti_nO_{2n+2}$ (19).

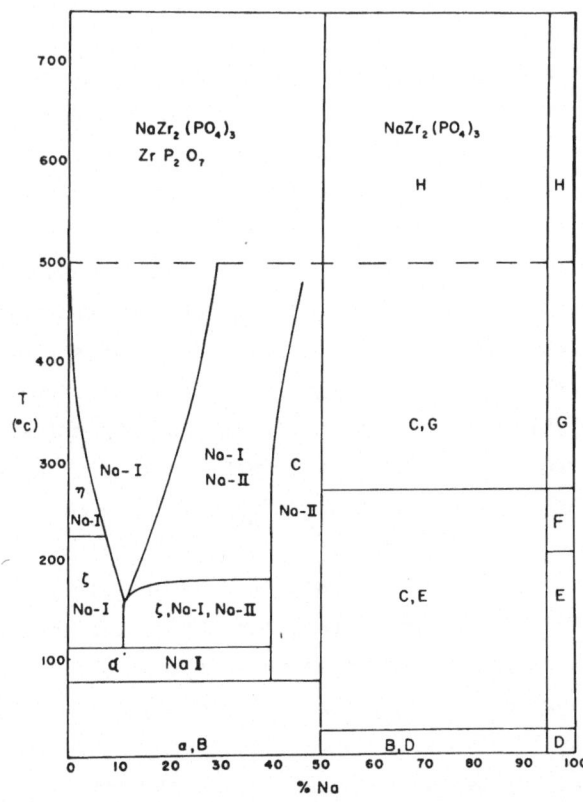

Fig. 3. Phases formed by sodium exchanged α-zirconium phosphate.

Almost all of these compounds and their potassium analogues can be converted to the protonated forms. It would be interesting to determine whether additional novel oxides would result from careful dehydration of these solid acids. But in any case, novel chemistry can be carried out with these compounds. We have prepared $Na_4Ti_9O_{20}$ by heating TiO_2 in NaOH. This nonatitanate shows promise as an exchange medium for the removal of radioactive strontium ion from nuclear waste streams (20). The exchange reaction with aqueous $SrCl_2$ yielded $Na_2(SrOH)_2Ti_9O_{20}$ whereas when the acid form was used the product was $Sr_2Ti_9O_{20}$. Undoubtedly the nature of the reaction is being controlled by the acidity or basicity of the exchanger. Furthermore when these products were heated, the reactions yielded totally different results as shown below.

$$Sr_2Ti_9O_{20} \xrightarrow{600°C} 2SrTiO_3 + 7TiO_2 \qquad (2)$$

$$Na_2[Sr(OH)]_2Ti_9O_{20} \xrightarrow{870°C} Na_2Ti_6O_{13} + \text{Unknown phase} + H_2O \quad (3)$$

The unknown phase in eq. (3) may well be $Sr_2Ti_3O_8$. An understanding of these reactions is essential to utilizing sodium titanates in nuclear processes.

Another example, using ion exchange, involved reactions of MnO_2. We prepared this compound electrolytically from an $MnSO_4$ solution in 1M H_2SO_4. The product obtained depends upon the temperature of the solution. At 10°C the electrolytic product was approximately $(Mn_{0.89}Mn_{0.11}O_{1.69})_2OH \cdot 2H_2O$ (21).⁻ The hydroxyl proton could be exchanged by Li^+, whereupon heating to 500°C yielded the spinel $LiMn_2O_4$. Treatment of this compound with acid converted it to HMn_2O_4. The protonated form was able to exchange with Li^+ but not with Na^+ and K^+ and thus is specific for Li^+. In contrast the original electrolytically prepared exchanger does exchange with larger ions such as Na^+ and K^+, but in lesser amounts the larger the cation. Rather than going through an electrolytic preparation, we found that the spinel could be formed at 500°C from the reaction of Li_2CO_3 with MnO_2 or Mn_2O_3. However if Na_2CO_3 is used in a similar reaction, then $Na_4Mn_{14}O_{27}$ is formed. Boiling this phase with water yielded birnessite, $Na_4Mn_{14}O_{27} \cdot 9H_2O$ (21) with expansion of the layers from 5.54Å to 7.04Å to accommodate the water. In this hydrated condition it is easy to exchange protons for the sodium ions but the exact composition of the new protonated phase has not been determined. Through a series of ion exchange reactions followed by heating it may well be possible to form a whole host of new metal manganites. We are using such reactions to prepare compounds with interesting magnetic and electrical properties.

Some time ago we attempted to grow single crystals of $Zr(NaPO_4)_2$ for crystal structure determinations using a hydrothermal procedure. To our surprise we obtained $NaZr_2(PO_4)_3$, at temperatures in the range of 250-300°C. This compound which we have mentioned earlier, is the end member of a family of superion conductors of general formula $Na_{1+x}Zr_2Si_xP_{3-x}O_{12}$ termed NASICON (23). Building on our hydrothermal procedure we mixed zirconium phosphate with an Na_4SiO_4 solution and heated the mixture at 300°C in a sealed tube. The resultant product, $Na_4Zr_2Si_{2.25}P_{1.8}O_{15}$, when heated above 1000°C loses P_2O_5 to yield a nonstoichiometric form of NASICON (24). The composition of this NASICON as determined by X-ray fluorescence was $Na_{3.3}Zr_{1.65}Si_{1.9}P_{1.1}O_{11.5}$. The reason for this strange stoichiometry became apparent from the results of a neutron diffraction study on a powder sample (25). The starting model used was the stoichiometric one proposed by Hong (23). However subsequent refinement of occupancy factors led to the formula $Na_{2.88}Zr_{1.68}Si_{1.84}P_{1.16}O_{12}$ which contains an insufficient amount of cationic charge. Especially noticeable is the low sodium ion content. We therefore added a contribution for Na^+ at the zirconium atom site and re-refined the data after each added contribution. The best results were obtained for 0.32 moles of Na^+ and this gave the formula $Na_{3.2}Zr_{1.68}P_{1.84}P_{1.16}O_{11.54}$ which accords very well with the results obtained by elemental analysis. The oxygen content was adjusted to balance the positive charge rather than through the refinement procedure. The close agreement between analysis and neutron defined compositions indicates that very little if any amorphous material was present.

In an effort to gain additional information on the stoichiometry of NASICONS, the sol-gel method was employed to prepare both stoichiometric and nonstoichiometric NASICONS. The nonstoichiometric preparation duplicated the composition of the NASICON obtained by the hydrothermal route. However, it gave slightly larger unit cell dimensions and this structure, determined by Rietveld refinement of neutron diffraction data (26), revealed that all the Na^+ were in the cavity sites. This then requires that some of the Zr^{4+} lattice sites are actually empty. Additional interesting facts arose from a neutron diffraction study of the stoichiometric NASICON prepared by the solgel method. Its composition based on occupancy refinement was $Na_{3.0}Zr_{1.93}Si_{1.9}P_{1.1}O_{11.91}$, and further, the phosphorus and silicon atoms were segregated into different crystallographic sites (26). In contrast, when the NASICON was prepared by a high temperature reaction of a mixture of oxides and phosphates, the Si and P atoms were randomly distributed over the available sites. Thus, the end product in this system appears to depend upon the method of preparation. While it is tempting to suggest that perhaps the equilibrium compositions have not been attained in these preparations, we feel that such is not the case, and that true nonstoichiometric or defect structures are the rule in this system. All the compositions can in fact be represented by the general formula

$$Na_{1+x+4y}Zr_{2-y-z}Si_xP_{3-x}O_{12-2z} \tag{4}$$

This formula allows for a deficiency of z moles of ZrO_2 and y moles of Zr^{4+} compensated by 4y moles of Na^+. It is now incumbent upon the synthetic chemist to find the limits of y and z.

We see from the examples described above that, while we have not been able to control our synthetic procedures so as to design specific structures or build in desired properties, we have been able to obtain new and interesting compounds. Further work along these lines needs to be pushed in many directions to begin to understand the underlying principles involved. However, there is one more area I would like to discuss in which a great deal of progress has been made and which gives promise of being able to design materials with specific structure-property relationships.

We have recently been engaged in a study of a new class of compounds which are organic derivatives of zirconium phosphate. These compounds were first described by Alberti, et al. (27) who prepared the phenyl derivative shown schematically in Fig. 4. This compound was synthesized by allowing phenylphosphonic acid to react with a soluble zirconium salt. The concept was enlarged by Dines and his coworkers who used phosphonic acids to bridge

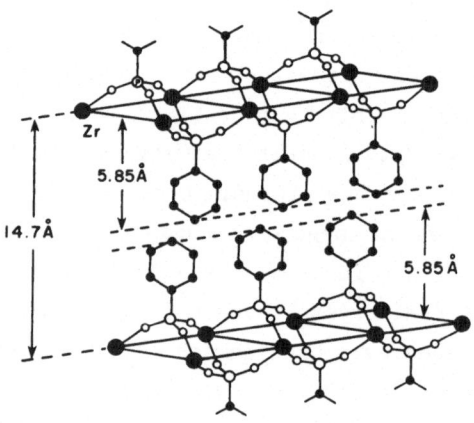

Fig. 4. Idealized crystal structure of zirconium bis(benzenephosphonate),
● - Zr; O - P; o - O; ● - C.

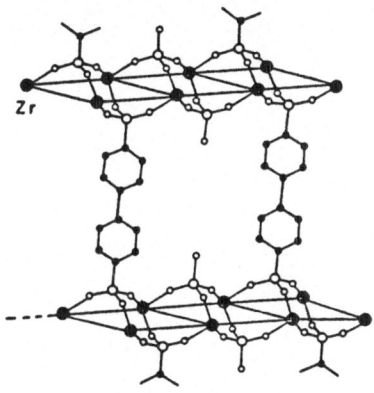

Fig. 5.   Pillared α-zirconium phosphate.   Pillars are separated by phosphate
or phosphite groups.
● - Zr; O - P; o - oxygen.

across layers (28).   By mixing in a certain amount of H_3PO_4 or H_3PO_3 with the
organic acid they were able to randomly space the pillars to achieve
derivatives with a mixed pore structure (Fig. 5).   We have repeated this work
and confirmed that, while surface areas are high, a variety of pore sizes are
created when the compound is precipitated rapidly.   This is shown in Fig. 6
where we have plotted the pore size distribution for a rapidly precipitated
triphenyl pillared phosphate.   On the other hand when precipitation is
carried out slowly, the pore size distribution is more regular as shown in
Fig. 7.   Regularity may be achieved by adding HF to the zirconium compound to

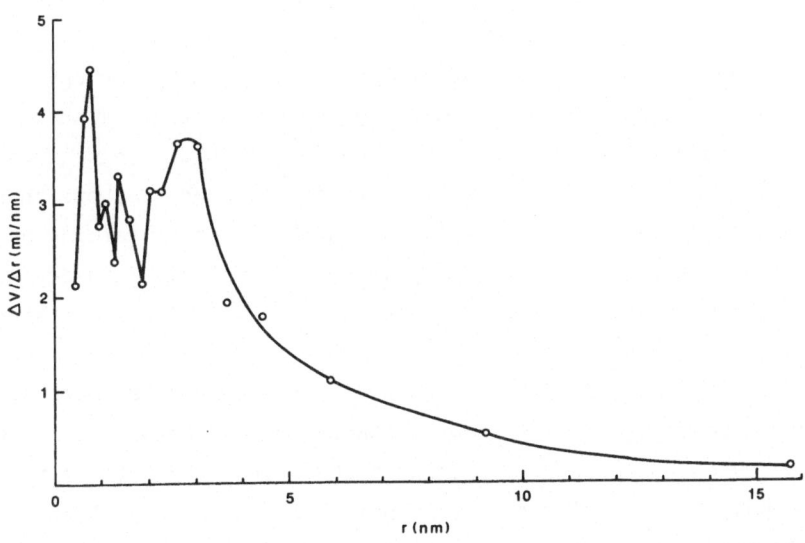

Fig. 6.   Pore size distribution in amorphous $[Zr(HPO_4)_{1.3}(O_3P-ph-ph-ph-PO_3)_{4.33}]$.

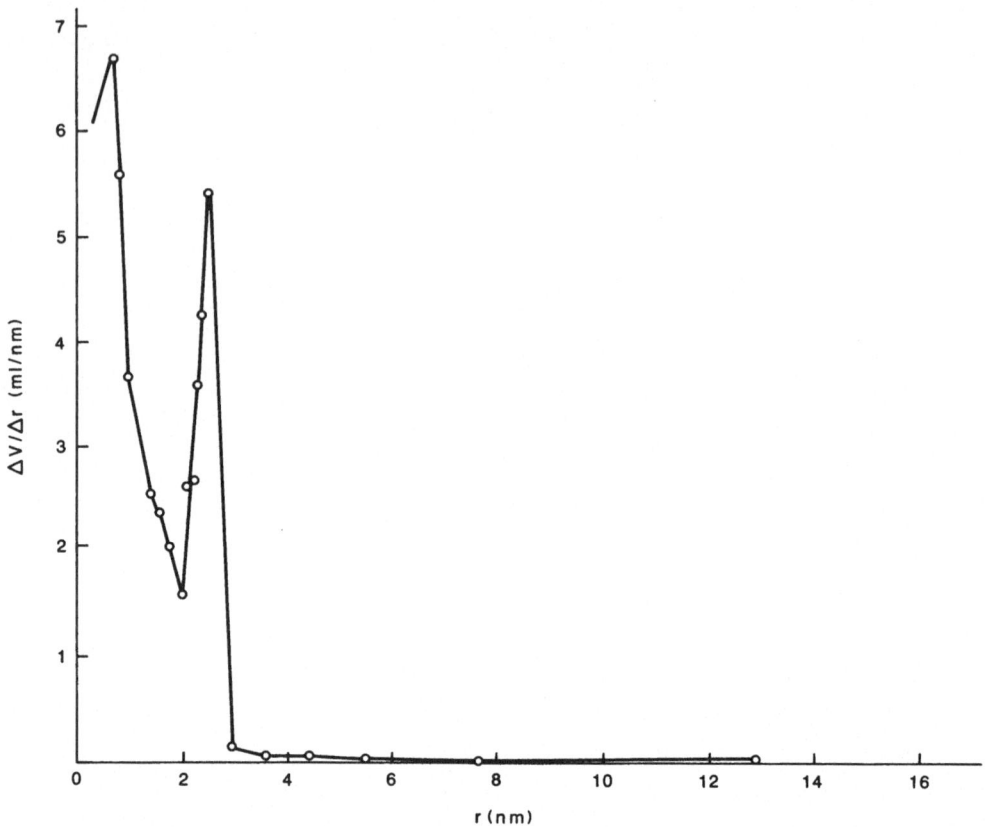

Fig. 7. Pore size distribution in crystalline $Zr(O_3PC_6H_4-C_6H_4PO_3)_{0.5}(HPO_4)$.

form $ZrF_6^=$. This anion keeps the zirconium in solution in the presence of
the phosphonic acid. However on heating to 60° or above, precipitation of
the desired product slowly takes place. The solid is much more crystalline
as shown by the sharper, more detailed X-ray powder pattern. Although the
use of HF promotes regularity it also has the undesirable effect of causing
segregation. That is, there is a tendency to form two phases, one being
totally inorganic [$Zr(HPO_4)_2$ or $Zr(HPO_3)_2$] and the other containing all of
the organic groups along with some phosphate or phosphite. In order to
obviate this behavior we have followed Dine's procedure of using mixtures of
ethylphosphate and the diphosphonic acid. The ethyl ester mixes in all
proportions with the phosphonic acid and does not appear to segregate. Thus,
the pillars can be spaced at different distances by choice of the ratio of
the ester to diphosphonic acid. Heating to 200°C splits out the ethyl group
leaving phosphate behind. The resultant products would appear to be promising candidates for a variety of separations and as catalysts or catalyst
supports. They are stable to about 350°C in the absence of air or strong

130

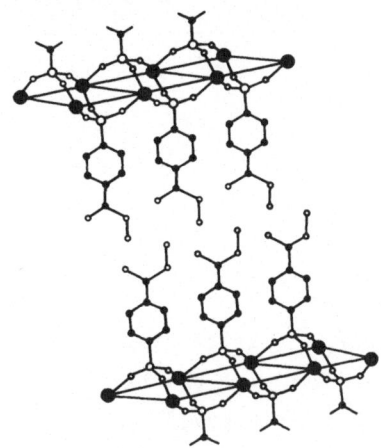

Fig. 8. Idealized crystal structure of P-carboxyphenylenephosphonate.

oxidizing agents.

Another class of interesting compounds results when the organic groups are functionalized. In our own work we prepared the phenyl carboxylate shown schematically in Fig. 8. When this compound is titrated to the endpoint with NaOH, it forms a clear, colloidal dispersion which is a powerful complexing agent for polyvalent cations. Other interesting derivatives resulted from the sulfonation of the phosphate-phosphonates as shown in Fig. 9. Derivative A is unpillared and forms colloidal dispersions in water since it swells to large interlayer distances. It prefers large divalent species as can be seen from a consideration of some K_d values (Table 1). The sequestering of large

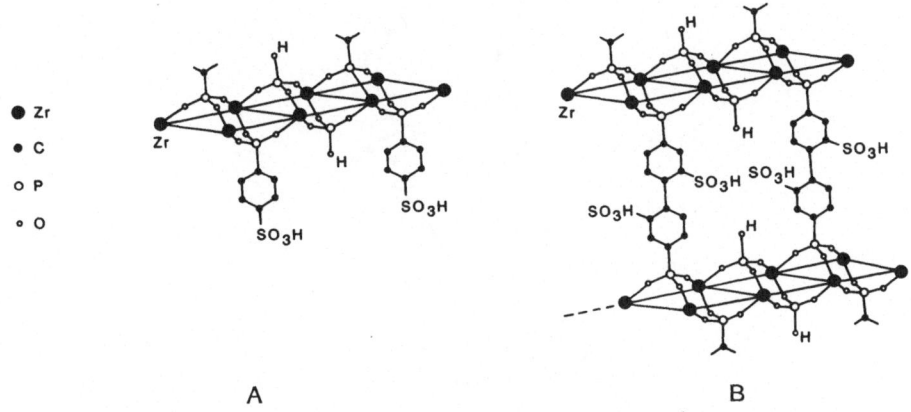

Fig. 9. Idealized crystal structure of layered zirconium phosphonates containing pendant sulphonic acid groups.

Table 1. K_d Values for $Zr(O_3PC_6H_5SO_3H)(HPO_4)$

Li^+	Na^+	K^+	Ca^{2+}	Ba^{2+}
52	108	375	24	20400

Initial Solution concentration $10^{-3}M$.

complexes, such as $[Ru(bipy)_3]^{2+}$, is so rapid that mixtures of two such complexes may be incorporated between the layers with a random distribution. This fact allowed us to space $[Ru(bipy)_3]^{2+}$ and methylviologen between the layers so that a photochemical electron transfer could take place. Since it is possible to change the distance between donor and acceptor species by increasing the number of phosphite groups, this system lends itself to a study of rates of electron transfer reactions and consequent photochemical water splitting reactions. Derivative B, in which the layers are bridged, shows sieving behavior, large complexes being excluded.

As a final example we shall mention the polyether derivatives of zirconium phosphate. We have prepared a series of these compounds of varying

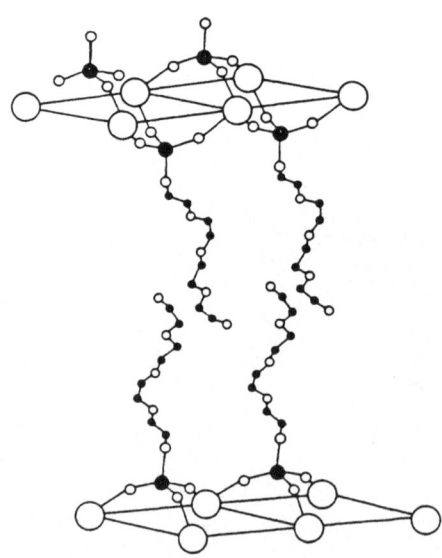

Fig. 10. Schematic depiction of zirconium tetraethyleneoxidephosphate derivative $\{Zr[O_3PO(CH_2CH_2O)_4H]_2\}$. Polyether chains are present on both sides of the layer.

polyether chain length (29). A schematic representation of the tetrameric derivative is shown in Fig. 10. Polyethers are known to incorporate electrolytes and become ionic conductors (30) and our derivatives are no exception. When the polyether chain length exceeds a degree of polymerization of 10, the layers exfoliate to form colloidal dispersions in water. This results from the highly hydrophilic nature of the polyether. Such dispersions may be used to prepare membranes or to coat inert substrates. Heating to 200°C removes the organic leaving a coating of zirconium phosphate on the substrate. Rapid surface exchange and ion separations are then possible.

We are attempting to obtain interlayered compounds by mixing colloidal dispersions of two different group IV metals and then reconstituting the layers by removal of the polyether chains. Some success in this direction has been achieved.

It is clear from the foregoing brief discussion that the application of ion exchange, sol-gel and hydrothermal techniques and the manipulation of layered compounds are fruitful areas for chemists to investigate. We need to understand the mechanisms of these processes for better control of the processes and design of new materials.

ACKNOWLEDGEMENT

This work was supported by the U. S. Army Research Office, Grant No. DAAG29-85-K-0124, for which grateful acknowledgement is made.

REFERENCES

1. "Trends and Opportunities in Materials Research," NSF 84-17; Washington, D. C., 1984.
2. Clearfield, A.; Stynes, J. A. J. Inorg. Nucl. Chem. (1964) 26, 117.
3. Clearfield, A.; Oskarsson, A.; Oskarsson, C. Ion Exchange and Membranes (1972) 1, 91.
4. Clearfield, A.; Oskarsson, A.; Kullberg, L. J. Phys. Chem. (1974) 78, 1150.
5. Clearfield, A.; Berman, J. J. Inorg. Nucl. Chem. (1981) 43, 2141.
6. Alberti, G.; Casciola, M.; Costantino, U.; Levi, G.; Ricciardi, G. J. Inorg. Nucl. Chem. (1978) 40, 533.
7. Alberti, G.; Casciola, M.; Costantino, U.; Levi, G. J. Membr. Sci. (1978) 3, 179.
8. For a more detailed description of the ion exchange properties see Clearfield, A. in "Inorganic Ion Exchange Materials," Clearfield, A.; Ed.; CRC Press: Boca Raton, Fl., 1982.
9. Clearfield, A.; Duax, W. L.; Medina, A. S.; Smith, G. D.; Thomas, J. R. J. Phys. Chem. (1969) 73, 3424.
10. Clearfield, A. "Annual Review of Materials Science," Huggins, R. A.; Ed.; Vol. 14; Annual Reviews Inc.: Palo Alto, California, 1984; pp. 205-229.
11. Alamo, J.; Roy, R. J. Am. Ceram. Soc. (1984) 67, C-78.

12. Roy, R.; Agrawal, D. K.; Alamo, J.; Roy, R. A. Mat. Res. Bull. (1984) 19, 471.
13. Oota, T.; Yamai, I. J. Am. Ceram. Soc. (1986) 69, 1
14. Clearfield, A.; Pack, S. P. Mat. Res. Bull. (1983) 18, 1343.
15. Clearfield, A.; Smith, G. D. Inorg. Chem. (1969) 8, 431.
16. See U. Costantino in Ref. 8, Ch. 3
17. Clearfield, A.; Tindwa, R. M. Inorg. Nucl. Chem. Let. (1979) 15, 251.
18. Marchand, R.; Brohan, M.; Tournoux, M. Mat. Res. Bull. (1980) 15, 1129.
19. Werthmann, R.; Hoppe, R. Z. Anorg. Allg. Chem. (1984) 519, 117.
20. Lehto, J.; Miettinen, J. in "Inorganic Ion Exchangers and Adsorbents for Chemical Processing in the Nuclear Fuel Cycle," IAEA-TECDOC-337, Vienna, 1985; pp. 9-17.
21. Shen, X.-M.; Clearfield, A. J. Solid State Chem. in press.
22. Clearfield, A.; Jirustithipong, P.; Cotman, R. N.; Pack, S. P. Mat. Res. Bull. (1980) 15, 1603.
23. Hong, H. Y.-P. Mat. Res. Bull. (1976) 11, 173.
24. Clearfield, A.; Jerus, P.; Cotman, R. N. Solid State Ionics (1981) 5, 301.
25. Rudolf, P. R.; Subramanian, M. A.; Clearfield, A.; Jorgensen, J. D. Mat. Res. Bull. (1985) 20, 643.
26. Clearfield, A.; Thomas, J. R. Inorg. and Nucl. Letters (1969) 5, 775.
27. Alberti, G.; Costantino, U.; Allulli, S.; Tomassini, N. J. Inorg. Nucl. Chem. (1978) 40, 1113.
28. Dines, M. B.; Di Giacomo, P. D.; Callahan, K. P.; Griffith, P. C.; Lane, R. H.; Cooksey, R. E. in "Chemically Modified Surfaces in Catalysis and Electrocatalysis," ACS Symp. Ser. 192, Washington, D.C., 1982.
29. Ortiz-Avila, C. Y.; Clearfield, A. Inorg. Chem. (1985) 24, 1773.
30. Shriver, D.; Papke, B.; Ratner, M.; Dupon, R.; Wong, T.; Brodwin, M. Solid State Ionics (1981) 5, 83.

MATERIALS DESIGN BY MEANS OF DISCHARGE PLASMAS

S. Veprek

Institute of Inorganic Chemistry,
University of Zürich, Winterthurerstr. 190
CH-8057 Zurich, Switzerland

ABSTRACT

Plasma chemistry of heterogeneous systems is still in its
infancy. Thus, most of the preparative work is being done by
the empirical approach of trial and error in tailoring the
materials properties. This requires a large number of
experimental points to be known of the multidimensional
experimental matrix. The present paper will concentrate on the
question of **controlled material design** which can be achieved
with the present understanding of the complicated plasmachemical phenomena under non-thermal conditions far away from
thermodynamical equilibrium. A question of this kind obviously
addresses the issue of non-equilibrium dissipative systems,
i.e. the domain of the thermodynamics of irreversible
processes. The master equation for a typical plasmachemical
system contains many terms with insufficiently known
elementary constants (cross sections) and it cannot be solved.
All problems can be "solved" and some answers "given" to most
of the questions when using "mathematical modelling". This
consists of trying to find a set of unknown elementary
constants which allow one to fit the measured curve. Although
one can gain some usefull information and a feeling of how to
design an appropriate experiment (e.g.ref./1/), this approach
has brought only a very limited improvement of our
understanding of the chemistry taking place in a given system.
Therefore we shall concentrate on several selected systems in
order to illustrate some general rules which govern the
chemical transformations in heterogeneous inorganic systems.
Whenever possible the systems to be discussed as illustration
will be chosen with respect to their potential applications in
industry as well as in basic solid state research.

1. INTRODUCTION

The low pressure plasmas are produced by electron impact excitation and ionization when electrical current is passing a gas at a pressure between about $1 \cdot 10^{-2}$ and a few millibar. Because of the non-thermal excitation, the degree of dissociation of molecular gases and the concentration of reactive radicals can reach significant values at the relatively low temperatures of several hundred degrees centigrade. Under such conditions the fluxes of the reactive species are of the order of 10^{17} to 10^{20} cm^{-2} s^{-1}. As the reaction probabilities for the deposition or for the formation of volatile products are relatively high, deposition and etch rates of several hundred Angströms per second can be achieved /2,3/. Therefore, plasmas are new attractive **synthetic tools of increasing importance.** In this presentation emphasis will be on the preparative aspects of plasma chemistry.

In general the plasmas can be used in three different ways /3,4/:

1) As "catalysts" to decrease the reaction temperature (or to increase the reaction rate);
2) To provide conditions where "a unique kind of chemistry" takes place which has no counterparts under conditions of thermodynamic equilibrium;
3) To modify the structural properties of the solid.

Plasma treatment of solid surfaces is becoming important for the preparation of wear and corrosion protective surface layers on metallic alloys, for electronic devices and for improvement of the surface properties of organic polymers. In these cases the reaction rate is controlled by the diffusion in the solid which, however, can be significantly enhanced under plasma conditions. The equilibrium phase diagrams generally do not apply to the plasma conditions where the stability range of compound phases in a solid-gas system can be extended to lower gas pressures and higher temperatures.

Plasma Induced and Plasma Assisted Chemical Vapour Deposition (PCVD) is increasingly being used for the preparation of thin films and bulk materials for various applications and for basic solid state research /4/. Among the most important applications of P CVD to be mentioned are thin films of electronic and optical materials, and wear and corrosion protective coatings of carbides, nitrides, borides and oxides.

The preparation of amorphous and nanocrystalline (nc) materials by means of P CVD has contributed significantly to our understanding of solid state science. Examples will be given including oxides, nitrides, III-V compounds and elemental semiconductors.

The more recent trend in P CVD of preparing inorganic materials from metalorganic volatile precursors will be discussed briefly. Epitaxial growth of silicon and gallium arsenide will also be discussed.

2. HETEROGENEOUS CHEMISTRY IN THE FOURTH STATE OF MATTER ?

The plasmas represent the fourth state of matter. Under thermodynamic equilibrium this state is reached only at high temperatures of $\geq 1 \cdot 10^4$ K where neither solids, nor chemical compounds exist. Thus, materials preparation and/or treatment in such "thermal plasmas" can be performed only under conditions of short contact time of the 'cold' solids with the hot plasma, or in the quenching zone of a fast streaming plasma jet. Typical processes utilizing these kinds of plasmas are plasma spraying of protective coatings, spheroidization and synthesis of finely dispersed refractories /5-7/. The fast quenching rate which is obtained enables one to also prepare metastable phases /5,8/.

The **non-isothermal**, "cold" plasmas are produced by electron impact excitation and ionization when electric current passes a diluted gas at a pressure of 10^{-2} to several torr. The degree of ionization of $\leq 10^{-5}$ is rather small and the kinetic temperature of the molecules and radicals does not exceed several hundred centigrade. However, the concentration of reactive species such as atoms and radicals can reach high values due to the non-equilibrium nature of the excitation processes and to relatively small recombination rates. In weak glow discharges ion-molecule reactions can dominate because of their high rate constants at ambient temperatures /2/.

The non-equilibrium nature of the glow discharge plasmas arises due to a relatively weak energy coupling of the electrons with the heavy molecules, atoms and radicals in the gas phase /9/. Thus, surfaces exposed to these kinds of plasmas only reach temperatures which are close to the kinetic temperature of the particular gas. However, they experience high fluxes of reactive species which can never be achieved under thermal equilibrium /2/.

Let us, for illustration, consider a simple chemical evaporation ("etching") with formation of thermodynamically stable products, eq. (1).

$$a \cdot A(s) + b \cdot B_2(g) \quad = \quad c \cdot C(g) \tag{1}$$

Under thermodynamic equilibrium the reaction on the solid proceeds with molecular gaseous reactants B_2 and it needs a certain activation energy of let us say 20 to 40 kcal/mole. Under plasma conditions a degree of dissociation of several at. % can easily be obtained resulting in an atomic flux towards the surface of the order of 10^{20} cm^{-2} s^{-1}. With a typical reaction probability for the atoms between 0.001 an 1, the etch rate can reach several hundred Angströms per second or more.

The rate determining process during deposition can be either the sticking rate of reactive monomers or the rate of their formation at the surface. In the former case the monomers (radicals) are formed by fragmentation of molecular precursors in the gas phase and their steady state concentration can reach several at. %. With a sticking probability of 0.01 to 1 the resulting growth rates can reach several hundred Angströms per second. Such deposition rates are found for the deposition of polymers (see e.g. /10/) and amorphous phosphorus /2,3/ at a temperature below 200 °C, as well as for epitaxial growth of silicon above 700 °C /11/ and GaAs above 400 °C /12/.

As a rule, the upper limit of the deposition rate in glow discharges is determined by the rate of surface processes during the formation of the amorphous or crystalline solid network. These processes can significantly limit the deposition rate at which a material of the desirable quality can be prepared. This is valid, for example for the deposition of amorphous silicon, a-Si, where the deposition rate of a material with a quality sufficient for solar cell applications is limited to about 10 A sec⁻¹. Materials deposited at significantly higher rates have not achieved the necessary quality·thus far /13-15/ although their other properties are quite satisfactory. The problem lies in the relatively low absolute value of the photoconductivity, a quantity which is directly related to the product of the life time and the mobility of photogenerated carriers, and which is strongly dependant on the structural quality of the material.

The understanding and control of the surface processes during the film growth is one of the most important and challenging tasks for future research. Today we are only beginning to see what is important and what is not. This applies, for example, to the question of the role of the substrate bias during the deposition. At a negative bias of several tens of volts the bombardment of the surface with low energy negative ions can, under appropriate circumstances, facilitate conditions where the decomposition of the molecular precursors is limited to the surface and, simultaneously, the polymerization in the gas phase is avoided. Obviously these are the conditions necessary to obtain structurally well ordered deposits with good physical and chemical properties. The above mentioned problem with the high rate of deposition of a-Si is related to these phenomena. Because of the importance of thin film silicon and the available data on the plasma chemistry of this system, particular attention will be devoted to the control of the deposition and of the properties of a-Si and nc-Si in sect. 10.1.

3. THE DESIGN TOOLS

There is a variety of experimental arrangements used for material deposition and preparation in non-isothermal plasmas. The most typical and frequently used ones will be briefly described and references given to relevant papers.

Figure 1a shows the "radial flow parallel plate reactor" which is most widely used for industrial applications /4/. It operates typically at a frequency of 13.6 MHz. Above a certain pressure which depends on the dimensions of the reactor the plasma is confined between the plate electrodes. If the upper electrode is coupled to the R.F. generator via a blocking capacitor (i.e. the parallel resistance to this condensor shown in Fig. 1a is "infinitive") the electrode experiences a large negative D.C. bias, and it is called the "cathode".

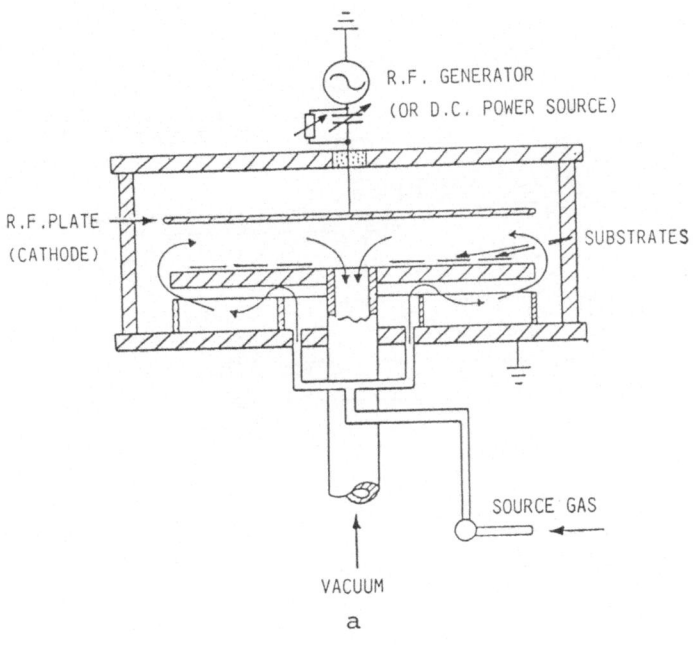

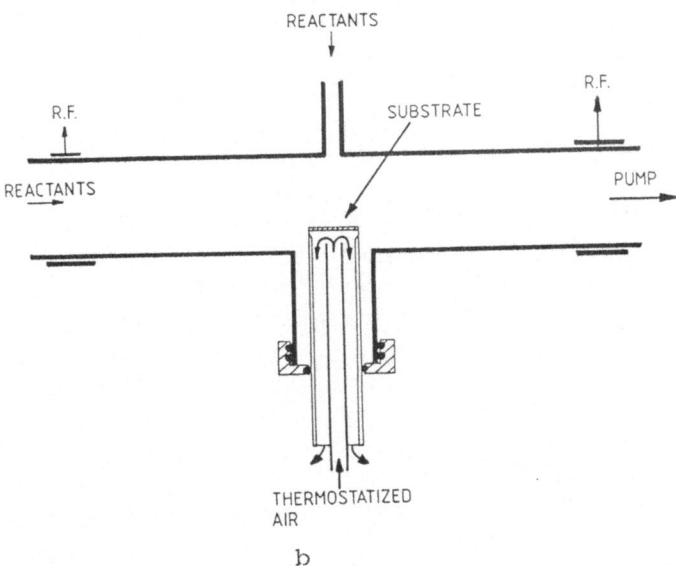

Fig. 1. a: Radial flow parallel plate reactor; b: Discharge tube with the substrate located in the plasma column.

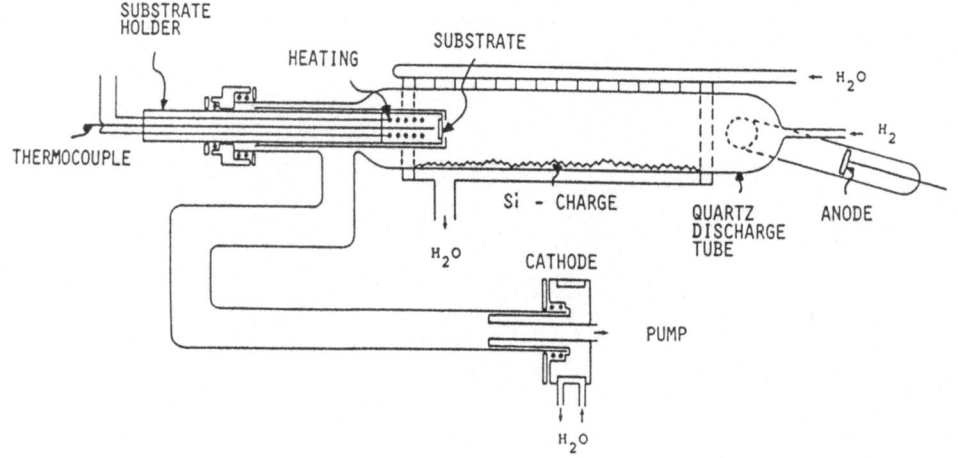

Fig. 2. D.C. discharge tube with a substrate whose temperature
and bias can be exactly controled.

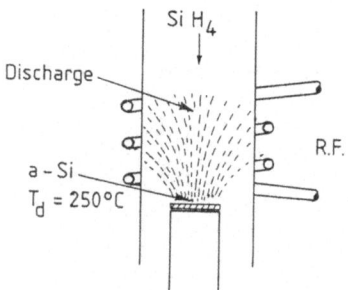

Fig. 3. The so called "inductively coupled" discharge.

For a given reactor design the D.C. bias increases with decreasing area of the cathode and with decreasing pressure. Thus, the parallel plate reactor can be operated either in a sputtering mode (R.F. or D.C.) in a total pressure of several millitorr, or in the deposition mode in a pressure range between about 0.1 and several torr where the bias decreases to a relatively low value. If for preparative purposes an even lower bias is needed, one preferentially uses higher excitation frequency in the range of 80 MHz.

The arrangement shown in Fig. 1b offers the advantage of better control of the substrate bias. This is obtained because of the small area of the substrate and the substrate holder made of electrically insulating materials avoids any capacitive coupling. Therefore, thermostatized air or oil have to be used instead of ohmic heating. A further, significant advantage of this arrangement is the fact that the substrate is placed in the plasma column of the R.F. discharge whereas in the former case it is located in the electrode regions where only a limited control of the plasma parameters and particle energies is possible.

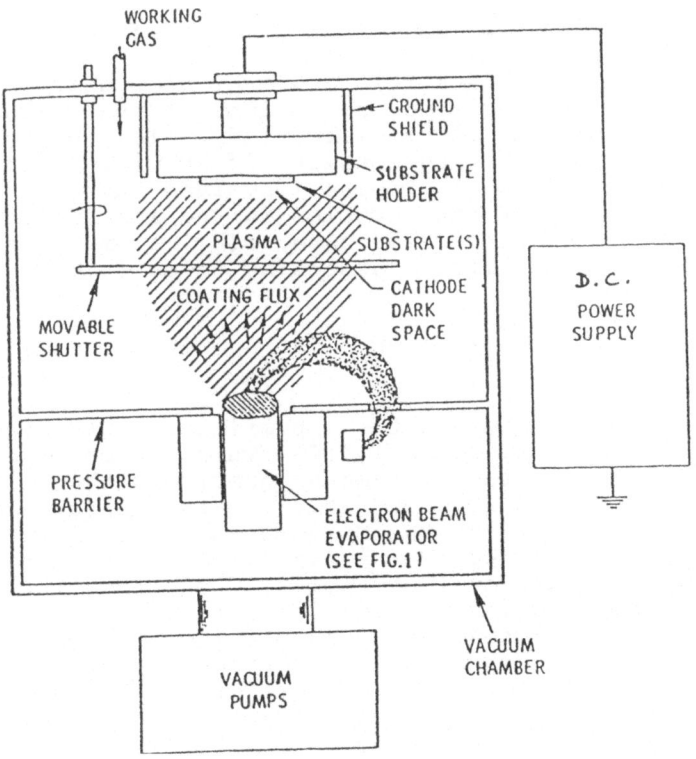

Fig. 4. Experimental arrangement for ion plating and activated reactive evaporation.

141

The best control of the plasma parameters and of the bias
provides a D.C. discharge with a small area substrate. The
substrate bias can exactly be measured and even a positive
bias with respect to the plasma can be reached /3,16/. The
disadvantage is the small substrate area and the limitation of
the bias control to electrically conducting materials only.
Examples of the control of the materials properties of nc-Si
via the substrate bias are summarized in ref. /16/.

Very often used is also the so called "inductively coupled"
R.F. discharge shown in Fig. 3. Its advantage is the
possibility of exciting a relatively weak discharge. The
disadvantage lies in the spatial inhomogenity of the plasma.
The excitation occurs capacitively in most cases (see /4/ and
literature given therein).

Ion plating and Activated Reactive Evaporation /17/ are
important techniques utilizing glow discharge activation
during thermal evaporation of solids. Typical equipment is
shown in Fig. 4. The great advantage of this technique is its
versatility as regards the material to be deposited. The
disadvantage lies mainly in the directionality of the
deposition /4/.

4. ARE THEORETICAL PREDICTIONS OF PLASMA CHEMICAL PROCESSES IN

GLOW DISCHARGES AVAILABLE AS DESIGN TOOLS ?

Because of the non-isothermal nature of glow discharge plasmas
the most powerfull tool for predictions of the course of a
chemical reaction, equilibrium thermodynamics, cannot be used.
The exact kinetic approach consists of solving a set of
balance equations for all species involved in their respective
excited states. The first problem is the insufficient
availability of cross section data, $\sigma(E)$, from which the rate
constants can be calculated according to eq.(2).

$$ k = c \cdot \int_{c_1}^{\infty} E^{1/2} \cdot \sigma(E) \cdot f(E) \cdot dE \tag{2} $$

The second problem lies in the approximation of the electron
energy distribution function, $f(E)$. This can be in principle
obtained by solving the Boltzmann equation with the inherent
problem of the cross section data. The frequently used
approximation of $f(E)$ by the Maxwell-Boltzmann distribution is
in most cases incorrect. Particularly the high energy tail of
the actual distribution in the plasma is strongly reduced in
the plasma compared to the M.-B. function. This results in an
error of the calculated rate constant k of factor 10 or more
/18/. More recent work of several groups has shown that the
problem is even more serious because of the intense
vibrational excitation which introduces strong deviations from
the M.-B. distribution function even at low electron energies
/19-20/.

A further complication arises due to the R.F. field compared
to a D.C. discharge. As already pointed out by the present
author some time ago, the equivalence of the D.C. and R.F.
plasmas in the discharge column applies only at high

frequencies. Simple arguments based on the high frequency conductivity of plasmas led me to a prediction that the equivalence applies only at frequencies of 20 to 100 MHz /23/. More recent calculations of Winkler, et al. confirmed this estimate /21,22,24/. In particular their calculations show that the frequently used "effective field approximation" applies only at frequencies above about 30 MHz /24/. In view of these problems it is not surprising that even the calculation of dissociation of simple molecular gases represents a tremendous problem which has not been satisfactory solved so far. This justifies my sober assessment of the theoretical approach to plasma chemistry in non-isothermal plasmas.

Indeed, there is only a limited number of examples of plasma chemical transformations in which theoretical calculations have led to conclusive results (e.g. the ozone synthesis /25/ and refs. therein). In general, homogeneous gas phase systems are simpler than the heterogeneous ones and D.C. glow discharges are simpler than the R.F. ones. Nevertheless, simple phenomenological kinetic models can help in identifying the crucial processes and in understanding of the system. Worth mentioning is for example the modelling of silica deposition for preparation of optical fibres /1/ and of the Si/H-system to be discussed further below.

5. WAYS TOWARDS MATERIALS DESIGN

Empirical tailoring of the materials properties can be done by a simple black box approach (see e.g. /26/). However, enormous experimental work is needed in order to complete the whole data matrix. With 8 experimental parameters and only three values for each of them the total number of possible combinations amounts to $3^8 = 6561$ deposition experiments. With three experiments per day it would require more than eight years to complete the work /26/. Evidently, the black box approach is not only uneconomical but it also does not improve our understanding of the system.

The materials design in non-isothermal plasmas requires a sufficient knowledge of discharge physics which is, unfortunately, neglected in many current papers. Furthermore, the knowledge of the present status of crystal growth, kinetics of nucleation and film growth within the framework of the conventional thermal CVD is a necessary prerequisite for successfull P CVD work. For example it is not of crucial importance for the deposition of good quality films to study the kinetics of silane polymerization in the gas phase but it is necessary to understand the conditions under which it can be avoided. Only such an approach based on the presently available knowledge of glow discharge physics and conventional CVD can enable one to design an efficient experimental matrix, to reduce the amount of the necessary work and to increase our understanding. Simple theoretical considerations allow one to assess which processes are possible and which cannot proceed under plasma conditions /9/.

In the following part of the paper selected examples will be given to illustrate the basic guidelines for materials design by low pressure plasmas.

6. MODIFICATION OF NEAR SURFACE LAYERS: "DIFFUSION COATINGS"

In many applications the complex requirements on materials properties cannot be met by a single alloy or compound. A typical example is high mechanical strength and elasticity combined with high wear resistance. Materials which display a high wear resistance are in general brittle and do not meet the requirements regarding mechanical properties. A similar situation is found with respect to corrosion resistance. It should be emphasized that these tribological and corrosion processes result in very high losses in the national economies reaching many billions of US $ per year.

As both, the wear and corrosion resistance are determined essentially by the properties of the surface layer with a thickness of several microns, the above mentioned complex requirements can be met by a combination of a bulk material with the desirable mechanical properties and a surface wear or corrosion resistance coating. The coatings can be prepared either by CVD (or PVD) or by diffusion of appropriate reactants into the surface region of the workpiece.

Nitriding, carburizing and boriding are the processes currently used in steel technology. In all of them the plasma technique is becoming increasingly used. Let us, for illustration, compare the nitriding technologies /27/.

There are three technologies currently used for industrial nitriding:

1. Molten salt bath
2. Thermal gas nitriding in ammonia and its mixture with other gases
3. "Ion nitriding" in a glow discharge (Fig. 5.).

Although the costs for ion nitriding are somewhat higher than for the other techniques they are more than compensated for by the superior properties of the ion nitrided surfaces. This is essentially due to the possibility to prepare a single phase compound layer (either pure γ' - Fe_4N or ε - $Fe_{2-3}N$ phases) at the surface followed by a diffusion α - Fe/N layer. Furthermore, the important advantage of the plasma nitriding is the lower process temperature /27/.

The maximum achievable hardness of the nitrided steels reaches 1300 kp mm^{-2} compared to 200 - 300 kp mm^{-2} for the untreated material (Fig. 20). Of particular interest is the improved corrosion resistance of nitrided iron and of some steels. Mittemeier and Colijn /28/ explained it by the formation of a magnetite corrosion layer during high temperature oxidation of nitrided steel. Because of a smaller Piling-Bedworth relation (P.B. relation is the ratio of the molar volumina of the oxide to that of the metal) of magnetite as compared to hematite, which is the natural product of bare iron or steel, the magnetite layer displays a better adherence to the metal and acts as a diffusion barrier against corrosion (P.B. ratio for hematite on iron is 2.09, for magnetite 1.80; P.B. of alumina on aluminum is 1.41).

However, a critical researcher cannot accept this as a true

explanation. The formation of magnetite was also observed in our laboratory during room temperature corrosion experiments of nitrided iron in a humid atmosphere. The problem is obviously more complex and requires further studies.

The effect of plasma nitriding and carburizing on the corrosion behaviour of pure iron is illustrated by Fig. 6 /29/. The samples were kept at 45 °C in air 1 cm above saturated solution of NaCl in water. A similar experiment was run with pure water and similar results were obtained. The samples were periodically abraded with a brush and weighted. The weight losses are plotted in the figure. At the end of the experiment the samples were analyzed by X-Ray Diffraction and, afterwards, the remaining corrosion layer (and partially also the uncorroded metal) were etched in diluted HCl for 30 seconds. The corresponding weight losses are shown by open circles. The significant improvement of the corrosion behaviour of the nitrided samples is evident. The carburized sample did not show any sign of corrosion observable by XRD or by visual inspection, and the weight losses were essentially within the errors of the measurements at such a long time scale. Recently, Yoneda, et al. reported on the improved wear resistance and hardness of plasma carburized steels, but they have not done any experiments on the corrosion behaviour /30/.

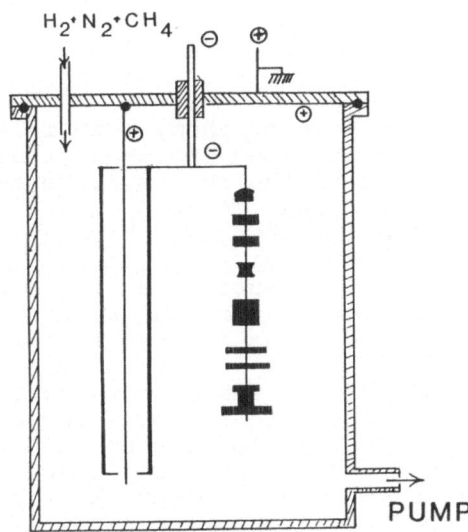

Fig. 5. Apparatus for industrial ion nitriding, carburizing and boriding.

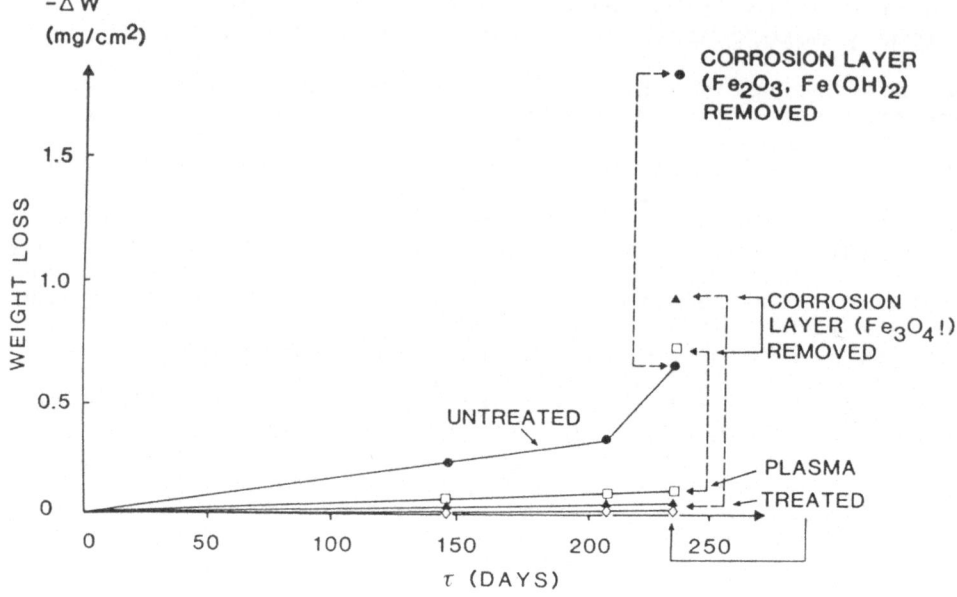

Fig. 6. Corrosion resistance of nitrided and carburized iron.

A relatively new approach to the modification of surfaces is ion implantation in an energy range of 10 to 1000 keV /31-33/. The advantage lies in the large flexibility with respect to the combination of the target material and the implanted ions. By a proper choice of the implantation energy and -temperature the implanted profile can be controlled to a certain degree. The drawback is the relatively low implantation depth, high costs of the equipment and the directionality of the process.

What are the fundamental effects which the plasma plays in these processes?

There are essentially two of them, both related to an enhanced concentration of the adsorbed (or implanted) reactive species at the surface and/or in the near surface region. This question has been discussed in some detail in refs. /3,4/ and only a brief outline will be given here.

When a molecular gas is dissociated in a non-isothermal plasma at a relatively low temperature the suface coverage of a solid exposed to the flux of atoms increases significantly as compared to the equilibrium with the molecular gas at thermodynamic equilibrium. The driving force for the diffusion of the adsorbed species into the bulk increases accordingly. It is obvious that the rate of the build up of the diffusion layer increases due to this enhanced concentration gradient. The thinner the layer (i.e. shorter the treatment time) the higher the enhancement becomes.

At long treatment times the gradient decreases due to the long diffusion length. However, there is still a significant difference between the equilibrium chemistry and that under plasma conditions /4/. As the steady state surface coverage can be higher under plasma conditions, compounds with

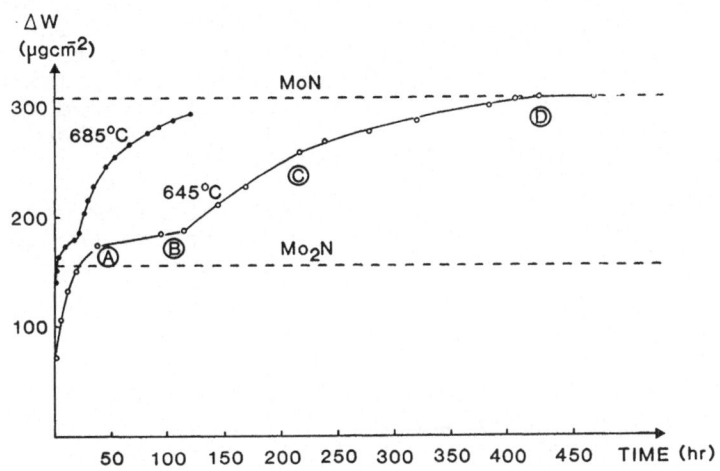

Fig. 7. Nitriding of a 5 μm thick molybdenum sheet in a glow discharge at about 1.25 mbar /34/.

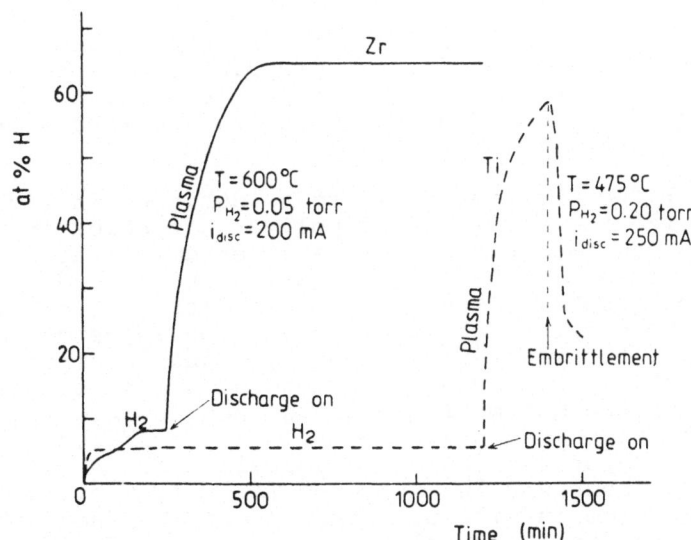

Fig. 8. Effect of glow discharge on the hydriding of zirconium and titanium /35/.

a higher content of the gaseous component (i.e. metastable phases) can be prepared. This is illustrated by the formation of nitrides, Fig. 7, and hydrides, Fig. 8.

The δ-MoN phase is unstable above 600 °C, but it can be formed and stabilized in the nitrogen discharge up to 850 °C if a sufficient degree of dissociation is provided. In Fig. 7 several regions of diffusion controlled nitride formation are seen.

The formation of hydrides proceeds much faster because of the high diffusivity of hydrogen in the metals (Fig. 8). In the case of titanium, catastrophic embrittlement occurs resulting in sample disintegration. The final and extrapolated maximum concentrations for Zr and Ti, respectively, correspond to an equilibrium pressure of molecular hydrogen of more than 100 atm (see /34/ and references given there).

Similar problems arise during the low temperature oxidation of silicon in glow discharges. This problem is beyond the scope of the present paper and we refer to a recent review /36/.

7. PLASMA INDUCED CHEMICAL VAPOUR DEPOSITION

As this subject has been reviewed recently by the author /2-4/ only some of the fundamental aspects of the materials design problem will be discussed here with emphasis on the control of materials' properties.

As mentioned in section 1 three effects of the plasma during the deposition can be distinguished.

1. The application of glow discharge to a heterogeneous system in which the desirable reaction is thermodynamically allowed results in an increase of the deposition rate at a given temperature ("catalytic effect").

The favourable decrease of the process temperature is, however, accompanied by some undesirable effects resulting in a degradation of the properties of the deposit. All of them are related to the build up of the solid (amorphous or crystalline) network.

First of all the experiment has to be designed in such a way as to minimize the formation of polymeric species in the gas phase since the incorporation of such species into the solid network requires much higher activation energy and -entropy than that of monomers. Even if this condition is met the deposition temperature and rate have to be adjusted in order to facilitate the monomer decomposition and incorporation into the solid matrix. Otherwise high concentrations of residual components of the monomer(s) will be incorporated into the deposit resulting in degradation of its properties (e.g. OH in silica films deposited from organometallic compounds or from silane below about 250 °C, hydrogen in silicon nitride and in silicon, chlorine in silica, titanium nitride, and other materials deposited from chlorides, etc). For further discussion regarding the control of the incorporated gases see Ref. /4/.

2. In an intense discharge high concentration of atoms and
   reactive radicals is present and it can lead to a quite new
   kind of pseudoequilibria ("Partial Chemical Equilibria",
   PCE).

Examples of such systems were discussed in Ref. /2-4/, and the
recently studied silicon/hydrogen system is just a prototype
of this "thermodynamic effect". Only if such a PCE state is
established in a given system, can chemical transport of the
solid be utilized for a controlled growth of films and
crystals /2/.

3. The number of exothermic processes taking place at the
   surface of a growing film or a crystal under plasma
   conditions influences strongly the formation of the solid
   network /3,37/.

Either formation of the stable or of the metastable phases can
be facilitated depending on the conditions. The crucial role
comes from the substrate bias. At low negative bias, when the
surface experiences bombardment by ions with energies below
the threshold for lattice displacement, stable phases are
formed as a rule. On the other hand, high energy bombardment
favours the formation of metastable phases. Examples are given
in Ref. 37 and the particular case of the growth of a-Si and
nc-Si was discussed in Ref. /16/. In our more recent papers a
detailed study of the kinetics and mechanisms of the silane
decomposition and of the silicon deposition has been done
/38,39/.

8.   PROPERTIES OF THE PLASMA DEPOSITED FILMS

Most of the surprising and unusual properties of the plasma
deposited solids are directly related to their structural
properties. Let us briefly summarize several examples.

A surprisingly high stability of a-B, a-Si, a-P (see /3,37/
and references given there) and of nc-Si (/16,40/ and ref.
there) against oxidation in air is related to a well ordered
solid network of the deposit together with a saturation of
dangling bonds by chemisorbed hydrogen.

However, considering the extremely high stability of nc-Si as
compared with single crystal Si surfaces which is illustrated
by Fig. 9 the question arises as what is the atomic structure
of the nc-Si surface deposited via chemical transport in the
plasma /16,40/. None of the known reconstructions of single
crystal silicon surface can explain such a behaviour and the
question remains open. A quite similar and suprising stability
was found for a-P which did not show any sign of oxidation
even after several hours of exposure to air at room temperature /3,37/.

The excellent electronic properties of plasma deposited a-Si
and a-Ge, in particular the possibility of their
substitutional doping are due to a low density of states in
the mobility gap. This, in turn is related to the structural
order, saturation of dangling bonds by hydrogen and to
relaxation of stress in the amorphous network. The pioneering

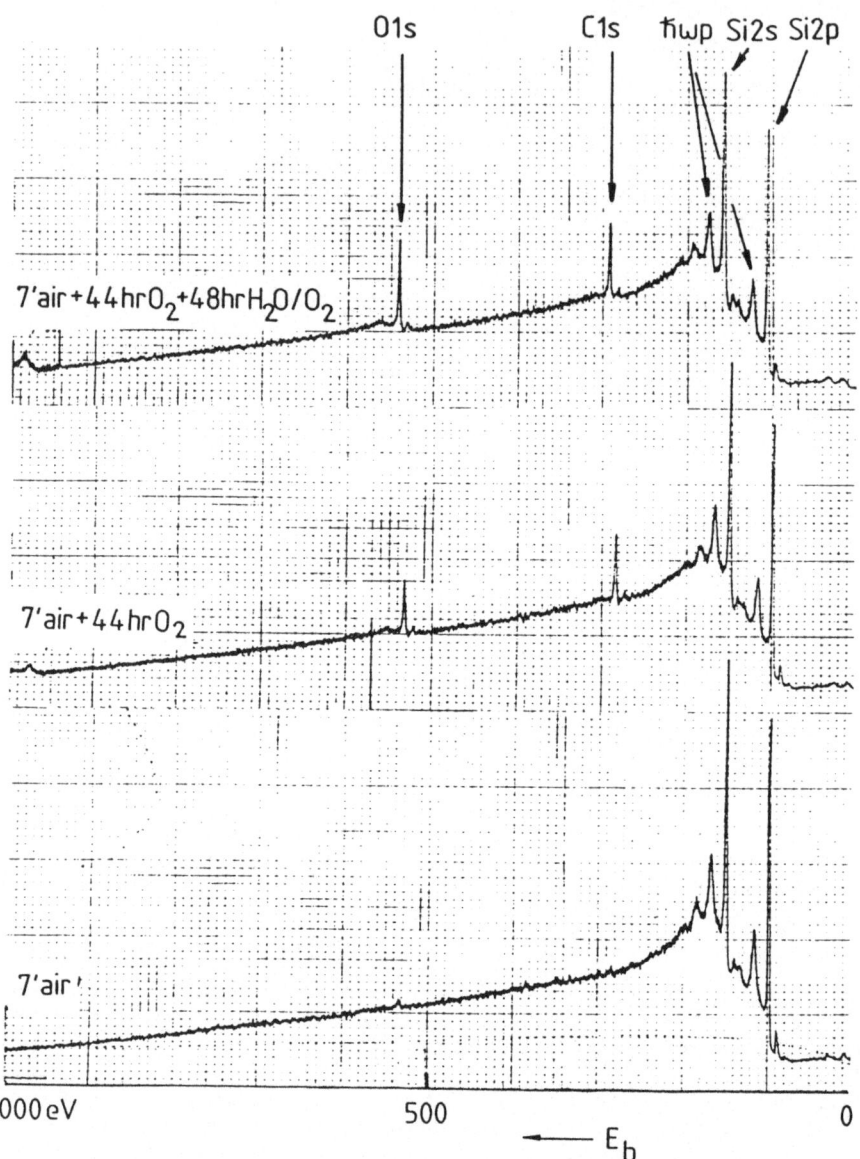

Fig. 9. X-ray photoelectron spectra of nanocrystalline silicon prepared by chemical transport in hydrogen plasma and exposed to air for various periods of time.

work of Chittick, et al. /41/ and the first demonstration of the possibility of substitutional doping of these materials by Spear and Le Comber /42/ opened up new areas of solid state research /43/.

9. CASE STUDIES IN P CVD

9.1. Nanocrystalline Silicon

The preparation of nanocrystalline silicon, nc-Si (sometimes also called "microcrystalline", μc-Si) was reported by Veprek and Marecek in 1968 /44/. This work has been continuing at Zürich /16,38,39,40,45-50/. In 1980, Matsuda et al. reported on the preparation of "μc-Si" in an intense glow discharge in silane diluted with hydrogen /51/. The later work of the Japanese groups has been reviewed recently by Osaka and Imura /52/.

Both, a-Si and nc-Si can be deposited from silane in a glow discharge. The parameters controlling the formation of either the amorphous or crystalline network were studied in Refs. /46,49,50/. Figure 10 shows the dependence of steady state concentration of silane on the dwell time in the glow discharge under conditions which are typical for the deposition of thin silicon films /39,50a/. One can see that with increasing dwell time the silane concentration decreases down to a constant value of about 0.3 mol. %, which is about five orders of magnitude higher than the thermodynamic equilibrium value /39/. The same concentration is reached if solid silicon is allowed to react with hydrogen (see the lower curve in Fig. 10). Thus, this state corresponds to the PCE under plasma conditions and it can be utilized to grow nc-Si via chemical transport /3,44/. Alternatively, nc-Si can be prepared if silane is diluted to ≤ 1 mol. % with hydrogen and the deposition is done under conditions close to PCE /3,46,49/.

However, severeal research groups have reported that the films deposited in this way consist of an agglomerate of small crystals imbedded in an amorphous matrix /52/. Figure 11 shows the relevant part of an XRD pattern of nc-Si deposited via chemical transport at various values of substrate bias /48/. A detailed analysis of these patterns /46/ shows that films deposited at a bias V_b ≥ 600 Volts consists of nanocrystalline material without any noticeable amount of the amorphous matrix. The amorphization due to ion bombardment occurs only at a more negative bias (for further details see /16,46,48/) .

Why is it apparently not possible to obtain such a one component material from a mixture of silane with hydrogen? The reason probably lies in a poor control of the deposition process, in particular of the steady state concentration of silane in the gas phase. In the case of chemical transport this concentration is automatically adjusted to the PCE value and, consequently, the deposition is controled by the silane decomposition at the surface. This is illustrated by Fig. 12 /38/ which shows clearly that the deposition of silicon is controlled by ion impact induced fragmentation of silane species at the surface of the growing films. The formation of dimers and higher polymeric species via the fast insertion reaction $SiH_2 + SiH_4 = Si_2H_6$ is surpressed under these conditions /39/.

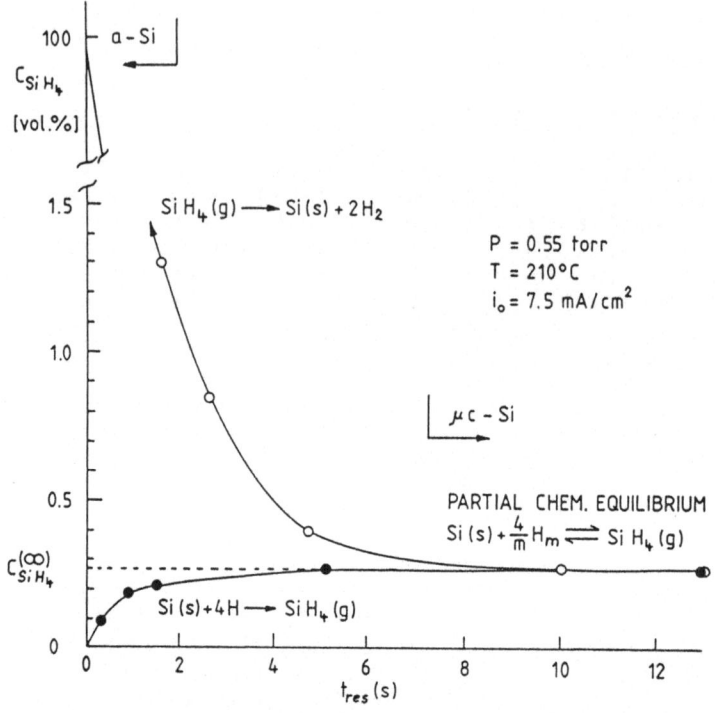

Fig. 10. Dependence of the steady state concentration of silane on the dwell time in the discharge.

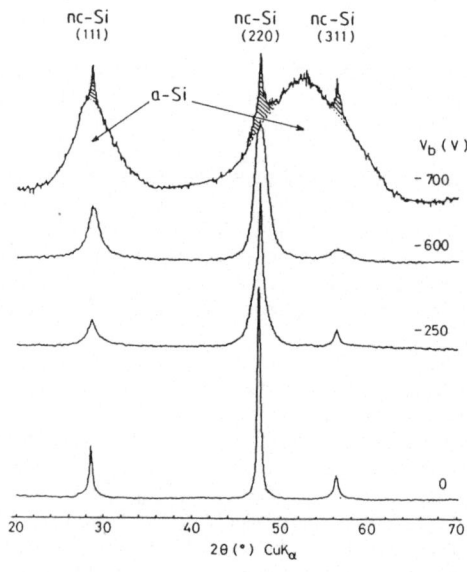

Fig. 11. X-ray diffraction pattern of nanocrystalline silicon deposited via chemical transport in hydrogen plasma at various substrate bias.

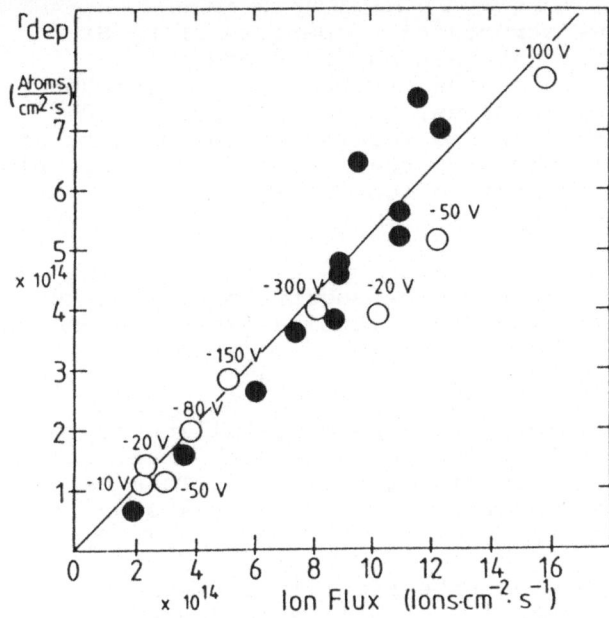

Fig. 12. Dependence of the deposition rate of nc-Si on the ion flux towards the surface. o - floating potential, o - negative substrate bias, T_{dep} = 260 °C.

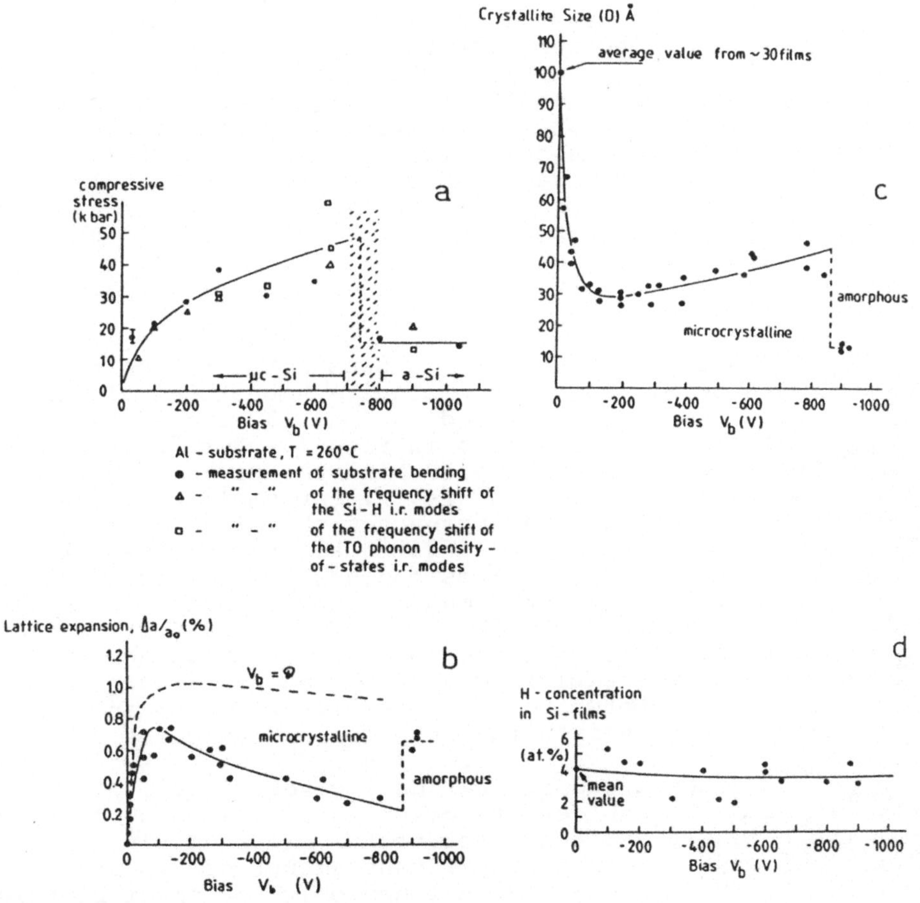

Fig. 13. Effect of the substrate bias on various properties of nanocrystalline silicon (ref. /16/).

The precise control of the structural properties of nc-Si allows one to study several interesting questions of a fundamental importance, such as the crystalline-to-amorphous transition /46/, optical properties, Raman scattering and electronic properties (see /49/ and references to the work of other groups given there). Here, we shall briefly discuss only the effect of compressive stress on some of the physical properties. This is induced by bombardment of the surface of the growing film with energetic ions /16/.

Figure 13 summarizes the effect of the substrate bias on the compressive stress, crystallite size, lattice expansion and on the hydrogen content /16/. For a detailed discussion we refer to the previous papers /16,48/. It is seen that a large compressive stress of up to 40 kbar can be build up in the nc-Si films (the material remains purely nanocrystalline up to a bias of -600 Volts, Fig. 11). Films deposited at floating potential can have a certain tensile stress /46/. This has an interesting effect on the vibrational properties as seen from the Raman scattering /45c,53/.

For films deposited at floating potential, i.e. having a small tensile stress /56/, the central phonon line in Raman scattering at Γ'_{25} point of the Brillouin zone is observed for crystallite sizes down to about 30 A. This is the lowest limit of stability of the crystalline lattice in silicon /45b/. It means, that the selection rule for light scattering remains preserved at crystallite sizes which are orders of magnitude smaller than the wave length of the light. As long as the phonons are localized at the individual crystallites the longest phonon wave length is limited by the crystallite size, D, to $\lambda \approx$ D. Thus, the Raman selection rule for single crystal, $K_{ph} \approx 0$, is for nc-Si modified to $K_{ph} \approx 1/\pi \cdot D$. The experimental data confirmed this rule and they have shown that also the dispersion relation for the optical phonons within this range of K_{ph} and Ω_{ph} remains unchanged as compared to the single crystal /45c/.

Figure 14 shows the effect of the substrate bias on the Raman frequency in nc-Si /53/. These results have to be discussed in context with the data presented in Fig. 13. Therefore, we replotted the stress data also in Fig. 14. For "stress free" films deposited at floating potential the Raman frequency should follow the broken line /45c/. For small compressive bias the experimental data approaches this line, i.e. the phonons are localized at the individual crystallites. When the compressive stress in the films reaches a value of about 25 kbar ($V_b \approx$ -150 Volt) the Raman frequency begins to increase and for stress larger than 30 kbar ($V_b \leq$ -300 Volt) it reaches a constant value close to that for single crystal. This dependence was explained by improved coupling of the phonon vibrations between the crystallites due to an increasing connectivity within the grain boundaries /53/. These boundaries are very narrow /48/. The Raman scattering data discussed here and further data given in Ref. /16/ together with the XRD results given in Fig. 11 show clearly that there is a significant structural order within the grain boundaries and only an insignificant amount of "amorphous component". These results were found only for nc-Si prepared via chemical transport and **they illustrate the exciting possibility which this technique offers for materials design.**

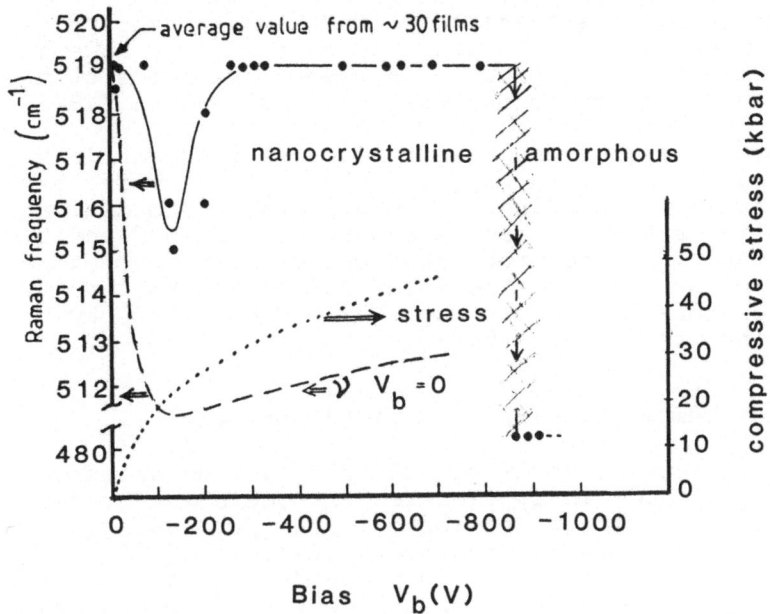

Fig. 14. Delocalization of phonons in nc-Si due to increasing
compressive stress in the film as seen by Raman
scattering.

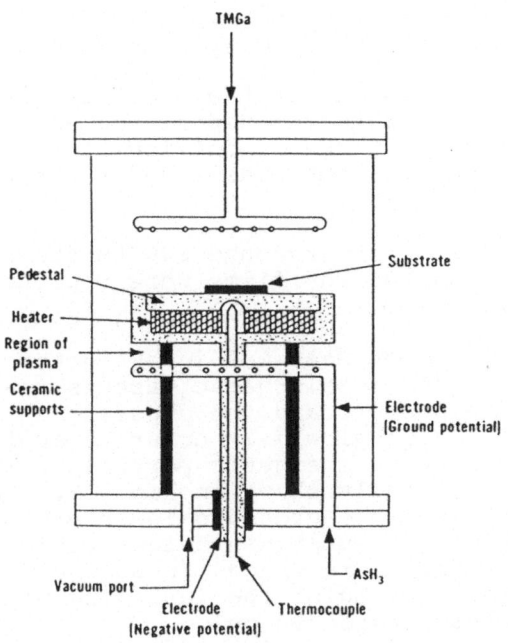

Fig. 15. Cross section view of the deposition apparatus for
plasma enhanced MO CVD of GaAs (from ref. /12/).

In a more recent paper a detailed study of the dependence of the electronic properties of nc-Si on the structural properties has been done. Also in this case the properties and the connectivity of the grain boundaries play an essential role. However, this problem is beyond the scope of the present paper and we refer to the original publication /48/.

The compressive stress also influences the stability of the films with respect to oxygen incorporation during exposure to air /45a,47/. Depending on the preparation conditions, the oxygen incorporation can be completely avoided and the films made stable. On the other hand one can introduce a certain amount of oxygen into the films /47/ and make them sensitive to reversible gas adsorption /45a/. Such films may be used as selective gas sensors /54/.

9.2. Low Temperature Epitaxy of Si and GaAs

Epitaxial deposition of silicon and gallium arsenide is of great importance in fabrication of modern integrated circuits and optical devices. The conventional processing by CVD requires high temperatures (e.g. 1050 - 1200 °C for silicon and 600 - 800 °C for gallium arsenide) which results in undesirable effects such as autodoping, outdiffusion of dopants and formation of thermally generated defects. In order to avoid, or at least to diminish these effects lowering of the deposition temperature is highly desirable. Already some, years ago several research groups reported on experiments using a glow discharge plasma for this purpose /55/. More recently this problem has attained more attention /11,12/.

The recent work on plasma epitaxy of Si was summarized by Reif /11b/. Two effects of the plasma were identified:

1. Improved cleaning of the substrate by a glow discharge applied prior to the deposition and
2. Enhancement of the deposition rate as compared with the deposition without plasma under otherwise identical conditions.

The deposition rate reached 200 to 450 A/min at a temperature of 600 to 900 °C. The films were reported to be of a good quality /11/.

The deposition of GaAs was investigated by Pande and Seabaugh /12/. They used metalorganic gaseous reactants as sources for gallium and arsenic and the deposition was done in the post discharge ("afterglow") in order to avoid defect formation due to bombardment by energetic particles. Their experimental set up is schematically shown in Fig. 15. The discharge was excited between the grounded lower electrode (anode) and the rear part of the substrate holder (cathode), and it remained confined within this space. In such a way, the discharge provided activation of the inflowing monomers but it did not reach the substrate region.

At a temperature above 400 °C the deposition rate was proportional to the flow rate of the reactants and it reached up to 2000 A/min. Detailed study of the film quality revealed that this process is a viable alternative to the conventional

MO-CVD and that it can operate at significantly lower temperatures.

9.3. Fabrication of Optical Wave Guides

Optical wave guides are presently being introduced into modern telecomunication as well as into many autoregulation and control systems which are susceptible to electric perturbations. This is the case for control systems of high voltage breakers, steering systems of airplanes flying through a thunderstorm, etc. The principle of optical wave guiding fibres and their various forms are illustrated by Fig. 16 (from ref./56,57/).

The most important requirements regarding the quality of the optical fibres are low attenuation losses and a low dispersion. The former determines the losses of the light signal and, therefore the maximum useful distance at which the signal can propagate. The dispersion corresponds to the broadening of a narrow pulse after passing a certain distance (typically 1 km fibre length) and it limits the density of information which can be transfered through a single fibre per unit time (see /56,57/ and references given there). The present status and trends in the development, fabrication and application of optical fibres was summarized by Khoe and Lydtin /58/.

The fabrication of the fibres involves the following steps:

1. Preparation of the preform which is a rod of doped silica glass with an appropriate radial profile of the index of refraction.
2. Pulling of the thin fibre from the preform.
3. Encapsulating of the fibre and its integration into a mechanically stable cable.

There are several techniques used for the preparation of the preform. All of them are similar in the sense that the radial index profile is obtained by CVD of silica controllably doped with germania, boron oxide and fluorine. Most of the methods use a CVD activated either thermally or by a thermal plasma torch, and, therefore, they operate at high temperatures and pressures (see /56-58/). Although very low attenuation can be reached by these techniques the high temperature and pressure used bring limitations regarding the exact control of a complicated graded radial index profile which is necessary for achieving the low dispersion.

This problem is avoided by the low pressure P CVD method developed at the Philips Research Laboratory at Aachen by Lydtin and co-workers /56-58/. The schematic view of the deposition unit which is now already used in industrial fabrication is shown in Fig. 17. The deposition takes place inside of a "cladding" silica tube which is rotated around its axes. The starting materials, silicon tetrachloride, oxygen and dopants such as germanium tetrachloride and C_2F_6 are fed into the tube by means of a computer controlled gas supply system. The total pressure is about 10-20 mbar and the temperature is maintained at 1150 °C. The microwave cavity

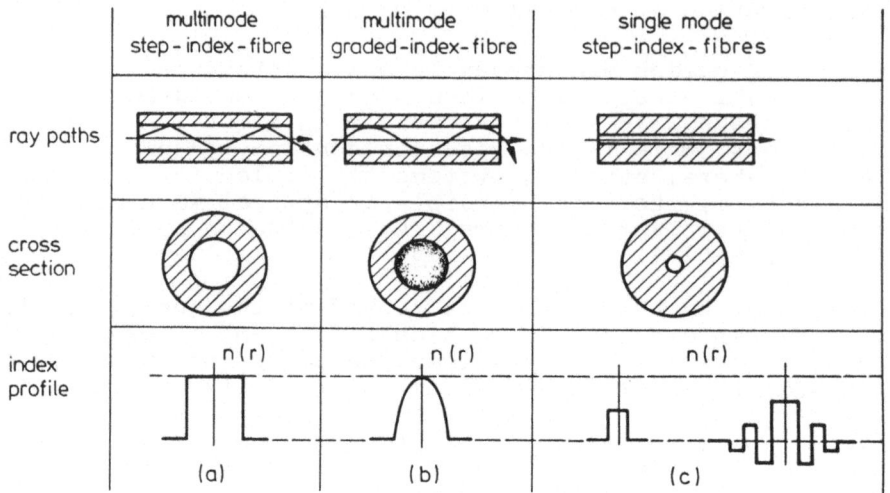

Fig. 16. Principles of optical wave guides: Ray path, cross
sections and index profiles of multimode step
index a), multimode graded index b) and single mode
c), structures (from ref. /56,57/).

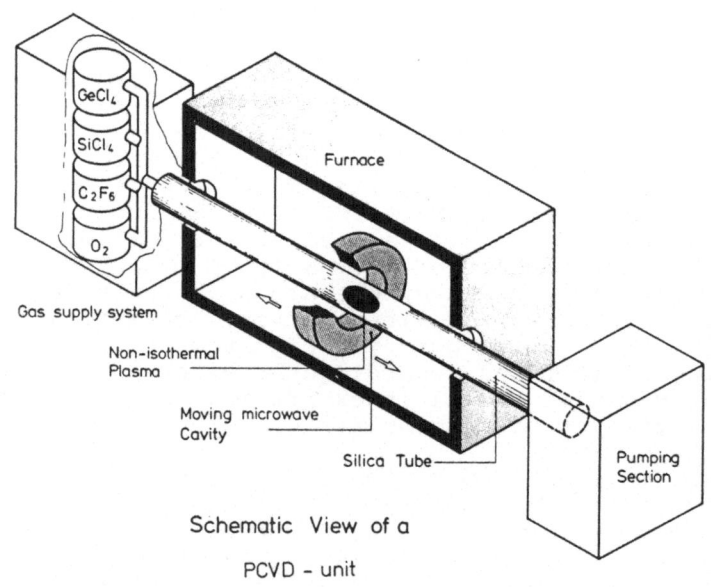

Schematic View of a

PCVD - unit

Fig. 17. Schematic view of a P CVD apparatus for the
fabrication of optical wave guides (from ref.
/56,57/).

(2.45 GHz, 3-4 kW) is moved to and fro along the tube and in each pass a compact layer of glass is deposited on the inner wall of the tube at a rate of 3-4 g/min. A stationary deposition profile of the deposition rate is shown in Fig. 18 /1/.

A phenomenological kinetic model of the deposition process based on the measured deposition profile and some others process parameters /1/ allows the researcher to optimize the fabrication. Figure 19 shows the measured radial profiles of the refractive index which were fabricated so far only by this technique /58/. An example of the fibre performance achieved so far is shown in Fig. 20 /57/. It is an impressive example of a precisely controlled design of materials with new properties.

9.4. Hard Coatings

The importance of wear and corrosion resistant coatings in many branches of industrial applications has been already mentioned in sect. 7 together with their preparation by ion implantation and diffusion. In this section we shall briefly discuss their preparation by P CVD.

The technology of conventional thermal CVD and PVD for protective coatings has been developed during the last three decades and it has achieved a high degree of sophistication /17/. However, these techniques have severe limitations, in particular regarding the high process temperature - typically in the range of 1000 °C. This brings about serious problems for example in coatings of steels which, if heated above about 550 °C, change their mechanical properties. Therefore, thermal post treatment of conventionally coated steels is necessary. Also the mechanical stress introduced into the coatings due to the different thermal dilatation coefficient of the coating and the substrate materials increases with increasing deposition temperature.

The plasma based coating processes are:

1. Thermal plasma spraying.
2. Activated Reactive Evaporation (ARE).
3. Ion plating (ion beam and ion cluster deposition).
4. Sputtering and reactive sputtering.
5. Low pressure CVD.

Several review articles and books were published recently on this subject /4,5,17,26,59/. The obvious advantage of all these techniques is the significantly lower deposition temperature as compared with the conventional CVD. Furthemore, the ion bombardment induced cleaning of the substrate surface prior to the deposition usually improves the coatings adherence. During the deposition, the ion bombardment allows one to adjust the stress in the coatings (Generally, a certain compressive stress is desirable; see also the above discussed nc-Si.)

The most serious problem are the residual gases incorporated in the films deposited at a low temperature. Detailed studies have shown that a temperature of 1150 °C is necessary in order

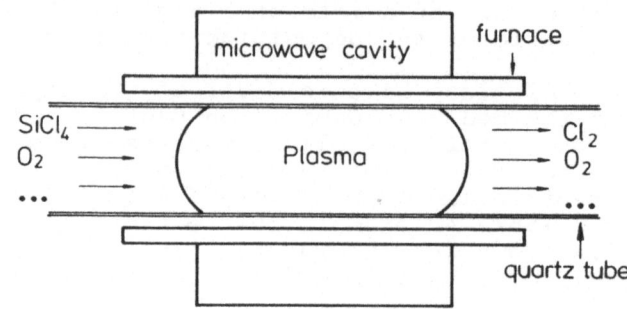

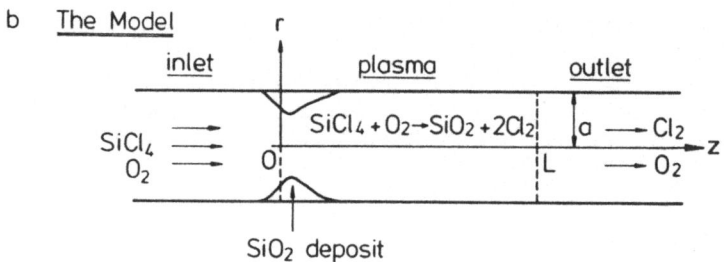

Fig. 18. A schematic view of the stationary deposition system
(a) and of the resulting deposition profile (from
ref. /1/).

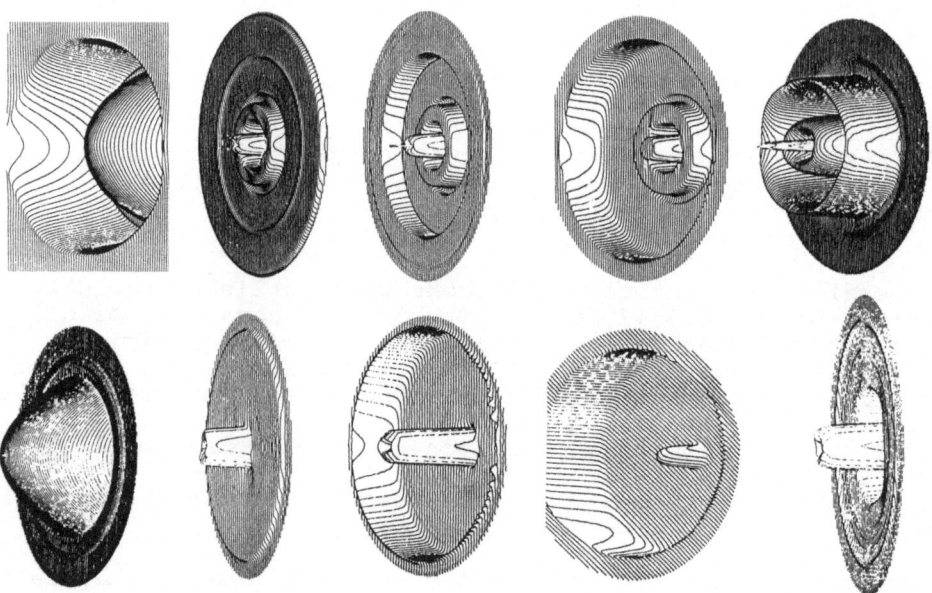

Fig. 19. Measured refractive index profiles prepared by means
of P CVD (from ref. /58/).

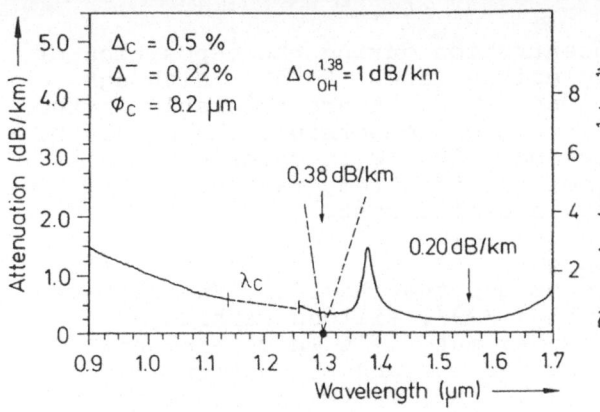

Fig. 20. The performance of optical wave guides prepared by means of P CVD (ref./56,57/).

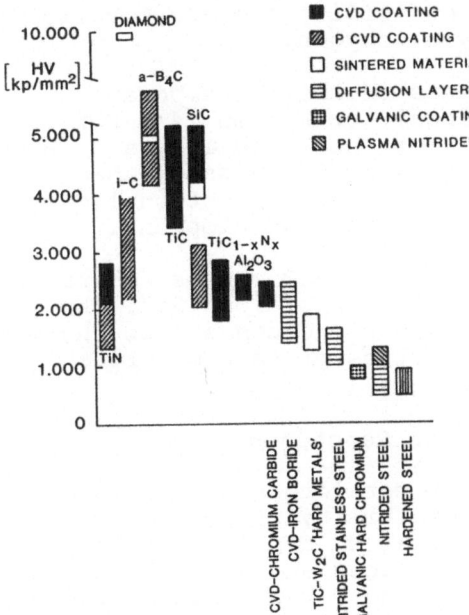

Fig. 21. Typical hardness of contemporary hard coatings.

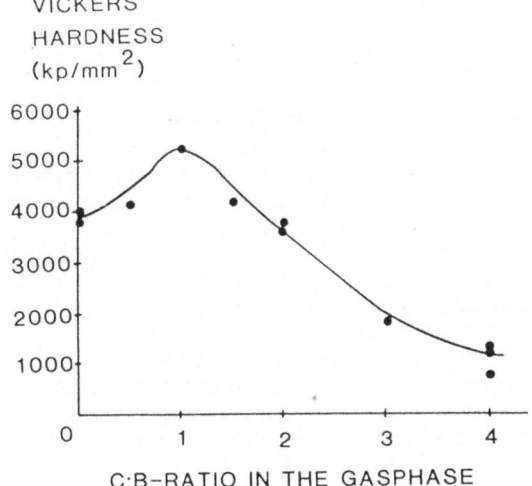

Fig. 22. Hardness of a-B$_x$C repared by P CVD vs. gas composition /29/.

to reduce the chlorine concentration during the deposition of the optical fibres to a neglegible level /57-59/. Although higher chlorine concentrations of ≤0.5 % are tolerable in the protective coatings, the deposition temperature can hardly be reduced significantly below about 500 °C in chlorine containing systems. Nevertheless, this temperature is compatible with many materials of interest.

The problem of incorporated gases can be significantly relaxed when using other starting compound than halides. Examples are the deposition of boron carbide /29/, silicon carbide, amorphous hard carbon, and other materials which elemental components form volatile hydrogen compounds. Another possibility offers Metal-Organic (MO) P CVD, as the majority of MO volatile compounds are very unstable and decompose at relatively low temperatures already without plasma. In this case the problem lies in the possibility of incorporation of carbon and, in case of oxyalkyls, oxygen.

Figure 21 summarizes the hardness of the contemporary materials prepared by conventional techniques /4/. For comparison, the hardness of plasma deposited TiN, a-C, B_xC, and a-SiC /60/ are indicated as well. It is seen, that these coatings are fairly compatible with the conventional ones in spite of the much shorter time of their development.

Figure 22 shows how the hardness of a-B_xC can be controlled by an appropriate adjustement of its composition /29/. The future development will show to what extent one can adjust the most optimum wear and lubricant properties of this material.

9.5. Deposition from Metalorganic Reactants

As already mentioned, the use of metalorganic volatile compounds for CVD offers the possibility of decreasing the deposition temperature. The thermal MO CVD is presently a fast growing field /61/. The plasma induced and -assisted MO CVD is emerging into a frontier area of the P CVD research /62-64/. The areas of applications of P MO CVD include:

1. Epitaxial deposition of semiconductors (see above).
2. Preparation of metal doped organic polymer films.
3. Preparation of protective coatings.
4. Deposition of metallic films.

The significant advantage of the P MO CVD is the low temperature at which the reactions occur. This is due to the unstable nature of most of the MO-s, such as alkyls, oxyalkyls and carbonyls. The drawback is the hazardeous nature of these compounds which are explosive when mixed with air, and extremely toxic. Also the removal of the carbon, hydrogen and oxygen residue from the growing film may be a serious problem, e.g. if metals should be deposited which form stable carbides.

Particularly exciting is the most recent development in a related field by Haag and Suhr who have demonstrated the possibility of the application of metalorganic for plasma etching /64/. It is expected that the plasma chemistry of metalorganic heterogeneous systems will be fast developing in the near future.

10. SUMMARY AND CONCLUSIONS

Large progress in the application of plasmas of electrical discharges for materials processing has been achieved during the last two decades. Thermal plasmas of arc discharges are being used for plasma spraying, spheroidization, preparation of finely dispersed refractories, in extractive metallurgy, for welding, cutting and for other applications. Much progress has been achieved also with respect to the understanding of the complex phenomena which take place during these processes. However, there are still many unresolved problems regarding the understanding of many fundamental processes and the economical aspects of the industrial applications of thermal plasmas.

The application of **non-isothermal plasmas of glow discharges** which are being produced at low pressures and temperatures offers unique possibilities for design of new materials. Also many well known materials can be prepared by means of these plasmas using new synthetic routes which results in materials with new, unusual properties. These preparation techniques are relatively inexpensive and they can meet most of the industrial requirements regarding the materials quality, throughout, environmental hazard and possibilities for process automation.

The most important areas of application of the glow discharge plasmas are the deposition of thin films of electronic materials for the fabrication of integrated circuits and, more recently, for wear and corrosion resistant coatings. The same applies for diffusion coating of oxides, nitrides, carbides, borides and metallic alloys. In this case also ion implantation at high energies is being successfully used. The main advantages of the plasma based processing in all these applications are the significant lowering of the process temperature, the enhancement of the reaction rate and the superior properties of the product materials. Low pressure plasmas have also been successfully used for the growth of single crystals and epitaxial thin films.

The possibility of preparing materials with new, unusual properties opened up new areas of the basic solid state research. The preparation of amorphous and nanocrystalline silicon should be mentioned as an example.

Most of the preparative work which has been done so far was based on the empirical approach. This was due to the large complexity of the physical and chemical phenomena taking place in these systems under conditions remote from thermodynamical equilibrium. Nevertheless, it has been demonstrated that the materials preparation by means of the plasma related techniques can well be planned by combining the present understanding of various research fields.

THE MATERIALS DESIGN BY MEANS OF DISCHARGE PLASMAS IS AN INTERDISCIPLINARY UNDERTAKING.

REFERENCES

/1/ F. Welling, J. Appl. Phys. 57(1985)4441

/2/ S. Veprek, in: Current Topics in Materials Science,
Vol. 4, ed. E. Kaldis, North-Holland Publ. Co.,
Amsterdam (1979),p.151

/3/ S. Veprek, Pure and Appl. Chem. 54(1982)1197

/4/ S. Veprek, Thin Solid Films 130(1985)135

/5/ P. Fauchais, E. Bourdin, J. F. Coudert and R. McPherson,
in: Plasma Chemistry Vol. IV, eds. S.Veprek and
M. Venugopalan, Topics in Current Chemistry 107,
Springer-Verlag, Berlin 1983

/6/ R. M. Young and E. Pfender, Plasma Chem. and Plasma
Processing 5(1985)1

/7/ D. Degout, F. Kassabji and P. Fauchais, Plasma Chem. and
Plasma Processing 4(1984)179

/8/ Xi Chen and E. Pfender, Plasma Chem. and Plasma
Processing 3(1983)97

/9/ S. Veprek, in: Topics in Current Chem. Vol 56,
Springer-Verlag, Berlin 1975, p.170

/10/ Y. Catherine and A. Zamouche, Plasma Chem. and Plasma
Processing 5(1985)353

/11a/ T. J. Donahue, W. R. Burger and R. Reif, Appl. Phys.
Lett. 44(1984)346

/11b/ R. Reif, J. Electrochem. Soc. 131(1984)2430

/12/ K. P. Ponde and A. C. Seabaugh, J. Electrochem. Soc.
131(1984)1359

/13/ S. J. Hudgens, A. G. Johncock and S. R. Ovshinsky, J.
Non-Cryst. Solids 77&78(1985)809

/14/ S. Kato and T. Aoki, J. Non-Cryst. Solids 77&78(1985)813

/15/ J. C. Anderson and S. Biswas, J. Non-Cryst. Solids
77&78(1985)817

/16/ S. Veprek, in: Poly-Micro-Crystalline and Amorphous
Semiconductors, eds. P.Pinard and S.Kalbitzer, Les
éditions de physique, Les Ullis, France (1984) p.425

/17/ R. F. Bunshah, J. M. Blocher, D. M. Mattox, et al,
Deposition Technologies for Films and Coatings;
Development and Applications, Noyes Publications, Park
Ridge, N.J., (1982)

/18/ A. von Engel, Ionized Gases, Clarendon, Oxford,
2nd.ed.(1965)

/19/ M. Capitelli and E. Molinari, in: Plasma Chemistry Vol II, eds. S. Veprek and M. Venugopalan, Topics in Current Chemistry 90, Springer-Verlag, Berlin 1980

/20/ C. Gorse, F. Paniccia, J. Bretagne and M. Capitelli, J. Appl. Phys. 59(1986)731

/21/ R. Winkler, J. Wilhelm and A. Hess, Ann. Phys. 42(1985)537, (see also Beitr. Plasmaphys. 24(1984)285)

/22/ R. Winkler, M. Capitelli, M. Dilonardo, C. Grose and J. Wilhelm, Plasma Chem. and Plasma Processing, to be publ.

/23/ S. Veprek, Pure and Appl. Chem., 48(1976)163

/24/ R. Winkler, M. Dilonardo, M. Capitelli and J. Wilhelm, to be published

/25/ B. Eliasson, M. Hirth and U. Kogelschatz, Proc. 7th Int. Symp. on Plasma Chem. ed. C. J. Timmermans, Eindhoven Univ. of Technol. (1985), The Netherlands, p. 339

/26/ E. Ehnke, in: Proc. Int. Conf. on Wear and Corrosoin Protection by Ion- and Plasma Assisted Coating Technol., ed. K. H. Kloos, Techical Univ. Darmstadt, (F.R.G.) 1983, p. 105

/27/ B. Edenhofer, Härterei Technische Mitteilungen (1974)29,105

/28/ E. J. Mittemeiher and P. F. Colijn, Härterei Technische Mitteilungen (1985)40,77

/29a/ S. Veprek and M. Jurcik-Rajman, in Ref. /25/, p. 90

/29b/ M. Jurcik-Rajman and S. Veprek, To be published

/30/ Y. Yoneda, S. Takami and W. Scheuermann, Härterei Technische Mitteilungen (1985)40,80

/31/ G. K. Huber, O. W. Holland, C. R. Clayton and C. W. White, Ion Implantation and Ion Beam Processing of Materials, North-Holland, New York (1984)

/32/ O. Auciello and R. Kelly: Ion Bombardment Modification of Surfaces, Elsevier, Amsterdam (1984)

/33/ R. Leutenecker, in: Verschleiss - Schutzschichten und Anwendung der CVD/PVD-Verfahren, ed. H. K. Pulker, Tech. Akad. Esslingen (F.R.G.) 1985, p. 217

/34/ E. Wirz, Ph D Thesis, University of Zürich (1979)

/35/ K. Yamashita, J. K. Gimzewski and S. Veprek, J. Nucl. Materials 28&129(1984)705

/36/ S. Gourrier and M. Bacal, Plasma Chem. and Plasma Processing 1(1981)217

/37/ M. Venugopalan and S. Veprek, in: Plasma Chemistry Vol. IV Ref. /5/, p.1

/38/ S. Veprek, K. Ensslen, M. Konuma and F.-A. Sarott, in Ref. /25/ p. 27

/39/ K. Ensslen and S. Veprek, Plasma Chem. and Plasma Processing (submitted)

/40/ J. K. Gimzewski and S. Veprek, Solid State Commun. 47(1983)747

/41a/ P. C. Chittick, J. H. Alexander and H. F. Sterling, J. Electrochem. Soc. 116(1969)77

/41b/ H. F. Sterling and R. C. G. Swann, Solid State Electron. 8(1985)653

/42a/ W. E. Spear and P. Le Comber, Solid State Commun. 17(1975)1193

/42b/ W. E. Spear and P. Le Comber, Phil. Mag. B 33(1976)935

/43/ Proc. 11th Int. Conf. on Amorphous and Liquid Semiconductors, Rome (1985) J. Non-Cryst. Solids 77&78(1985)

/44/ S. Veprek and V. Marecek, Solid St. Electron. 11(1968)683

/45a/ S. Veprek, Z. Iqbal, R. O. Kühne, P. Capezzuto, F.-A. Sarott and J. K. Gimzewski, J. Phys.C: Solid State Phys. 16(1983)6241

/45b/ Z. Iqbal, F.-A. Sarott and S. Veprek, ibid, 16(1983)2005

/45c/ Z. Iqbal and S. Veprek, ibid, 14(1982)295

/46/ S. Veprek, Z. Iqbal and F.-A. Sarott, Phil. Mag. B (1982)45,137

/47/ H. Curtins and S. Veprek, Solid State Commun. 57(1986)1980

/48/ M. Konuma, H. Curtins, F.-A. Sarott and S. Veprek, Phil. Mag. B, submitted

/49/ S. Veprek, Z. Iqbal, H. R. Oswald, F.-A. Sarott and J. J. Wagner, J. Physique C4(1982)251

/50a/ J. J. Wagner and S. Veprek, Plasma Chem. and Plasma Processing 2(1982)9

/50b/ J. J. Wagner and S. Veprek, ibid 3(1983)219

/51/ A. Matsuda, S. Yamasaki, K. Nakagawa, et al, Jap. J. Appl. Phys. 19(1980)L305

/52/ Y. Osaka and T. Imura, in: Amorphous Semiconductor Technologies & Devies, ed. Y. Hamakawa, OHMSHA Ltd, Tokyo and North-Holland, Amsterdam (1984) p.80

/53/ F.-A. Sarott, Z. Iqbal and S. Veprek, Solid St. Commun. **42**(1982)465

/54/ F. Mattenberger and S. Veprek, Chemtronics submitted

/55/ W. G. Townsend and M. E. Udin, Solid St. Electron **16**(1973)207

/56/ P. Geittner, J. W4. El. Chem. Soc. Proc. **84&85**(1984)479

/57/ P. Bachmann, Pure and Appl. Chem. **57**(1985)1299

/58/ G. D. Khoe and H. Lydtin, to be published

/59/ H. K. Pulker, ed., Verschleiss - Schutzschichten unter Anwendung der CVD/PVD-Verfahren

/60/ S. Veprek and W. Portmann, to be published

/61a/ R. Morancho, Proc. 4th Europ. Conf. on Chem. Vap. Deposition ed. J. Bloem, et al., Philips Cent. Manuf. Technol. Conf. Eindhoven, Netherlands (1983), p.36

/62/ H. Suhr, A. Etspüller, E. Feurer, H. Grünwald, C. Haag and C. Oehr in Ref. /25/, p.53

/63/ R. K. Sadhir and H. E. Saunders J. Vac. Sci. Technol **A3**(1985)2093

/64/ Ch. Haag and H. Suhr, Plasma Chem. and Plasma Processing (1986) in press

ADVANCED CERAMIC MATERIALS AND

PROCESSES

Roy W. Rice

W. R. Grace & Co.
Research Division
Columbia, MD

I. INTRODUCTION

Ceramics have been the focus of increased interest in recent years.
This interest is reflected in part by governmental programs to stimulate
advanced development and uses of ceramics, e.g. in the U.S., Japan, and
Germany. This interest is also shown in the wide press coverage,
including terms such as "ceramic fever" and "the new ceramic age". More
concrete are the interests and commitment shown by a number of large,
especially chemically based, corporations, in ceramics not only in the
U.S. and Japan, but also several other countries, such as England,
France, Germany and Sweden. This conference and the resultant procedings
are another manifestation of this interest.

The first of two basic questions to ask is how large is the potential for ceramics, i.e. how much of the current "hype" has a real
foundation. The second question, which becomes pertinent only if there
is a sufficiently encouraging answer to question 1, and then would become
the critical question, is how to realize the potential of ceramics. Both
of these questions are large and complex, with the complexity compounded
by the potential inter-relationship between them. Thus, these questions
can only be partially addressed in this paper, which will focus on one of
the most critical elements of question 2. However, it is useful to first
outline some aspects of other answers to these questions.

While in reality only time can determine the answer to question 1,
the key aspects of the physical basis behind all of this interest in
ceramic materials can be outlined. Two factors are most fundamental:
The first in both the broad ranges, and especially the extremes, of
physical properties that ceramics represent. Of broad significance are
electrical and thermal conductivities, hardness, stiffness, and chemical
stability, providing not only broader ranges, but more extreme values
than obtainable in any other materials. This diversity of physical
properties is related to the broad diversity of ceramic materials, which
will in part be discussed later. The second basic factor is the
elements that are the constituents of ceramics of most common interest,
namely Al, Ca, Mg, Si, B, C, O and N. Besides being light in weight,
these elements are amongst the most abundant, hence minimizing scarcity
issues and providing basic potential for limited costs.

The basic potential of ceramics has been recognized for quite some time. What is novel is the increasing indications that many of the applications implied by this potential are now beginning to be realized, or will potentially be realized in the foreseeable future. Clearly, the significant expansion expected for the largest present and future ceramic application, electronics, is of vital importance. Also of great importance is potentially large scale application of ceramics as critical components in turbine or piston engines to greatly improve their capabilities. However, of equal or greater importance is the expected diversity and scope of many less spectacular, but collectively very important, applications. Less critical uses in engines already underway, e.g. use of glow plugs and pre-combustion cups in diesel engines, and more recently, the introduction of ceramic exhaust port liners in piston engines are part of these other uses, and others, e.g. valve lifters and bearings are likely to be added later. Other wear applications include better cutting edges, pump parts, and metal extrusion or forming dies. Filtration applications should also be considered, ranging from the removal of inclusions from molten metals for casting to filtration of food and biological products. Finally, other new applications, e.g. for fuel cells for chemical production, could be quite significant in the future.

Now consider question 2 above, i.e. how to get there from here. This involves two basic elements: (1) materials technology, and (2) design technology, with a close inter-relationship between these two. The development and inter-relationship of fracture mechanics, statistical analysis of failure, and finite element analysis are major factors in expanding design capabilities. The most important and basic aspects of materials technology are the selection and processing of materials, which are the primary focus of this paper. While all aspects of this cannot be covered in this paper, the goal is to illustrate the significant ceramic opportunities to further realize their potential by improving their processing, as well as developing new and improved ceramic materials.

II. CERAMIC PROCESSING

A. Overview of Ceramic Processing

The majority of ceramic processing is based upon densification of powder compacts. It is expected that such powder-based processes will continue to be the single, most widely used method of processing of ceramics for the foreseeable, and possibly indefinite, future. The extensive emphasis on powder processing may give the impression that powder-based processing, especially by sintering, is effectively the only method of any consequence for modern, high technology ceramics. However, as following sections of this paper illustrate, this is indeed a misleading view, and in the future may be even more misleading.

This author contends that in contrast to a single focus on powder consolidation and subsequent sintering, a variety of processes are important in the present status of ceramics, and are likely to become more so (Fig. 1). Most of these processes, which are also amenable to producing powders, can directly produce finished ceramic products. While an important focus of this paper is on such alternate approaches to processing, it is useful to initially consider some of the new opportunities in conventional, i.e. powder, processing of ceramics.

Consider first the sintering aspect of conventional processing. Much of the focus on sintering has been on making improved powders for sintering, especially by increased application of chemistry. This is an important approach and is again in part reflected in this conference.

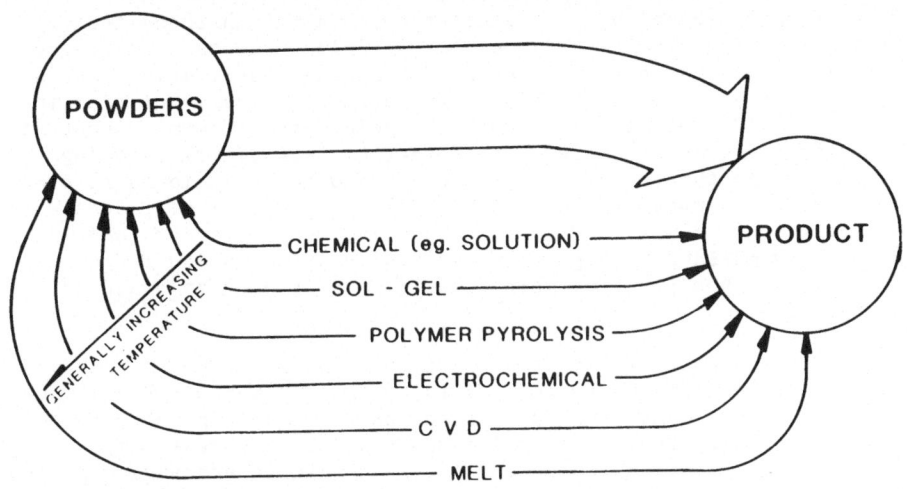

Fig. 1.   Schematic outline of ceramic processing methods. The common
view of ceramic processing would not include some of the
processes shown here and would considerably de-emphasize, if
not totally neglect, many of the direct routes to ceramic
products, i.e. those not developing a final ceramic via powder
processing.

However, there are also interesting developments regarding the sintering
process itself. For example, recent experiments have shown intriguing
possibilities for sintering under various plasma conditions (1-3)* The
potential of very rapid (in seconds to minutes), and very efficient
sintering is suggested, e.g. by heating of components only and not the
surrounding furnace structure. However, a great deal still needs to be
done to further define key parameters of such a process (e.g. accuracy
and control, and uniformity that can be achieved). While there will
probably be material as well as size and shape limitations on the
process, this and other developments can provide expanded opportunity and
diversity.

Another opportunity in sintering is the control of sintering atmospheres, e.g. as shown by recent research of Readey and colleagues (4-6).
This appears of particular importance in developing bodies of less than
full density with designed microstructures. While a great deal of
attention is appropriately placed on achieving dense structures, there
are significant applications for designed microstructures of less than
full density and atmosphere control can be an important factor in designing
such microstructures.

Two other aspects of special sintering conditions are worthy of
note. The first is hot pressing. While this is a long-established
process, it is typically viewed as being very limited in its application

_____

*Because of the diversity of subjects covered, it is not possible to
provide comprehensive referencing. Instead, a few key or representative
references are generally given.

because of cost, shape and size limitations. Although these factors still restrict the applicability of hot pressing, substantial progress has been made so that hot pressing is now a more widely used process and will become more so for high-technology ceramics. Thus, hot pressing of components 6"-12" square or larger is becoming more common. Constraints still exist in terms of shape, but increasing versatility is being achieved through the use of powder media and other techniques to provide a quasi-isostatic pressing condition. For example, ogive, radome shapes have been successfully hot pressed. Finally, with regard to cost, advances in simultaneously pressing several components and reducing cycle times have made hot pressing more competitive for high value added products.

The other aspect of the densification process worthy of note is the increasing use of hot isostatic pressing (HIP'g). While HIP'g of powder compacts is still limited because of the cost and the encumbrance of canning technology, the successful demonstration of glasses as canning materials (e.g. for Si_3N_4 components) is encouraging and suggests opportunity for further development. Also worthy of consideration is the use of HIP'g to reduce or eliminate residual porosity in components sintered to closed porosity. The very successful production use of HIP'g of metal bonded carbide products to reduce failures by eliminating isolated pores and pore clusters is extremely encouraging (7-8). While there are doubts as to whether this will generally work as well on ceramic materials, there are good indications that it will have significant applicability.

It is useful to consider some general aspects of improved powder processing that will not be covered in the subsequent sections on specific processes. As noted earlier, substantial attention has been paid to use of more sophisticated chemical techniques to produce powders with improved control of particle size, and possibly control of morphology, greater purity, and a broader range of compositions. Much attention has also recently been focused on obtaining monosized spherical particles. While this is clearly useful for study purposes and may have some applications, e.g. for sintering of thin layers in small components for electronic ceramics, it should not be viewed as an overall generic need of ceramics for a variety of reasons. For example, recent studies indicate advantages of bimodal over monomodal particle distributions (9). Further, different processes may have different optimum particle size distributions e.g., monomodal particles are not optimum for injection molding, as pointed out by Aksay (10). More basic, with regard to the issue of uniform particle size, is the question of the character of the resultant body. Uniform spherical particles do show better sintering both in theory and in practice than do particles of more variable shape and size. However, there are basic questions as to the real practicality of this. Even if perfectly uniform particles can be produced, they still exhibit stacking defects (Fig. 2) which can then become factors in determining the ultimate grain size in a sintered body, which is also of importance. Further, in reality there will commonly be some variation in shape and size simply because of the number of particles involved, e.g. $\approx 10^{12}$ particles/cm^3 for 1μm particles and $\approx 10^{15}$ particles/cm^3 for 0.1 μm particles. Even a quite limited population of particles of different shape or size can have a substantial effect (e.g Fig. 2). In view of inherent stacking defects of uniform particles and frequent variation in particle shape or size, there can be advantages in terms of introducing some heterogeneity, i.e. multi-moldability, in particle size distributions (Fig. 2), e.g to better control the final microstructure. Particles whose shapes are not amenable to uniform packing are also important; i.e. there are important applications for particles with clear, morphological, as opposed to simple spherical shapes. An important example of

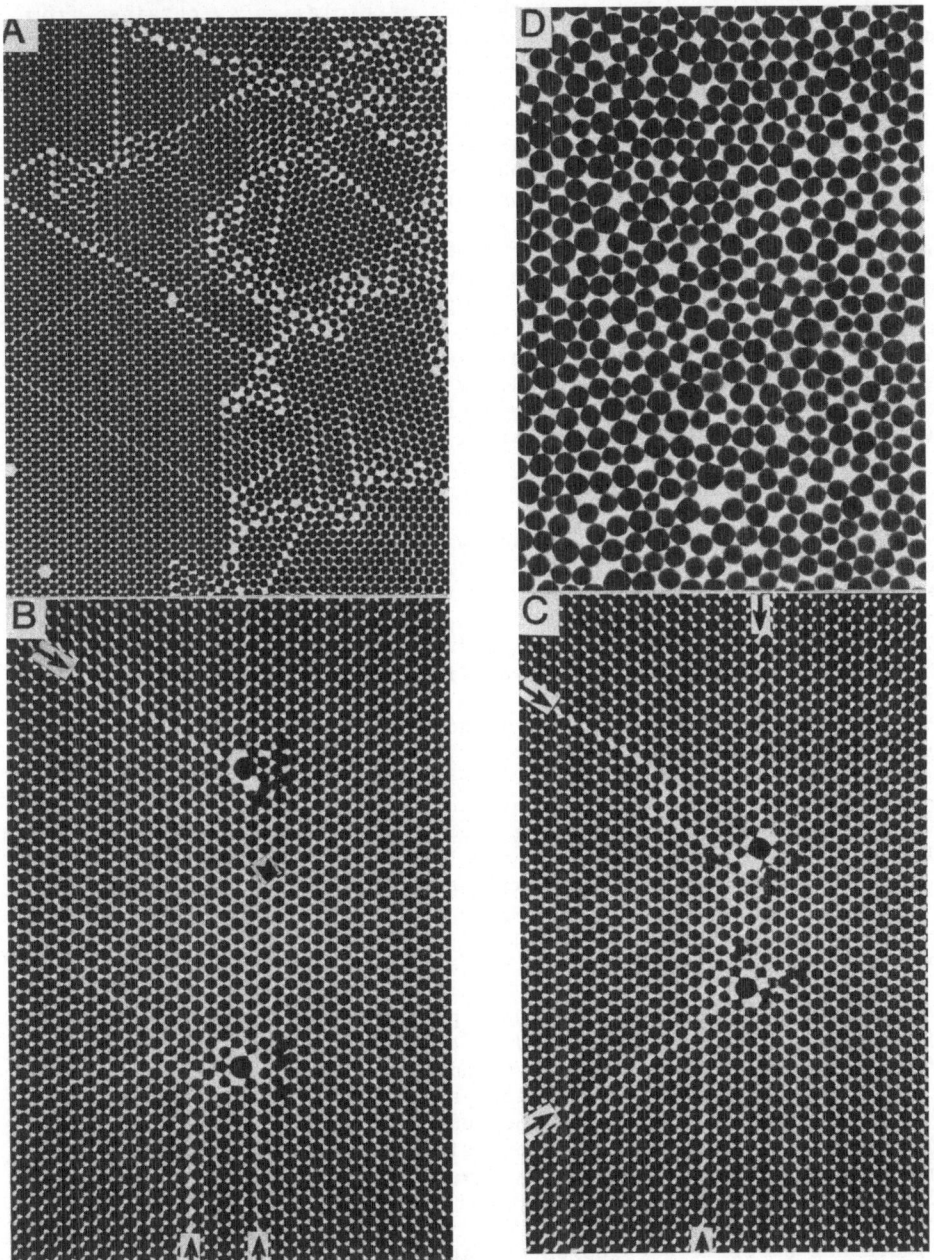

Fig. 2.   Two-dimensional stacking imperfections of spherical particles.
(A) point and line stacking imperfections with spherical
particles that are uniform in both size and shape. (B) and (C)
line (arrows) and extensive local point defects caused by the
presence of two larger spherical particles in an array of
otherwise uniform spherical particles. Note the local
distortions around the larger particles begin to interact when
they are of the order a dozen smaller particles in separation.
(D) imperfect stacking with the absence of linear stacking
defects and the much greater homogeneity of point and local
stacking defects due to some limited variation in both
sphericity and particle size.

this is the use of plate-like or lath-like particles, e.g. obtained from fused salt preparations, to obtain significant preferred orientations to take advantage of important anisotropies, e.g. for ferroelectric or magnetic applications.

With regard to composition, several points are worth noting. Increased purity, which has been a basic goal of much research, can be important. This is particularly true for non-oxide powders where residual carbon left from carbothermic reduction, or excess oxygen due to incomplete carbothermic reduction or subsequent contact of the powders with the air occurs can be serious problems. There are important opportunities for more sophisticated chemical preparation, especially for non-oxide powders. However, the predominant focus on higher purity is not always justified. More effort is needed on developing powders or additives to be used with existing powders , to allow suitable, if not superior, properties to be obtained with lower purity and lower cost materials than simply arbitrarily seeking higher purity, usually at higher cost.

B. Other Ceramic Processes

While sintering of ceramic powders, especially those derived by solution chemistry approaches, have attracted much attention, there are a variety of other approaches for both making powders, as well as directly making ceramic bodies (as noted in the previous section). While all of the alternate methods cannot be explored in this paper, three of the more developed and promising methods are outlined in this section.

The most widely used alternate method is melt processing. Large tonnage quantities of fused oxide grains, such as Al_2O_3 and ZrO_2, are produced for the refractories industry each year. Tonnage quantities of other fused oxide grains are made for other speciality purposes (e.g. MgO for insulation in electrical heating elements). In addition to these fused grains, tonnage quantities of monolithic fused materials are also made by casting, with some individual castings weighing in excess of 1 ton. At the high tech end are a few, to many, tons/year of single crystals such as Al_2O_3, $MgAl_2O_4$ and ZrO_2 grown each year by a variety of methods for both technical and jewelry uses.

Most of the above applications are predominantly viewed as "low tech", and hence typically neglected in considering high tech ceramics. However, analogies with advanced developments in melt processing of metals, actual development in ceramics, and the promise of ceramic composites all suggest realistic opportunities for medium and high tech applications. For example, melt processing of powders and newer concepts of casting (e.g. reho-casting) are suggested by developments occurring in the field of metallurgy.

There are specific developments that have already occurred, or are occurring now, in ceramics utilizing melt processing. It has been demonstrated that melting of multiconstituent materials results in greater homogeneity than does conventional preparation of powders for plasma spraying (11, 12) (e.g. by mixing and calcining). Skull melting and subsequent crushing of resultant ingots to produce multiconstituent powders (e.g. $MgAl_2O_4$, cordierite and various ZrO_2 compositions) have been in production for several years. Some organizations are now investigating the production of powders by atomizing a molten stream, which should save substantially on production costs. Investigations are underway to utilize such melt derived powders for fabricating bulk ceramic bodies. There are also other uses of melt derived particles. Thus, while sintering of Al_2O_3 in the past has been the predominant

174

method of producing sand milling media, substantial in-roads are being
made in the market by disassociated zircon particles made by melting and
solidification (Fig. 3).

Melting of ceramic compositions that yield toughened micro-structures
is of particular interest. Since these are typically two-, or multi-phase
bodies, they also help limit the scale of micro-structures developed,
which has often limited use of melt processed ceramics. Eutectic
structures, e.g. Al_2O_3-ZrO_2 (Fig. 4) are of particular potential for melt
processing. There are also interesting possibilities for combining
directional solidification of eutectic compositions with such unique melt
shaping techniques as the edged film-fed growth process, well known for
making shaped sapphire crystals (13, 14). The excellent precipitate
structures and mechanical properties obtained from skull melted partially
stabilized ZrO_2 crystals (15-17) are also very encouraging.

The potential for low costs using melt processing is clearly
indicated by the cost of fused grains being sold for approx. $2/lb.,
despite the added costs of crushing and grinding, over those of melting
itself. In terms of actual fused cast bodies, there is substantial size
capability (as indicated by the large weights of individual castings).
There is also increasing application of melt processed materials to
sizeable and complex wear shapes, e.g. components in cyclones.

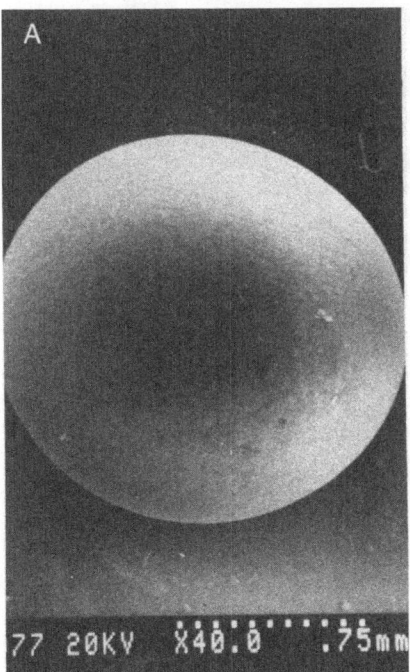
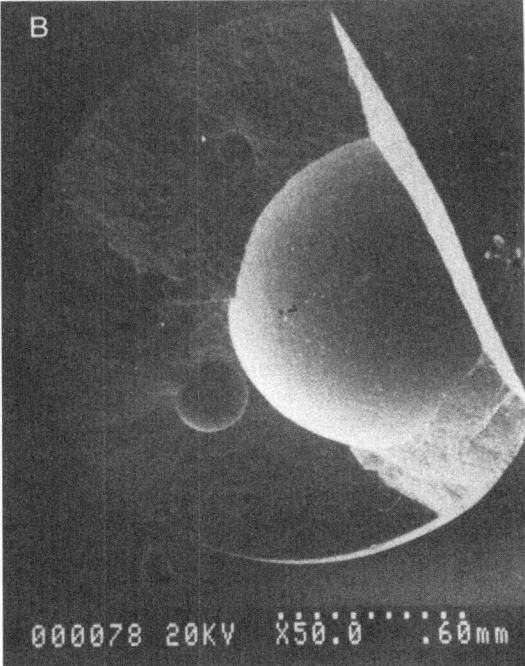

Fig. 3.   Examples of solidified molten ceramic droplets (A) substantial
           degree of spherical uniformity that can be obtained by solidi-
           fication of a molten droplet, in this case for use as sand
           milling media. (B) a fracture of a particle similar to that in
           (A) illustrating the large central cavity (and a few satellite
           pores) that are frequently, but not always, found in such
           solidified spherical droplets.

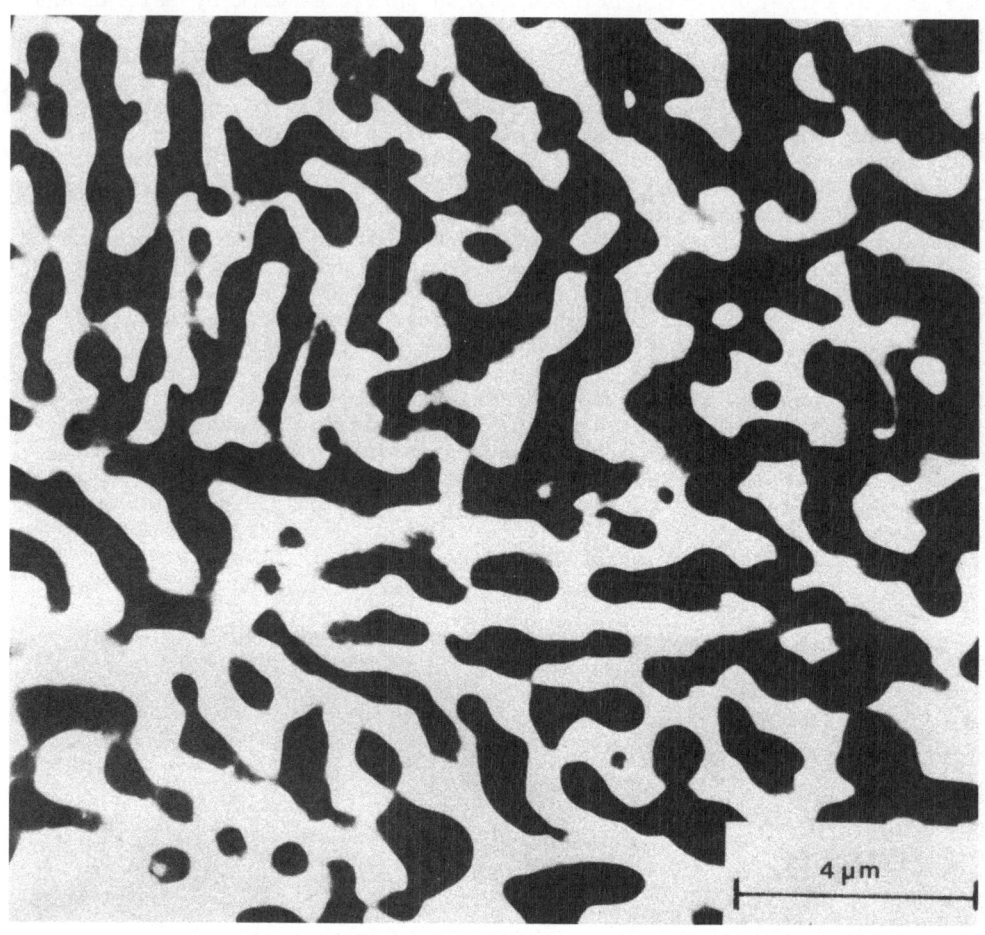

4 µm

Fig. 4. Examples of a eutectic structure from melt processing. The
Al_2O_3-ZrO_2 eutectic structure shown can be achieved only by
melt processing and results in a more homogeneous distribution
of the phases than does powder processing. Photo courtesy of
Dr. R. Ingel of the U.S. Naval Research Lab.

The basic issues in melt processing have been obtaining fine microstructures and controlling porosity from soidification shrinkage (which
can be substantial, e.g. 10-25% (Fig. 5.) Both composition and control
of solidification can be important in controlling porosity. As will be
shown later, ceramic composites compositions offers opportunity for
synergism between the microstructural control needs of melt processing
and homogeniety needs of composites.

The second method of processing considered in this section is
chemical vapor deposition (CVD). This is a recognized process for making
powders. Tonnage quantities of oxides, such as TiO_2 and Al_2O_3, are
produced by heterogeneous gas phase nucleation (i.e. CVD) using metal
chlorides and water. The large scale, fine particle size, and low cost
of these all attest to the potential of this method. CVD has also
demonstrated potential and some applicability for non-oxide powder (e.g.
Si_3N_4). In addition to powders, CVD is commonly associated with forming
of other thin structures, e.g. filaments for reinforcement, and coatings.

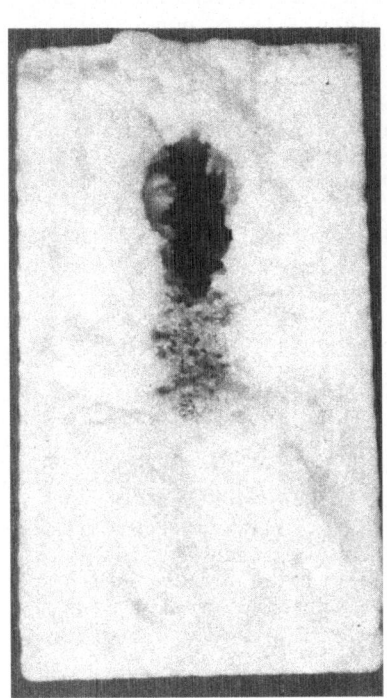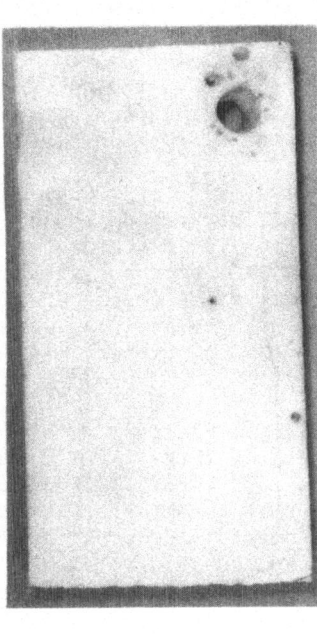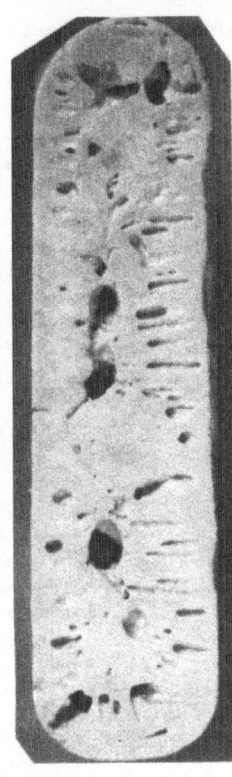

Fig. 5. Examples of porosity in macroscopic bodies made by fusion casting. Left), Cross-section of a fusion cast Al_2O_3 refractory brick (nearly 2" wide). The large shrinkage cavity near the top of the brick (due to casting two bricks back to back and the dark area underneath this showing more distributed porosity illustrates the combination of good management of the cooling process and favorable solidification characteristics to control the location of most of the porosity. Center) Cross section of another fusion cast refractory brick (about 1.5" wide) showing examples of isolated macropores scattered well away from the central shrinkage cavity. Right) Poor control of solidification, resulting in broad distribution of macro pores (piece about 0.2" wide) Note in all cases, there is microscopic porosity distributed both in a sporadic as well as in a pattern related to the overall cooling trend.

CVD has also been successfully applied to a variety of bulk, free-standing bodies and has further potential for such applications (18). Besides the extensive application of pyrolytic graphite (i.e. carbon made by CVD) to make sizeable and demanding bodies such as rocket motor exit nozzles and missile re-entry nose tips, there are a number of important applications for other ceramic materials. These include CVD of BN for semi-conductor applications (e.g. processing of GaAs). Other applications include IR windows, e.g. plates of ZnSe are made in sizes up to 48" x 36" x 1" (120 x 90 x 2.5 cm) for high-power laser windows because of the high purity and resultant low optical absorption that can be achieved. Similarly, large IR domes, which are sections of hemispherical shells, having a base diameter of about 10" and a wall thickness of approx. 1/4-1/2" in thickness, have been successfully made by CVD. These

IR domes are competitive in cost with those made by hot pressing, but are superior in optical quality. Potential for applicability to non-oxides is shown by making SiC in thicknesses of 1/2 - 3/4" and a variety of shapes (e.g. Fig. 6,7).

One of the advantages of CVD is its near net shape capability and the versatility of the shapes that can be achieved (e.g. Fig. 6). CVD of bulk shapes also has potential for low cost. Methyltrichlorosilane, which directly yields SiC in combination with hydrogen, costs approx. $1/lb., and theoretically yields approx. 1/3 lb. SiC per pound of methyltrichlorosilane. In practice, yields would be less, but even only achieving 1 lb. of SiC from 10 lbs. of methyltrichlorosilane would still produce a near net-shape SiC component with a raw materials cost around $10. Although energy and set-up costs can add a substantial amount to this, resultant costs can be competitive with, or less than, those of conventionally processed SiC. Further, with appropriate engineering, costs can be reduced.

Basic issues which have limited use of CVD are residual stresses, keeping fine microstructures (Fig. 7, 8) and maintaining uniformity and reliability in producing CVD materials. However, these issues can vary substantially from material to material. Additionally, these typically have been most troublesome in non-oxide materials where there is the largest driving force for use of CVD, since these materials can be made

Fig 6.   Examples of a diversity of free standing shapes processible by a CVD. Top row: a thick wall and a thin wall cylinder. Middle row: a carbon form (left) and resultant CVD ceramic paddlewheel-type rotor (middle) after deposition in, and removal of the carbon form, miniature thin wall ceramic heat exchanger (right). Bottom row, left to right: metal component model, resultant carbon form, and resultant CVD component. All samples are CVD SiC courtesy of Mr. Engdal of Synterials.

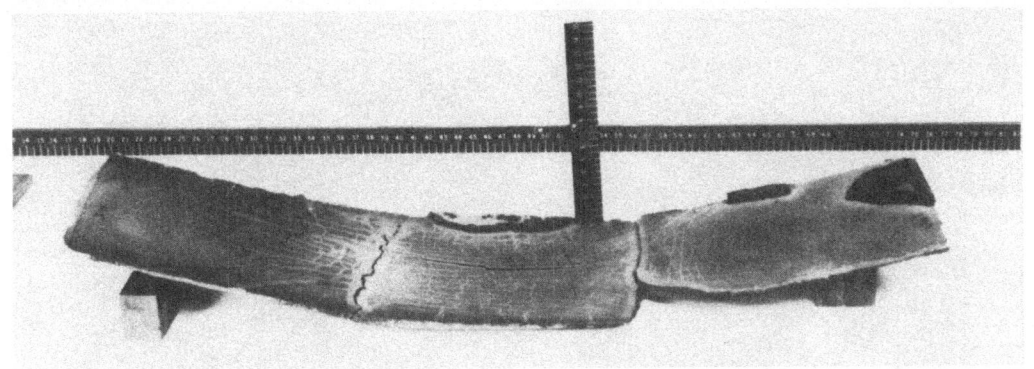

Fig. 7   An extreme case of residual stresses; Large, thick (>1/2")
          plate of SiC. Both the gross bowing from the original flat
          substrate and the complete and partial cracking of the plate
          illustrate an extreme case of stress generated in CVD.

by CVD without the use of additives, and hence, without degradation and
high temperature properties. As will be discussed later, application of
CVD to making composites presents important opportunities to minimize
these problems, as well as provide advantages in composite processing.

The newest method of processing discussed in this section is
preparing ceramics by pyrolysis of appropriate precursor polymers.
While this topic has recently been reviewed (19, 20), it is useful to
outline its important aspects. The well established application of this
process is the production of carbon fibers, bulk glassy carbon bodies and
carbon-carbon composites (Fig. 9). Various appropriate polymers are
first formed into the shape of fibers, various laboratory ware, or
infiltrated into fiber preforms for composites, and subsequently
pyrolyzed (i.e. slowly heated in a non-oxidizing atmosphere to remove
mainly the hydrogen constituents of the polymers, leaving the carbon
skeleton as the remaining constituent of the resultant carbon body). It
is now established that various polymers can yield bodies consisting of
predominantly SiC, or Si_3N_4 upon pyrolysis. Other ceramic materials
should also be feasible, such as B_4C or BN-based bodies. The term "based
bodies" is used here, because in general it will not be possible to
obtain exact stoichiometry, therefore resultant bodies after pyrolysis
will contain a few to several percent of other constituents, e.g. excess
carbon, silicon, boron, etc. However, the content of these can be at
least partly controlled by the atmosphere of firing, e.g. excess silicon
or boron may be converted to silicon nitride or boron nitride by
processing in a nitrogen or ammonia atmosphere.

Known and potential applications of polymer pyrolysis can be divided
into those that have one dimension that is small, e.g. well less than
1mm, and those in which the smallest dimension is of the order of 1mm or
more, i.e. bulk bodies. The most significant application falling in the
first category is fibers. The development of an SiC-based fiber by
pyrolysis of an appropriate polycarbosilane polymer based on developments
by Yagima (21), is having significant impact on composite development,
especially ceramic fiber composites that will be discussed later. This
process is in direct analogy with the formation of carbon fibers by
pyrolysis of previous rayon, and more recently, polyacrylonitrile or
pitch.

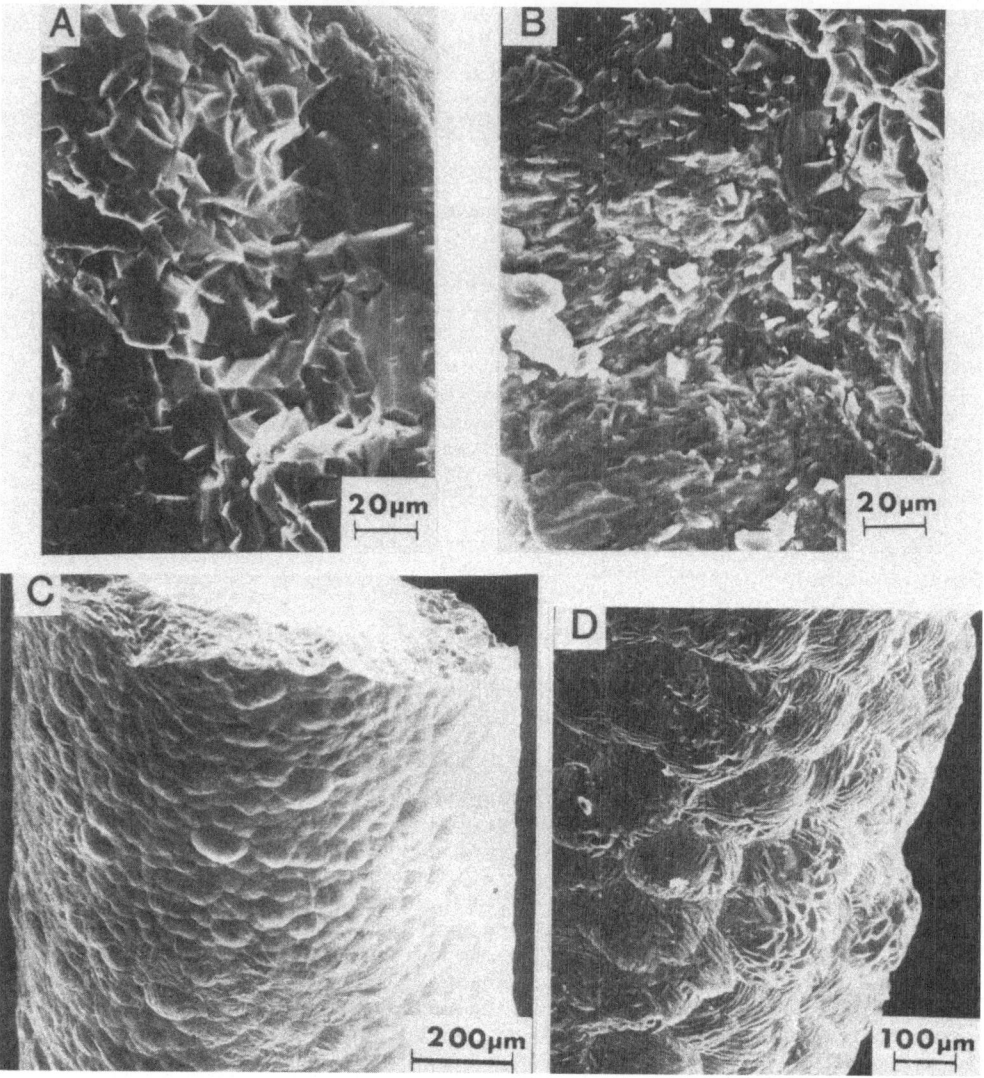

Fig. 8. Examples of less desirable micro-structures that can result
from chemical vapor deposition. (A) blocky, coarse microstructure that is sometimes obtained, often at high deposition
rates. (B) Coarse and columnar grains (toward left of photo),
and much more extreme columnar structures are frequently
associated with growth cones, which typically lead to surface
roughness as illustrated in C) (moderate case) and D), a more
serious case. Such growth cones can often be a source of
mechanical failure.

Another application of potential importance that has received more
limited attention in this category is that of coatings. While coatings
on bulk surfaces are an important possibility, more work appears to have
been done on use of such polymers for coatings on wires for high temperature electrical insulation. At least one Japanese firm has investigated
the use of polycarbosilanes to provide a high temperature electrical
insulation. Similarly, the author and colleagues demonstrated that
dipping a tungsten filament in a polysilane solution, and then curing (by
Joule heating of the filament) at temperatures of 270°C for about 15
minutes resulted in an insulating layer that prevented the high temperature

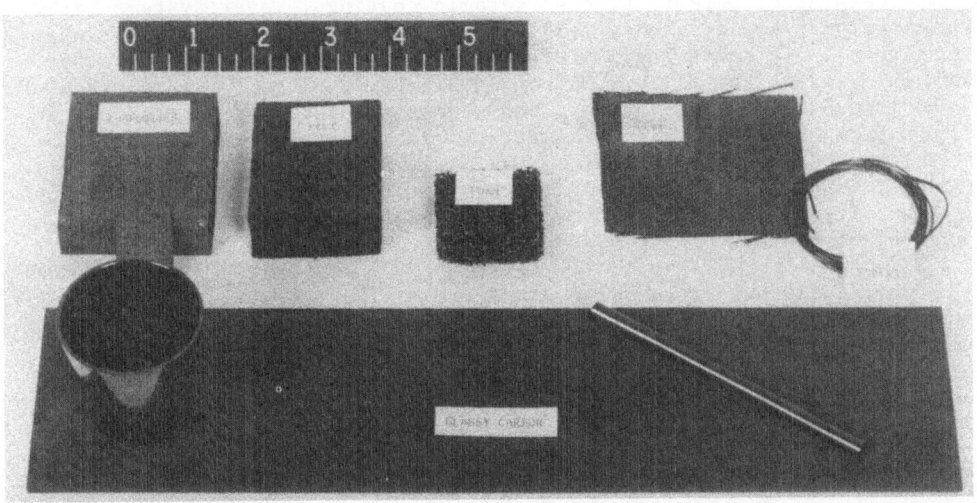

Fig. 9.   Carbon bodies from polymer pyrolysis. This photo illustrates a variety of carbon products made by polymer pyrolysis including glassy carbon (plate crucible and rod) graphite fibers and carbon-carbon composites, e.g. see ref. 19. The analogy of all of these is feasible in other ceramics and, in fact, fibers and ceramic fiber composites have already been made.

shorting out of the ignitor coil that occurred without such a coating. Although it appears to not yet have been investigated in any depth, an important application for such polymer pyrolysis of mainly film-type structures is for electronic packaging. While there may be some opportunity for producing bulk bodies of other ceramics by polymer pyrolysis in direct analogy with glassy carbon, the opportunities appear to be limited because of the greater changes in density from the polymer to the ceramic (19, 20). However, there is a significant opportunity for using polymer pyrolysis for processing ceramic fiber composites, as will be discussed later.

There are three major issues in terms of further development of polymer pyrolysis as a means of preparing ceramics. One major issue is cost. Estimates indicate that the costs in volume production of promising polymers may be of the order of $50/lb. Such costs will allow a range of application, but will require use of either limited amounts of the polymer, substantial value added, or both. The basic technological issue is the amount of shrinkage that occurs, which again is related to the change in volume, but also to the yield of actual ceramic from the polymer (i.e. the weight of the resultant ceramic relative to the starting polymer weight). Limited cross-sectional dimensions of parts, and particularly the use of fillers, as well as use in making matrices for composites are all important factors in controlling shrinkage.

III. PROCESSING OF SPECIALIZED MATERIALS, e.g. COATINGS

The preceding discussion, as have most discussions of ceramic processing, focused mostly on the processing of dense, bulk ceramic bodies as an end goal. There are, however, a variety of specialized applications that either require modification of accepted processing methods, or new approaches. Most significant amongst these methods are the processing of ceramic fibers, ceramic coatings, layer systems (i.e. for electronic applications, mainly electronic packaging and multilayer capacitors), and special structures (e.g. foams, honeycombs, and other open structures) for filters, heat exchangers, fuel cells, etc. Each of these is a broad topic by itself and cannot be discussed extensively in this paper. We will instead illustrate such special form processing by one of the oldest applications of ceramics, namely as coatings. This is not only important, showing significant growth and future potential, but is also often one of the ceramic technologies most neglected by the research community.

While coatings often serve a variety of functions, some of which were discussed briefly earlier, two of the most extensive are for wear resistance and thermal protection (22,23). Ceramic coatings have been utilized for several years in combustor transition ducts in jet engines on commercial aircraft. Coatings are also being introduced commercially in large Marine diesel engines (e.g. Fig. 10) to allow the use of lower

Fig. 10. Example of large-scale application of ceramic coatings. This photo shows the application of ceramic coatings to a piston for a large marine diesel engine being applied by plasma spraying using a hand-held torch. More production oriented trends would be to use more automated equipment with computer control.

grade fuels and have been under investigation for use on turbine vanes
(Fig. 11). Coatings were also the predominant form of ceramics used in
the low heat rejection diesel engine successfully demonstrated for the
Army by Cummins (24).

While ceramic coatings on ceramics, mainly by glazing, continue to
be important, coating of ceramics on metals is particularly important.
The most widely used method of forming such ceramic coatings is via melt
spraying, most commonly plasma spraying. Chemical vapor deposition (CVD)
is also used, but is limited by temperature and chemical effects on the
metal substrates. Physical vapor deposition (PVD), sputtering and
various modifications of these are also used, mainly for thin coatings.
Sol-gel coating methods, as well as polymer pyrolysis, are also being
explored, as discussed earlier. Electrochemical methods have also been
demonstrated for producing various non-oxide coatings (25-29) (as well as
for producing some powders, e.g. of oxides). The processing of most
ceramic coatings does not involve sintering, and only melt spraying
involves the use of powders (but in a larger size range than those used
for sintering bulk ceramic bodies) again illustrating the need for a
diverse view of ceramic processing.

Uniformity and reproducibility are important needs for further
application of ceramic coatings, as they are with all forms of ceramics.
The use of melt homogenization of multiconstituent powders discussed
earlier is one step in this direction, and computer control of the
spraying process is another. Another basic issue is that of residual
stresses. While again, compositional homogeneity can be an important
factor, another factor is the mismatch in properties, especially thermal
expansion, between the coating and the substrate. This is frequently a
severe problem since ceramic coatings are most often put on metals, which
typically have thermal expansions of $15-25 \times 10^{-6} \, ^\circ C^{-1}$ while most ceramics
of interest have thermal expansions of $10 \times 10^{-6} \, ^\circ C^{-1}$ or less. Development of techniques to produce coatings with highly preferred orientations,

Fig. 11.   Ceramic Coating on Turbine Vane. Photo of vane (length ~ 4")
           from marine gas turbine engine after service. The ZrO_2 plasma
           sprayed coating has spalled some from the leading edge, but is
           still serviceable.

which appears feasible, for example, with vapor deposition methods, could provide new opportunities for matching coating expansion to the substrate, as well as providing other advantages. This would use the thermal expansion anisotropy that exists in many ceramic materials, very commonly those that have been less extensively addressed in the past, since we did not know how to utilize their high thermal expansion anisotropy. Thus $CaWO_4$, $LiNbO_3$, and α quartz have expansion coefficients of about 14, 21, and 22 x 10^{-6} $°C^{-1}$. $LiNbO_3$ also has a very low expansion in the third direction. (Table 1) Were it possible to make coatings of such types of materials so that the two high expansion coefficients were oriented in the plane of the coating, and the third expansion coefficient normal to it, it would produce a much closer match of expansion with the substrate. Having a low expansion perpendicular to the substrate, would give the advantage of maintaining closer tolerances, as well as improving the spall resistance of the coating, e.g. from thermal stresses.

IV. NEW MATERIALS

A. Composites

Ceramic composites have received a great deal of attention due to wider diversity and improved physical properties, or combinations of physical properties that they can produce, especially in terms of mechanical behavior. Both the improvement in physical properties of composites as well as how they are processed depends on the type of composite. There are particulate composites where a ceramic particulate phase is dispersed in a ceramic matrix, e.g. BN or graphite particles in matrices such as Al_2O_3, mullite; or tetragonal ZrO_2 particles dispersed in Al_2O_3, mullite, beta alumina, etc. Whisker or short fiber composites have whiskers (e.g. Si_3N_4 or SiC) or chopped fibers (e.g. SiC or C) dispersed in various matrices. Continuous ceramic fibers may contain unidirectional, cross-plyed, or multidimensionally oriented filaments or cloth. Both particulate and short fiber composites show increased toughness, but still exhibit catastrophic failure upon reaching their maximum stress. (Fig. 12) In contrast, continuous ceramic fiber composites typically do not exhibit catastrophic failure and hence retain substantial load carrying capability beyond their maximum stress capability, showing promising failure strains (e.g. greater than 1%).

Processing also depends on composite type. Particulate composites are more difficult to sinter than conventional monolithic ceramic bodies. Thus, while sintering is commonly used for particulate composites, such composites are frequently fabricated by hot pressing or HIPing. Hybrid methods, such as the preparation of melt derived composite powders for subsequent densification by sintering or hot pressing, are also being pursued as discussed earlier. Short fiber composites are more difficult to sinter than particulate composites, and continuous fiber composites are more difficult to sinter than short fiber composites. There has been some limited investigation of sintering of short fiber (mainly whisker) composites, but to date, fiber (and whisker) composites have been made mostly by hot pressing. Currently there are at least two applications of (Al_2O_3-SiC) whisker composites in production utilizing hot pressing. Other methods of fabrication of short fiber composites are being extensively investigated as discussed below for continuous fiber composites.

Fabrication of continuous fiber composites involves the broadest diversity of ceramic processing. Hot pressing has been used extensively for composites utilizing matrices based on SiO_2-containing glass systems (30) and for a variety of polycrystalline (oxide) matrices. Since fiber

Table 1. Examples of Thermal Expansion Anisotropy of Ceramic Oxides

Material	Melting Temp. (°C)	Thermal Expansion Coefficient ($10^{-6}°C^{-1}$)		
		αa	αb	αc
Al_2O_3	2050	8.6	8.6	9.5
TiO_2	1825	8.3	8.3	10.2
Cr_2O_3	2260	8.1	8.1	6.1
SiO_2 (α quartz)	1720	22	22	12
$LiAlSiO_4$ (β eucryptite)	1400	8.2	8.2	-17.6
$MgTiO_3$	1630	9.4	9.4	12.4
$LiNbO_3$	1410	21.2	21.2	0.1
Al_2TiO_5	1860	-3.0	11.8	21.8
$MgSnB_2O_6$		4.6	4.6	19.0
$MgTi_2O_5$	1650	2.3	10.8	15.9
$MgFeTaO_5$		1.9	9.1	13.7
$CaWO_4$	1580	13.7	13.7	21.5

costs and costs of fiber lay-up are typically higher than for making green bodies of conventional ceramics, hot pressing represents a smaller impact on the overall composite cost than it does in pressing conventional ceramics. Chemical vapor infiltration (CVI), i.e. using CVD to generate the matrix around fibers has been used to date primarily to generate SiC matrices (in addition to C matrices for long-standing development of C-C composites).

The predominant methods of making continuous ceramic fiber composites with non-oxide matrices have been by hot pressing or by CVI. There is considerable synergism between CVD and using it to form composites. First the heterogeneity of the composite structure helps limit the CVD microstructure, and the effect (as well as probably the level) of residual stresses. Second, the composite allows higher deposition rates since deposition occurs simultaneously throughout the composite on the large fiber surface area. Control of microstructure by the composite can also allow use of higher deposition rates.

Considerable progress and promise has been demonstrated utilizing polymer pyrolysis as a method of making ceramic fiber composites.(31) The predominant work is focused on Si-based matrices, e.g. SiC (Fig. 13). Again, there is a synergism between composite processing and polymer pyrolysis. The use of polymers as a matrix source allows application of conventional "fiber glass" type composite processing. Further, the presence of the fibers greatly reduces both the extent and the effect of shrinkage cracking or conversion of the polymer to a ceramic.

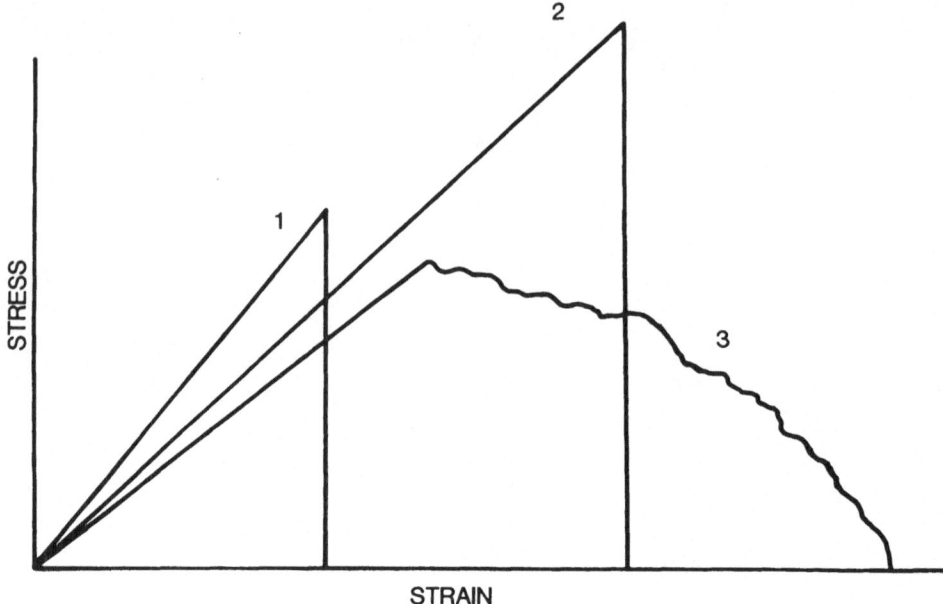

Fig. 12. Comparison of stress/strain curves for various ceramic bodies. Curve 1, for a conventional monolithic ceramic, illustrates the typical elastic loading until catastrophic failure occurs with total loss of load carrying capability. Curve 2, typical for a particulate or short fiber composite illustrates a lower slope (Young's modulus) and larger strain to failure (reflecting higher toughness), frequently occurring, but still with catastrophic failure and total loss of load carrying capability. Curve 3, behavior commonly found for continuous fiber composites showing substantial load carrying capability past the ultimate load and large strains to failure (37).

This is not to say that the above methods are perfect. Thus, as discussed in more detail elsewhere (32), hot pressing still presents challenges due to the temperatures needed and resultant fiber degradation versus lower temperatures and residual porosity. CVI can still present challenges of fiber degradation due to temperature − chemical environment effects (e.g. due to products such as HCl), as well as trapped porosity. Polymer pyrolysis still results in porosity and microcracking. However, the synergism is sufficiently effective that very useful materials can be produced at practical costs, with most of these materials not being producible by sintering.

Two other important aspects of ceramic composite processing demonstrate the diversity of processes that are needed. The first is making the fibers themselves. Sol-gel type processes are commonly utilized to produce oxide-based fibers, and polymer pyrolysis processes are used for most of the non-oxide based fibers. Large filaments, e.g. SiC, are being produced by CVD. Second is fiber coating. Substantial promise has been demonstrated in limited investigations of fiber coatings (Fig. 14). Most coatings have thus far been applied by CVD. However, some methods of liquid application, as well as applying coatings by reaction of the fibers in appropriate atmospheres are also known to be under investigation.

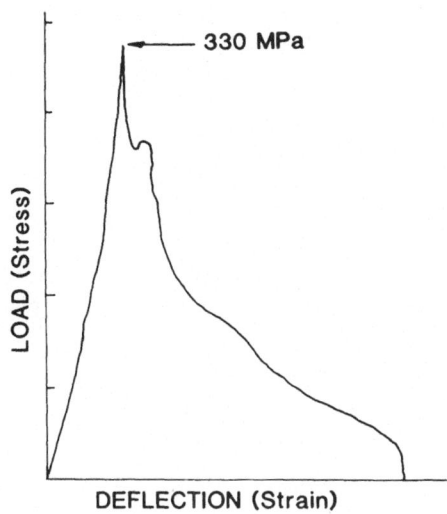

50 VOL. % FIBERS

MATRIX: 50 WT. % POLYSILANE
        & 50 WT. % CERAMIC FILLER

IMPREGNATE TOW

MOLD: 200° C, 500 psi

POLYMERIZE 270° C
        (UNDER PRESSURE)

PYROLYZE 1000° C
MATRIX YIELD 91%

FINAL ρ : 2.0 gm/cc

Fig. 13. Outline of processing procedure and resultant load deflection
          curve for a ceramic fiber composite made by polymer pyrolysis.
          Note although this composite still has considerable porosity, it
          has rather respectable strength (of the order of 50,000 psi)
          (31).

B. New Compounds

     Many tend to view ceramics as a field in which the primary needs and
opportunities were to further develop a well defined list of compounds.
With the exception of complex silicates, the field was viewed as one to
develop the known binary compounds of ceramics, i.e. compounds consisting
of two types of atoms by themselves or in various composite combinations.
While this is of great importance, the development of new compounds can
greatly extend ceramic opportunities.

     Some perspective can be obtained by recognizing that ceramics are
predominantly inorganic compounds. It is also readily recognized that
such compounds are not limited to those consisting of just two types of
atoms; they can consist of three, four and even greater numbers of atoms.
While the common ceramics, which are primarily binary compounds, represent
the simplest systems, this is only a small fraction of the potential
ceramic population. Even when it is recognized that many of the other
possible combinations of inorganic compounds in the periodic table would
not be of particular interest for the diversity of ceramic applications,
it can be readily seen that to date only a limited fraction of the total
population of ceramic compounds have been identified. Studies have been
performed on only a limited fraction of the total ceramics population, as
previously discussed (33, 34).

     There is substantial evidence that many other compounds do exist and
that many can be of considerable importance. Consider oxide materials,
where many more compounds, especially SiO_2-based compounds, exist in

187

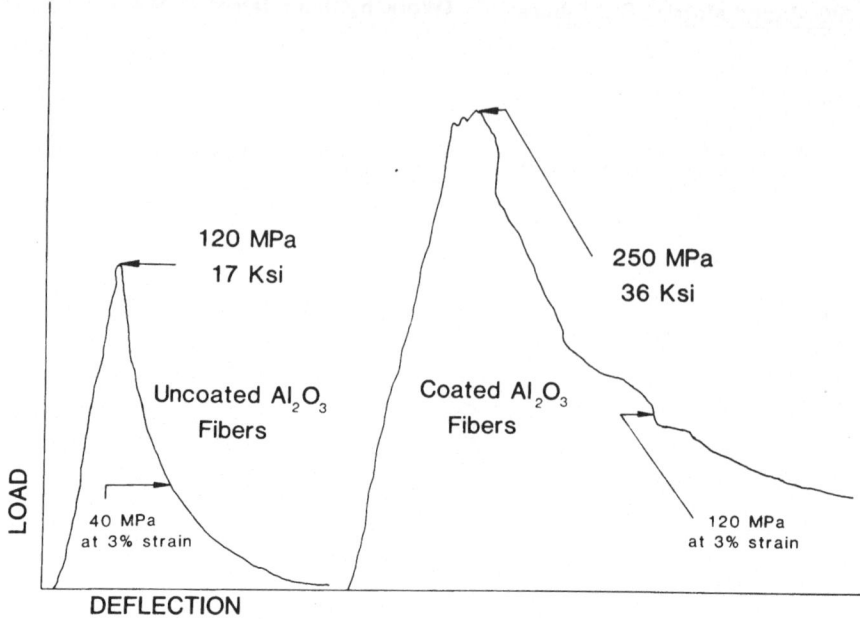

Fig. 14. Effects of fiber coating on continuous fiber composite
mechanical performance. Load deflection curves from room
temperature mechanical testing of two composites which are
essentially the same except the one on the right was made with
fibers coated with a thin layer of BN to limit the fiber-matrix
bonding thus resulting in both higher strength and higher
toughness (e.g. as reflected by area under the curve). Note the
increased fiber pullout and roughness of fracture with the
coated fibers. This composite was made by polymer pyrolysis
since it was one of the few methods available whereby an all-
ceramic composite could be fabricated within the temperature
limitations of the Al_2O_3 fibers used in this composite (31).

nature as a guideline and have led to greater development. Cordierite
($2Al_2O_3$ $2MgO$ $5SiO_2$) and mullite ($3Al_2O_3$ $2SiO_2$) are two examples utilized
for their respectively low and moderate expansion, combined with other
useful physical properties, e.g. moderate dielectric constant (cordierite)
and high creep resistance (mullite). However, many more complex oxides
that do not necessarily occur in nature have been of importance. $BaTiO_3$,
widely used in ceramic capacitors, and $PbTiO_3$-$PbZrO_3$ bodies, the dominant
materials used in piezoelectric transducers, are two important examples
of ceramics for electrical functions. Other examples are also found in
both the ionic conductors (e.g. beta aluminas), magnetic materials (e.g.
ferrites); all dominated by ternary or higher-order compounds.

The existence and potential of higher-order compounds, although less
explored in non-oxide materials, is also of great significance. We know
that a variety of complex compounds containing boron, nitrogen or carbon,
either singularly or in various combinations with themselves or other
anions, can exist (Table 2). These compounds can have interesting
properties, as indicated by the fact that the highest known melting
materials are not binary ceramics, but are 4TaC ZrC and 4TaC HfC (melting
temperatures ~4000°C).

188

Table 2
EXAMPLES OF BORIDE, CARBIDE AND NITRIDE CERAMICS

Typical Refractory Borides	Typical Refractory Carbides		Typical Refractory Nitrides	
AlB_{12}	Cr_3C_2	TaC	AlN	Si_3N_4
CaB_6	HfC	TiC	BN	TaN
TiB_2	NbC	WC	CrN	TiN
ZrB_2	SiC	ZrC	NbN	ZrN

SOME REPORTED MIXED NITRIDES[*]

LiMgN	Fe_3MgN	Ni_3ZnN	$MgSiN_2$
LiZnN	Fe_3ZnN	Ni_3AlN	$BeSiN_2$
Li_3ZnN	Fe_3AlN	Ni_3InN	(Nb,Ta)N
Li_2AlN_2	Fe_3GaN	$Zr_4Zn_2N_x$	$Ga_{.5}Al_{.5}N$
Li_3GaN_2	Fe_3InN	ZrTlN	Ti_2AlN
Li_5GeN_3	Fe_3GeN	Nb_4ZnN_x	Ti_3InN
Li_9CrN_5	Fe_3SnN	V_2GaN	Ti_3TlN
Li_7MnN_4	Co_3ZnN	Hf_2InN	Ti_2GaN
$LiTiN_3$	Co_3GaN	Hf_2SnN	Ti_2InN
$LiBN_2$	Co_3InN		$Ti_4Zn_2N_x$
Mn_3ZnN	Co_3GeN		
	Co_3SnN		

SOME REPORTED MIXED CARBIDES[*]

$4TaC \cdot 1HfC$	Cr_2AlC	Fe_3ZnC	Ti_3SiC
$4TaC \cdot 1ZrC$	V_2AlC	Fe_3AlC	Ta_2VC_2
$Dy_3AlC_{0.7}$	Nb_2AlC	Fe_3GeC	$SnMn_3C$
$W_{10}Co_4C_4$	Ta_2AlC	Fe_3SnC	Ti_2AlC
			Ti_2GeC

SOME REPORTED MIXED BORIDES

$YCrB_4$	$ZrIr_3B_4$	Nb_5Ge_3B	$AlMgB_{12}$
Y_2ReB_6	$AlBe_{0.7}B_{22}$	$CeCr_2B_6$	$LiAlB_{14}$
WCoB	$Mg_5Ni_{20}B_6$		

SOME REPORTED MIXED ANION COMPOUNDS

BeC_2B_{12}	LaB_2C_2	YB_2C_2	$Ga_{.28}B_{.85}N$
BeC_2B_2	LaB_2C_4	YB_2C	$Al_{1.05}B_{.27}N$
LaBC	Mo_2BC	$Er_5Si_3C_2$	$Si_3Al_4N_4C_3$
			$Si_3Al_5N_5C_3$

The search for new optical IR transmitting materials is a good example of seeking new compounds beyond simple binary compounds (35). Utilizing ternary or higher order compounds was first suggested for shorter wavelength window materials, and was used to identify oxide candidates. However, it has proved particularly useful in finding longer wave length materials based upon chalcoginites. The challenge has been to

---

[*]See for example, L. E. Toth "Transition Metal Carbides and Nitrites," pp. 55-66, Academic Press, N.Y., 1971.

find materials showing comparable optical transmission to that found in materials such as ZnS or ZnSe, but with higher hardness and hence better erosion resistance, and lower thermal expansion to exhibit greater thermal shock resistance. Several promising ternary sulfides have been identified which have suitable optical transmission, but which provide nearly double the hardness, with the same or less thermal expansion than ZnS or ZnSe. These are being actively explored for future generation long wave length IR windows for military and other applications. Besides finding new materials for use in bulk, such as the above IR windows, there are also many other applications for new materials such as coatings (e.g. to extend Table 1), and for new composite constituents.

The key need for most effective development of new compounds is to obtain better guidance as to what combinations of the periodic table are likely to produce the properties of interest. Specific and highly accurate predictive methods are not at hand, and are generally not likely to be for a substantial time. However, there are a number of existing and developing guidelines that may often be adequate. There are reasonable guidelines for indicating the approximate optical transmission of materials, what materials are likely to be electronic conductors, and those that are likely to be non-conductors. More demanding can be guidance in such properties as melting point, elastic stiffness, and especially thermal expansion. The latter is of great importance because of the dominant role this plays in thermal stress resistance or mismatch stresses in composites, and between coatings and substrates, as discussed earlier. There are both thermodynamic and empirical relations between these various properties that can be of use. Studies showing promising guidance on thermal expansion behavior of materials have recently been discussed (34).

V.   DISCUSSION AND SUMMARY

There are two overall aspects of ceramics and ceramic processing that are often lost sight of. These are first the diversity of ceramics as well as the processes by which we can, or should, be making them. The second is an overall systems point of view in selecting both the material and the process whereby we make it. All too often, we tend to be too proscribed in our view, and too focused on a single theme, often in a "band wagon" effect wherein a particular theme becomes overemphasized as "the" answer to improved ceramics. Two examples are the emphasis several years ago on sialons as being "the" answer to engine and other high tech ceramic applications, and a few years later, the extensive emphasis on non-destructive evaluation (NDE). This is not to say that either of these approaches did not have their merit, nor that they did not yield useful results. However, it is this author's opinion that in both cases there was an overexpectation of what would result from these efforts, and an overemphasis on them. Much of this results from the R&D community itself. However, some of this is also stimulated by the funding agencies wherein there may be some competition between them to fund "the break-through" in what then appears to be an exciting new approach, with much of this excitement based upon a rather limited view.

First, consider the case of sialons. There was a very significant surge of interest and effort in this area, motivated largely by the feeling that sialons offered "the" answer to the problems and limitations that were being experienced in using primarily Si_3N_4 for new, demanding applications, e.g. in heat engines. However, the output of the large surge of effort was, in fact, quite limited. We are now only beginning to see some useful, but modest, application of sialons today. What was unfortunate about this experience was that in a limited period of time, a large amount of effort was devoted, much inefficiently, into what turned out to be a very narrow spectrum of opportunities. We need to recognize the potential for the large expansion of ceramic materials. Even the field of oxynitrides, of which the sialons are one small element, is only a small fraction of the future opportunity of new ceramics. There may be an important payoff from some of this work, namely the developments of oxynitride glasses. However, it is contended that this development probably would be where it is today even if we had had a more balanced approach to our materials efforts, and hence much less effort in earlier development of sialons.

Consider next the issue of NDE. This is a very important area for ceramics and all materials technology. Overall, the technical efforts were more fruitful, but one of the primary problems was the perspective of people managing research. Many apparently thought this might be the "magic button" to solve the difficulties of applying ceramics to high technology applications. We have made progress on NDE and useful work continues. However, we have now come to the important, fundamental realization that (a) the ability to do NDE on critical ceramic components such that we can, with a very high degree of probability, assure success, is still a substantial distance off, and (b) no degree of successful NDE is useful unless we can produce components with a high rate of NDE acceptance. Thus, we went through a period where we tended to de-emphasize processing improvements because "it was much less glamorous", and tended to overemphasize NDE development as the "magic button" to solve the problems. We are now beginning to be more realistic on the amount of effort that we need to put into processing improvement. Further, there is a broader realization now that one of the important potential applications of NDE-type methods may be utilized in the processing steps to improve, and possibly control, processing. Although, again, we should not have excessive expectations, especially for a near-term payoff.

Consider next the diversity of processing versus the issue of dominance of sintering powders to make ceramics. As noted earlier, sintering is the major process of making ceramics, and has a number of important advantages. However, sintering also has basic limitations. Even with the most optimistic improvements of sintering, it seems unlikely that we can obtain ceramics with high levels of reliability for demanding applications, such as use in engines, without a number of aids to sintering or alternative approaches. Thus isolated pores or pore clusters that are typically left, even from the best sintering, can pose serious reliability challenges (34,36). Further, sintering has limited applicability to coatings and composites.

We need to use NDE to the maximum extent that is realistic as a means of improving our processing, and possibly even for in-process control for sintering as well as other processing. We also need to look at post-treatment, e.g. HIP'g, to further improve reliability, as well as still utilizing to the maximum extent that is practical and realistic NDE of the final product. At the same time, we also need to look at various alternatives, in particular utilizing the increasing diversity of processing approaches. Some of these approaches may emphasize the use of composites to reduce the defect sensitivity of resultant components, some may utilize processes which are less prone to produce defects of the character that would be detrimental to an approach. This author feels that a useful approach to take in evaluating the processes for particular applications is to ask the question, "What type of defects can realistically result from the different stages of processing proposed to make a particular component?" Here a key question is what are likely deviations from the norm, e.g. inhomogeneties in binder distribution. In turn, we need to ask, "To what extent will subsequent processing stages make defects more or less severe, or make no change in them?" Doing this, as well as evaluating the overall picture in terms of cost and quality from a systems standpoint is essential.

Finally, it is important again to stress the diversity of ceramic processing and materials. Sintering is, and will continue to be important. However, we need to look at a variety of different routes by which we get our powders, as well as a variety of routes that allow us to either densify these powders or obtain ceramics directly, without having gone through a powder stage, e.g. chemical vapor deposition and polymer pyrolysis. Similarly, for the longer term, we need to recognize the significant expansion that can be achieved in the diversity of ceramic materials available to us. This includes both an increasing array of composites, as well as significant opportunities for obtaining compounds that can be used as individual monolithic ceramics, or as constituents of composites, coatings, etc.

Maintaining and expanding the diversity of ceramic materials and processes should be an important goal. Identification and characterization of new compounds has many ramifications as does development of new composites. No single process or set of processes can meet all of the present needs of ceramic technology, and greater diversity will be needed in the future. The challenge is to understand and extend the advantages and disadvantages of each process so the best choices can be made. It is important that this diversity of ceramics, and especially processing, be recognized in establishing large R&D programs, e.g. centers of excellence, so they are not a means of entrenching a narrower approach. Integrating a variety of disciplines into the field of ceramics is also an important factor in developing this diversity.

REFERENCES

1. J. S. Kim and D. L. Johnson, Plasma Sintering of Alumina, Amer. Ceram. Soc. Bulletin. 62(5):620-622 (1983).
2. D. L. Johnson, W. B. Sanderson, J. M. Knowlton, and E. L. Kemer, Sintering of $\alpha-Al_2O_3$ in Gas Plasmas, Advances in Ceramics, 10:656-665 (1985).
3. E. L. Kemer and D. L. Johnson, Microwave Plasma Sintering of Alumina, Amer. Ceram. Soc. Bulletin, 64(8):1132-1136, (1985).
4. T. Quadir and D. W. Readey, Microstructure in Tin(IV) Oxide and Cadmium Oxide in Reducing Atmospheres, Mater. Sci. Res., 16:159-69, (1984).
5. J. Lee and D. W. Readey, Microstructure Development of Ferric Oxide in Hydrogen Chloride Vapor, Ibid., pp. 145-57.
6. D. W. Ready, J. Lee, and T. Quadir, Vapor Transport and Sintering of Ceramics, Ibid., pp. 115-36.
7. H. F. Fischmeister, Development and Present Status of the Science and Technology of Hard Materials, R. K. Viswanadham, D. J. Rowcliffe, and J. Gurland, eds., Science of Hard Materials, 1-42, Plenum Press, New York (1983).
8. L. M. Sheppard, Expanding HIP's Horizons, Advanced Materials & Processes, 37, (1985).
9. C. Han, I. A. Aksay, and O. J. Whittemore, Characterization of Structured Evolution by mercury porosity, pp. 339-348 in "Materials Sci. Res.," Vol. 19, Avd. in Mat. Char., Vol. 2, Ed R. L. Snyder, R. A. Conrad, and P. F. Johnson, ed., Plenum Press (1985).
10. Private communication, Dr. I. A. Aksay. University of Washington. 1986.

11. C. Ding, R. A. Vatorski, H. Herman, and D. Ott, "Oxide Powder for Plasma Spraying, The Relationship Between Powder. Char. & Coatings Prop.," J. Thin Solid Films, 118:467-475 (1984).

12. N. R. Shantor, H. Herman, S. P. Singhal, and C. C. Berndt, "Neutron & X-ray diffraction of plasma sprayed ZrO_2-Y_2O_3 thermal barrier coatings," J. Thin Solid Film, 119:159-171, (1985).

13. T. Surek, S. R. Coriell, and B. Chalmers, The Growth of Shaped Crystals from Melt, J. of Crystal Growth 50:21-32 (1980).

14. D. O. Bergin, Shaped Crystal Growth - A Selected Bibliography, J. of Crystal Growth 50:381-396 (1980).

15. R. P. Ingel, R. W. Rice, and D. Lewis, Room-Temperature Strength and Fracture of ZrO_2-Y_2O_3 Single Crystals, J. of the Amer. Ceram. Soc. 65:C-108-109 (1982).

16. R. P. Ingel, D. Lewis, B. A. Bender, and R. W. Rice, Temperature Dependence of Strength and Fracture Toughness of ZrO_2 Single Crystals, J. of Amer. Ceram. Soc. 65:C-150-152 (1982).

17. R. W. Rice, R. P. Ingel, Ba. A. Bender, J. R. Spann, and W. R. McDonough, Development and Extension of Partially-Stabilized Zirconia Single Crystal Technology, Cer. Eng. & Sci. Proc., 5(7-8), pp. 530-545 (1984).

18. H. O. Pierson, ed., Chemical Vapor Deposited Coatings, The American Ceramic Society, (1981).

19. R. W. Rice, "Ceramics from Polymer Pyrolysis, Opportunities and Needs - A Materials Perspective", Amer. Cerm. Soc. Bull., 62 (8), 889-892 (1983).

20. K. J. Wynne and R. W. Rice, "Ceramics via Polymer Pyrolysis", Ann. Rev. Mater. Sci., 14: 297-334 (1984).

21. S. Yajima, Y Hasegawa, K. Okamura and T. Matsuzawa, "Development of High Tensile Strength Silicon Carbide Fibre using an Organosilicon Polymer Precursor", Nature, 273 (5663) pp. 525-527 (June 1978).

22. J. W. Fairbanks, Advances in Heat Engine Performance and Durability Through Coating Applications, R. L. Clarke and J. W. Fairbanks, eds., "1st NATO Workship - Coatings for Heat Engines," Aquafredda Di Margta, Italy, (April 1984).

23. I. Kvernes, E. Lugscheider, and J. Fairbanks, "Potential of Ceramic Coating Systems Engineering Materials and Technical Aspects," 3rd Conf. of the European Materials Society, Advanced Materials Research and Development for Transportation, Symposium C - Ceramic Coatings for Heat Engines, November 26-29, 1985, Strasbourg.

24. W. Bryzik, "TACOM/Cummins Adiabatic Engine Program," Proceedings of the Twentieth Automotive Technology Development Contractors' Coordination Meeting, pp. 273-290 (1982).

25. K. E. Anthony and B. J. Welch, Electrodeposition of Zirconium Diboride from Oxides Dissolved in Fused Salts, Aust. J. Chem., 22, 1593-7 (1969).

26. D. R. Flinn, F. X. McCawley, G. R. Smith, and P. B. Needham, Jr., Electrodeposition of Erosion-Resistant Titanium Diboride Coatings, Report of Investigations 8332, United States Department of the Interior, (1979).

27. I. V. Zubeck, R. S. Feigelson, R. A. Huggins, and P. A. Pettit, The Growth of Lanthanum Hexaboride Single Crystals by Molten Salt Electrolysis, J. of Crystal Growth 34:85-91, (1976).

28. D. Elwell, R. S. Feidgelson, and M. M. Simkins, Electrodeposition of Silicon Carbide, Mat. Res. Bull., 17:697-706, (1982).

29. K. H. Stern and S. T. Gadomski, Electrodeposition of Tantalum Carbide Coatings from Molten Salts, J. Electrochem. Soc.:300 (1983).

30. K. M. Prewo and J. J. Brennan, "High-Strength Silicon Carbide Fibre-Reinforced Glass-Matrix Composites," J. Mater. Sci., 15:463-68 (1980).

31. J. Jamet, J. R. Spann, R. W. Rice, D. Lewis and W. S. Coblenz, "Ceramic-Fiber Composite Processing via Polymer-Filler Matrices," _Cerm. Eng and Sci. Proc._, 5 (7-8), pp. 443-474 (1984).

32. R. W. Rice and D. L. Lewis III, "Ceramic Fiber Composites Based Upon Refractory Poly-Crystalline Ceramic Matrices," to be published.

33. R. W. Rice, "Overview of the Naval Research Laboratory Ceramic Turbine Materials Program," _MCIC Report_, (1978).

34. R. W. Rice, "Mechanically Reliable Ceramics: Needs and Opportunities to Understand and Control Fracture," _J. Phys. Chem. Solids_, 45:1033-1050, (1984).

35. R. W. Rice, "Possible New Irdome Materials for Transmission to 4.5-5 Micrometers," _NRL Memorandum Report 3725_, (1978).

36. R. W. Rice and D. Lewis III, R. C. Bradt, A. G. Evans, D.P.H. Hasselman, and F. F. Lange, eds., "Limitation and Challenges in Applying Fracture Mechanics to Ceramics," _Fracture Mechanics of Ceramics_, 5:659-676, Plenum Publishing, (1983).

37. R. W. Rice, "Capabilities and Design Issues for Emerging Tough Ceramics," _Amer. Ceram. Soc. Bulletin_, 63(2):256-262, (1984).

RESEARCH ON HYDROGENATED AMORPHOUS SILICON

K. Weiser

Solid State Institute & Dept. of Electrical Engineering

Technion-Israel Institute of Technology, Haifa, Israel

INTRODUCTION AND HISTORICAL SURVEY

A review article of length suitable for these proceedings is hardly adequate to give more than a sketchy survey of activities in the field of amorphous silicon. Some excellent reviews are available on the subject (1,2) and the reader who wants more depth than can possibly be provided in this paper is referred to these references. The effort now devoted to an understanding of amorphous semiconductors, and more particularly to amorphous silicon, is an appreciable fraction of the total effort devoted to the study of all semiconductors. As one example we cite the fact that approximately 10% of all papers presented at the latest International Conference on the Physics of Semiconductors were devoted to amorphous materials, with the lion's share going to amorphous silicon (3). Furthermore, the field draws large attendances at biannual international conferences dedicated exclusively to the subject of amorphous semiconductors (4). In addition, more specialized meetings take place at frequent intervals which are devoted, for example, to photovoltaic applications.

To understand the interest in this field it is best to acquire some historical perspective. A landmark paper by Anderson(5) in 1958 addressed the question of what happens to the band structure of a solid when the long range order is disturbed. Fig. 1 illustrates the problem in one dimension. Here we see a linear array of potential wells whose depths vary randomly by an average amount Uo. In the absence of these potential fluctuations one would, of course, obtain energy bands in which all electrons are delocalized and therefore extend throughout the array if one neglects surface effects (6). If one of the bands is not

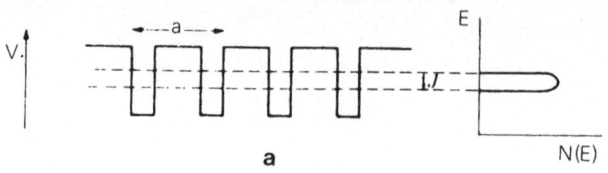

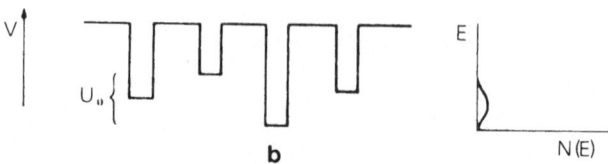

Fig. 1- One-dimensional array of potential wells and
corresponding density of states N(E). a) Periodic array.
b) Array with random potential U_o. From Ref. 10, p16.

completely filled electrical conduction is possible and the motion of
an electron will be impeded only by phonon collisions. If one of the
wells has a different depth from the others a carrier will be reflected
at this point and if a sufficient number of wells differ in depth
propagation of the carrier will cease. In three dimensions the situation
is less obvious since a carrier can always move around a disturbance.
Anderson (5) showed that the seriousness of the potential fluctuations
depends on the ratio of their mean value to the bandwidth in the absence
of the fluctuations. If the bandwidth is great it will take strong
fluctuations to destroy the band structure. For a narrow bandwidth, on
the other hand, the band structure is easily destroyed and all electrons
become localized and cannot move through the lattice except with phonon
cooperation. In the intermediate case where the potential fluctuations
are not too severe Mott has suggested (7) stat states near the band
edges become localized but those in the center of a band remain delocalized. This is believed to be the situation for amorphous silicon where
the potential fluctuations are largely due to changes in bond angle
and, to a lesser degree, changes in bond lengths (8) . Fig. 2 is a
two-dimensional representation of amorphous silicon, a-Si. (In this
paper we shall use the notation a-Si for amorphous silicon without
hydrogen and a-Si:H for amorphous silicon with H). Except for the
occasional "dangling bond" we see that every atom is connected to four
nearest neighbors just as in the crystalline case. A dangling bond DB is

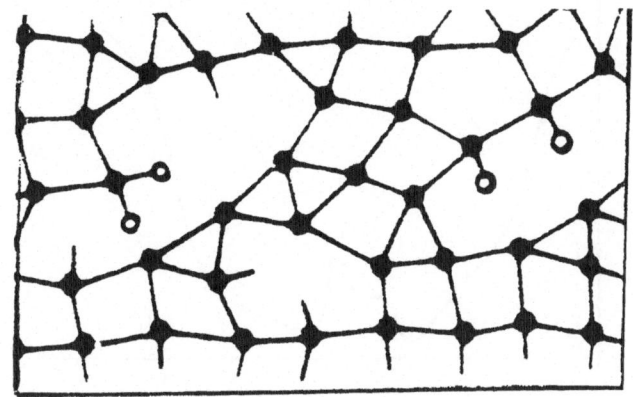

Fig. 2 - Two-dimensional representation of a-Si. Note that most atoms are four-fold coordinated but some have dangling bonds shown by empty circles.

an unpaired electron on a silicon atom which has less than four nearest neighbors, typically three. Such defects play a major role in understandigs the electronic properties of amorphous silicon and we shall refer to them frequently . On the basis of x-ray and neutron diffraction experiments (8) it is found that on the average bond distances do not vary more than about 1% and bond angles no more than about 9% from their crystalline values. On the basis of Mott's model one therefore expects the picture for the density of states between the conduction band and the valence bands shown in Fig. 3. The ordinate in Fig. 3 is drawn on an arbitrary scale though experiments and theory predict that the density of states near Ec and Ev are of order 10^{21} cm^{-3} eV^{-1} . (9). Ec and Ev are so-called mobility edges which separate localized from delocalized states. The energy range between Ec and Ev is usually called the mobility gap or just the bandgap. The abscissa scale in Fig. 3 is based on an estimate of an energy differ-ence of about 1.2 eV between Ec and Ev. The bumps in the density of states within the gap is due to highly localized defects, particularly DB's, which were described above. These defects are akin to deep impurity levels in crystalline semiconductors. (As discussed below the high density of midgap states in a-Si is responsible for the low resistivity of a-Si). The states above Ec or below Ev resemble conduction or valence band states only in the sense that an electron placed above Ec or a hole below Ev can propagate

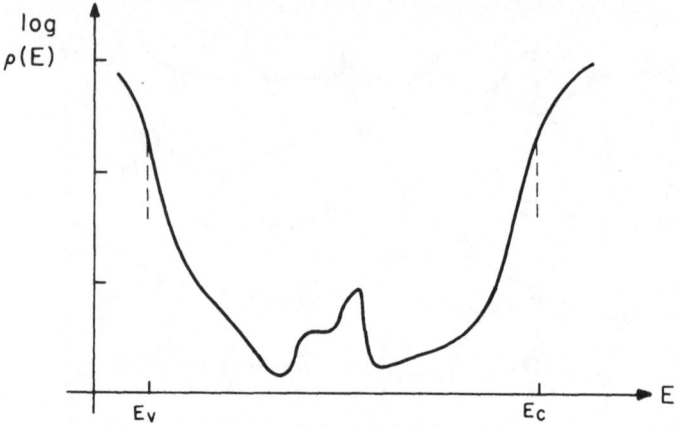

<u>Fig. 3</u> - Schematic density of states (E) vs energy showing conduction and valence band mobility edges E_c and E_v; see text.

through the system at $0^{\circ}K$. It must be remembered, however, that in the case of a perfet crystal the mean free path is virtually infinite at $0^{\circ}K$ whereas in the amorphous case the mean free path should be of the order of an interatomic distance. Mobilities of only about 20 cm^2/V sec are therefore anticipated[10].

In the late sixties or early seventies experimental investigations of these theoretical ideas concentrated largely on chalcogenides rather than on the Group IV elements Si and Ge (The term word chalcogenides refers to materials containing either Se or Te in compound form, such as As_2Se_3, as well as the elements themselves). The reason for the interest in these materials was due both to the desire to test the theory of the amorphous state and to their commercial applications. These applications were based on the use of chalcogenides in electrophotographic or "Xerographic" photocopiers [11]. At the time no-one thought of using amorphous silicon which, in its unhydrogenated form, has too low a resistivity for such a purpose. Since the process of electrophotography involves the creation of electron-hole pairs by photons, their separation by an electric field and the drift of carriers across the film much of the experimental and theoretical research in this period was centered on optical and transport properties of these amorphous semiconductors. There was also considerable interest in the fact that the conductivity of chalcogenides was thermally activated with an activation energy of typically 1 eV, e.g. for As_2Se_3 [12] while that of Group IV elements, e.g. Ge or Si, showed a much weaker temperature dependence which could not be fitted to an Arrhenius plot. Mott suggested (13) that in the latter case the Fermi level is near the middle of the gap within a band produced by the DB's whose density is so high that electrons can "hop" from a filled DB just below the Fermi level to an unfilled just above it with the help of phonons. Because of an intriguing process called variable range hopping the conductivity is not thermally activated in the Arrhenius sense of the word. If the concentration of dangling bonds is high enough the conductivity due to such a process can be quite high, an indeed, the conductivity of unhydrogenated amorphous silicon is typically 10^{-4}(ohm cm)$^{-1}$. In the case of amorphous chalcogenides one would also expect a large concentration of dangling bonds and it is therefore puzzling at first glance that a similar conduction process does not occur in this case. Street and Mott (14) explained the difference by postulating that in this case the DB's are not, on the average, singly occupied as in the case of Group IV elements, but that half of them are empty and the other half doubly occupied. One can show that under these circumstances conduction via the dangling bonds is negligible so that only carriers excited to the vicinity of the mobility edges conduct the current. This excitation requires approximately half the bandgap energy and the temperature dependence of the conductivity is therefore very steep. The resistivity of chalcogenides is therefore very high, greater than 10^{12} ohm cm at room temperature for a-Se.

To summarize, up to the early 1970's basic and applied research activities on amorphous semiconductors concentrated on chalcogenides with only a small fraction of the work being devoted to Group IV elements. Interest in a silicon picked up sharply around the end of that period when it was discovered by Spear and LeComber that hydrogenated amorphous silicon can be doped n- or p-type (15). This discovery and its subsequent application to the fabrication of p-n junctions (16) greatly increased the potential usefulness of amorphous silicon, particularly since a-Si:H is a good photoconductor. (We define a "good" photoconductor as one whose conductivity under ordinary room light is greater than its dark conductivity). The fact that hydrogen is responsible for the difference between a-Si:H and a-Si was not immediately obvious since a-Si:H was not produced by "doping" a-Si. Instead, it was first produced by glow discharge decomposition of silane (SiH_4) (17).

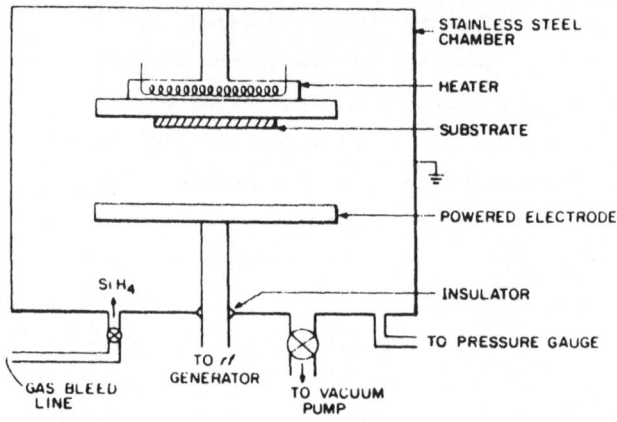

Fig. 4 - Diagram of reaction chamber for RF decomposition of silane. From Ref. 57, p214.

A typical reaction chamber is shown in Fig. 4. Silane gas is bled into the chamber and is decomposed by either an RF or a DC electric field between the electrodes. It is now known(18) that the decomposition is not into Si and H atoms but mainly into various SiH_x radicals from which the Si films grow at a substrate temperature of typically 250°C. (By contrast it should be remembered that the early work on a-Si was performed on films grown by simply evaporating Si and condensing the vapor (19)). The most immediate difference between glow discharge material and evaporated material was found in their resistivities and photoconductivity. Silicon prepared by evaporation and condensation typically has a resistivity of 10^4 ohm cm while that of glow discharge material is some six orders of magnitude higher. Similarly, while the photoconductivity of the former is practically nil the latter is a very good photoconductor in the sense defined above. At first it was thought that the difference between the two types of silicon was due to the presence of energetic species in the plasma present in the silane technique. It soon became clear, however, that the material produced by this technique contains hydrogen, by simply heating films in vacuum in a closed system and measuring the pressure rise (20). It also became clear from sputtering experiments(21) that the hydrogen is responsible for the difference in the electronic properties of the material. If Si is prepared by sputtering in an argon atmosphere only, one obtains material with properties similar to those of evaporated Si but if hydrogen is added to the argon then the properties are very similar to those of glow discharge material.

At this point the question naturally arises as to why hydrogen affects the electronic properties of amorphous silicon so drastically. There are two reasons for this phenomenon. The first reason is easy to understand and has been confirmed experimentally. Amorphous silicon withouthydrogen contains a large number of DB's, typically about 0.2% or of order of 10^{20} per cm^3 (22). (Among other effects these DB's are responsible for the high conductivity of a-Si, as already discussed).The DB concentrations are determined by electron spin resonance (ESR) measurements which measure the concentration of spins due to unpaired electrons. From the "g-values" seen in these experiments it is clear that the spins are due to localized electrons and they are therefore attributed to dangling bonds. In "good" hydrogenated amorphous silicon, on the other hand, the DB concentration can be as low as a few times 10^{15} cm^{-3} (23). Clearly, what has happened is that H atoms have formed chemical bonds with the unpaired electrons but one should only need less than one percent hydrogen to do so. In order, however, to produce "good" a-Si:H one needs a hydrogen content of about 10% and the question arises as to why so much "extra" hydrogen is needed. Surprisingly, no serious attempts seem to have been made until recently to explain this phenomenon. It has, of course, been realized that most of the hydrogen must have disrupted weak Si-Si bonds (24) but how this improves the properties of the material was not clear. Cody and collaborators(25) have suggested that hydrogen incorporation reduces the "disorder" in the material. With less disorder one might expect less band tailing, or a sharper drop in the density of states with energy in the regions adjoining the mobility edges. The effective mobility of photocarriers should therefore increase since fewer of them will be trapped. The model of Cody et al was recently put on a quantitative basis by Weiser and Shapiro(26). Their model is based on the idea that H incorporation allows the atoms surrounding the Si-H bonds to relax into less strained configurations than for pure a-Si. The lattice relaxes because the "over-constrained" condition (27) of a cell of about 8 A diameter is relieved when a Si-Si bond is replaced by two Si-H bonds. Such a process is equivalent to snapping a spring in a system of balls which have been displaced from their equilibrium positions by external forces. Weiser and Shapiro(26) put these ideas into quantitative form and, in particular , were able to explain why the effect on disorder sets in very gradually with H concentration, then accelerates and finally saturates. They were thus able to explain why considerably more than 1% H is needed to produce "good" hydrogenated amorphous silicon.

Our historical survey has covered the period from about 1960 to 1975 although we have already made occasional reference to current work. In the next Section we shall describe briefly research activities since 1975, which, generally, have followed a predictable course. By this we mean that anytime a "new" semiconductor material comes into prominence because it has more interesting properties than those of the material currently in use research and development activities fall into the following categories. In the first place one wants to understand the electronic properties of the material and to control them, which requires constant improvements of preparation techniques. In parallel with these efforts device research is carried out whose aim it is to understand the relations between device performance and materials properties. As device performance is better understood the direction of the basic research is usually affected. Occasionally, though this has not yet happened in the case of a-Si:H, entirely new and exciting phenomena are discovered which are directly connected with some specific property of the material and which simply could not have been observed in the then current material. A good example of such a discovery is the Gunn Effect which is found in GaAs and related substances but not in Ge or Si (28).

II. CURRENT RESEARCH ACTIVITIES IN HYDROGENATED AMORPHOUS SILICON

A. Basic Research

It would be grossly exaggerated to claim that a-Si:H has now joined the ranks of "standard" crystalline semiconductors such as Si or GaAs in regard to the type of basic research being carried out. In the case of the crystalline materials a tremendous amount is known about band structure, transport and recombination processes and defect states. As already alluded to in Section I, on the other hand, one's understanding of these properties for a-Si:H is still very sketchy. For this reason, while much of the current basic research on GaAs, for example, is being devoted to the physics of special structures such as "superlattices" most of the work on a-Si:H is still devoted to an understanding of basic properties.

Referring to Fig. 1 we return to some of the basic concepts embodied in it and discuss experimental techniques used to verify them, if possible.

First of all, the existence of mobility edges has never been verified experimentally with certainty. The best evidence so far seems to be based on the value of the pre-exponential factor in the Arrhenius type temperature dependence of the conductivity on temperature, $\sigma = \sigma_o \exp{-A/T}$. One can show(9) that if a mobility edge exists and if the temperature dependence of the bandgap and some other rather predictable parameters are known the value of σ_o can also be predicted. Mott (9) has claimed that the experimental values of σ_o are indeed in very good agreement with theory. Wagner and Overhof have, however, pointed out that the experimental values are often strongly influenced by long range potential fluctuations (29) so that one must be careful to come to firm conclusions with regard to the existence of a mobility edge on the basis of the value of the pre-exponential factor. Nevertheless, the existence of more or less sharp mobility edges is widely accepted, at least by experimentalists, because of its intuitive appeal and because without their existence it would be even more difficult to interpret transport and recombination experiments. As for the mobility of carriers in extended states it is difficult to obtain unambiguous information about its value from conventional Hall and resistivity measurements. The problem here lies with the interpretation of the Hall effect when the mean free path of carriers is small(30). One may expect, however, that a variant of the Hall-Shockley experiment might be applicable to amorphous materials. This is indeed the case although there are differences between time-of flight measurements carried out on amorphous materials and those on crystalline materials (31). In the former case, but only under favorable circumstances, one measures the mobility of carriers as an "effective" mobility rather than as a true microscopic mobility. To explain this point we first describe the experiment which is illustrated in Fig. 5. A pulse of strongly absorbed light is shone on the semi-transparent electrode and excites electron-hole pairs. Depending on the polarity either electrons or holes now drift across the film to the other electrode and one can show that during the transit time a current flows in the external circuit. In the ideal case where there is no trapping of carriers by deep traps this current is constant (insert of Fig. 5) and the effective mobility of carriers can be obtained from the transit time. Even in this case the effective mobility is nevertheless different from the true mobility since carriers interact with shallow traps, those which give rise to bandtail states, on the way to the other electrode. Rapid capture and release insure that all the carriers get across but the mobility measured, the "effective mobility" depends on the density of such shallow trap states as well as on the microscopic mobility. Hence information about trap states has to be available from other sources in order to deduce the true mobility of carriers. Reasonably reliable information of this kind is indeed available, as discussed below, and one then concludes that the mobility of electrons in extended states is approximately 10 cm^2/V sec at room temperature (32,33), in very good agreement with theoretical predictions. We again stress, however, that even in the best case where sharp current transientsare observed the true microscopic mobility deduced from such experiments is model dependent. (In many cases, particularly for holes, no sharp transients are observed because of deep trapping (32). In this case, carriers are trapped by midgap states and their release time is much longer than the transient time).

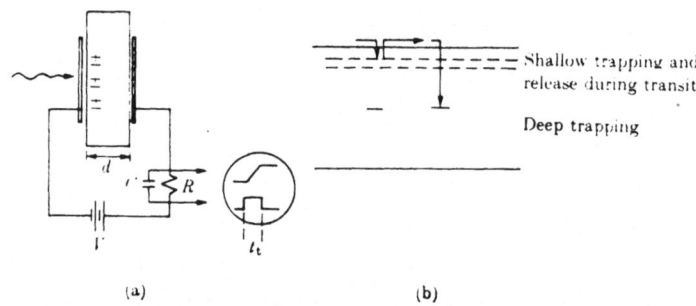

(a)   (b)

Fig. 5 - a) Principle of time-of-flight experiment with insert showing ideal current or voltage transient during transit time t_t. b) illustration of shallow and deep trapping as explained in text. From Ref. 10, p247.

We mentioned above that information about shallow traps, those which give rise to bandtail states is available. The main source of information comes from optical data and from photoconductivity and photo-induced absorption measured . A typical plot of the absorption constant alpha vs photon energy is shown in Fig. 6. At the highest energy, typically for alphas greater than 10^4 cm^{-1}, one generally obtains a relation $(\alpha h\nu)^{1/2}$ $\alpha(h\nu-E_o)$ from which an optical or Tauc gap E_o (34) is obtained. In this

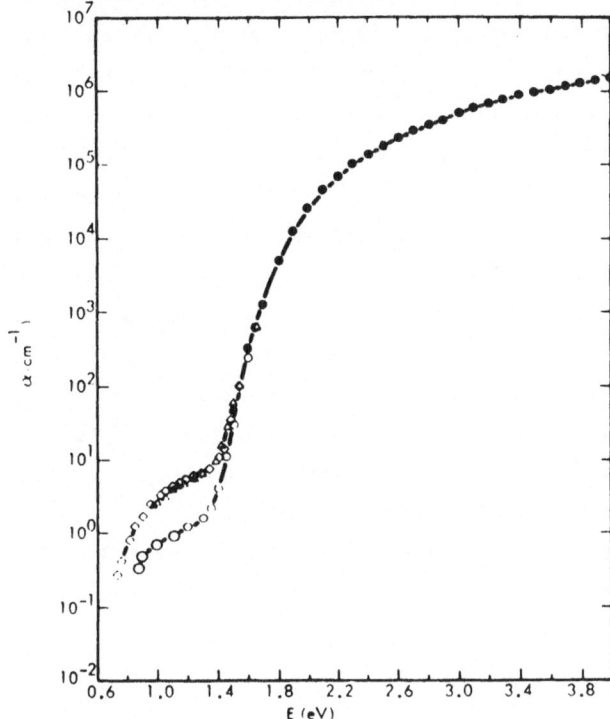

Fig. 6- Typical dependence of absorption constant alpha on photon energy E. From Ref. 34, p23.

range transitions are believed to occur between electrons in extended states in the valence band to extended states in the conduction band. Eventually, at energies such that the photon energies are too low to cause such transitions the dominant transitions will be between band-tail states. In this range it is found that the energy dependence of the absorption constant is given by the relation $\alpha = \alpha_o \exp(h\nu-E_1)/E_o$ where $alpha_o$ is a scaling factor and E_o is a characteristic energy called the Urbach energy (35) which determines the slope of the energy dependence on semi-logarithmic plot. Since ther is theoretical evidence that potential fluctuations give rise to exponentially decaying band-tail (36) an exponential depencence of alpha on photon energy is expected since the convolution of two exponential functions yields an exponential function. Thus, from optical data alone the energy dependence of the bandtail states can be deduced and it is gnerally accepted that such data yield E_o for the valence band-tail which is expected to be larger than for the conduction band only (37). These E_o are of the order of 30 to 50 meV's for good material (34), and this information is then used in the analysis of time-of-flight results in order to extract the mobility of carriers in extended states (32).

Continuing our discussion of optical data we see that for alphas below values of about several reciprocal centimeters the curve of Fig. 6 is no longer exponential. For such low values of alpha it therefore appears that the absorption is process is no longer due to bandtail to to bandtail transitions but rather due to transitions from localized midgap states to the conduction band. Indeed, it has been shown that the "excess" absorption in this energy range correlates with spin density(38) as determined by electron spin resonance measurements and one is therefore tempted to conclude that all midgap states are due to dangling bonds. Adler has suggested, however, (39) that other midgap states besides DB's such as two-fold co-ordinated Si atoms may also contribute to the midgap density of states. One might add that conventional optical absorption measurements in which one measures transmittance and reflectance of films becomes very difficult when the absorptance is small. In this case more indirect techniques, particularly photothermal deflection techniques are frequently used (40). The central idea is illustrated in Fig. 7.

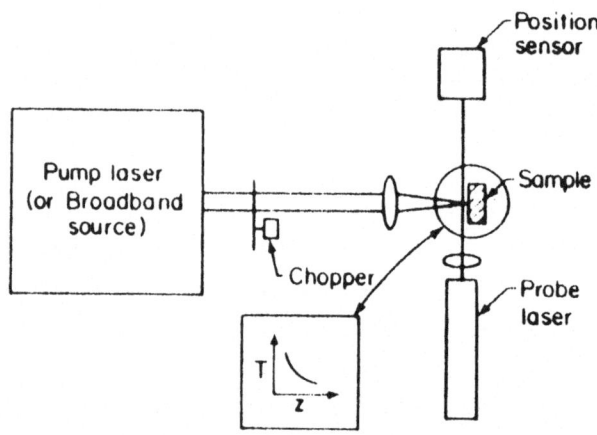

Fig. 7- Principle of photo-thermal deflection technique, as explained in text; in the insert T denotes temperature and z direction. From Ref. 40, p 86.

The absorbed light heats the film which in turn heats the layer of air above it and therefore changes its refractive index. A laser beam grazing the sample is therefore deflected and the degree of deflection is a sensitive measure of the absorption coefficient.

Enough has been said about optical measurements to point out that they are a valuable tool in mapping out the density of gap states but since there always two states involved in any optical transition and since one must assume something about matrix elements these techniques are only of semi-quantitative value in determining the density of states. Another valuable tool is Deep Level Spectroscopy but for lack of space we refer the interested reader to a review article on the subject for an understanding of the technique(41). Finally, one should mention that photo-emission experiments (42) are also used for studying the density of states but that this technique is particularly valuable for valence band and core states rather than for bandgap states. In short, mapping out of the density of states in amorphous semiconductors is not a trivial matter.

As a final example of basic research on a-Si:H we shall make a few remarks about the behavior of excess carriers. Suppose, for example, that a light pulse produces electron-hole pairs in extended states. In the case of a crystalline semiconductor the sequence of events which follows is well understood. The carriers lose energy quickly by optical phonon emission until further energy loss in impossible when they reach the band edges. Having done so they recombine through a variety of mechanisms(43). In the case of amorphous semiconductors the mobility edges may in some respect be equivalent to crystalline band edges but since the bandtails are contiuous in energy to the mobility edges there is no reason why carriers should not simply continue to lose energy upon reaching the mobility edges and recombine via bandtail states (44). Alternatively, they may be captured by bandtail states intially but have to be re-emitted into extended states in order to recombine (45,46). Which process dominates is not yet clear and may, indeed, depend on sample properties. At room temperature , at least, it appears from photoconductivity and photo-induced absorption measurements that bandtail to bandtail recombination predominates (37). At low temperature where there are hardly any carriers in the extended states the starting point of the recombination process is clearly in the bandtails. As seen from photoluminescence experiments this does not necessarily mean that the dominant recombination mechanism is from bandtail to bandtail since dangling bonds are also recombination centers (47,48).

Our final topic in this sub-section is not so much a research topic but is included mainly in order to satisfy the curiosity of the reader with regard to the determination of the hydrogen content of samples. Probably the most widely used technique is to measure optical absorption in the infrared region (from about 600 to 2100 cm^{-1}) which contains several absorption peaks due to various Si-H vibrations[49]. The most obvious one of these vibrational modes is a stretching mode which occurs at 2000 cm^{-1}. It is relatively easy to measure well defined absorption peaks of this kind for hydrogen concentration of the order of 10% but it is clearly necessary to know the oscillator strength of the vibration in order to translate the integrated absorption peak into a hydrogen content. In order to avoid having to rely on a theoretical estimate of this strength it is preferable to determine the H content by a direct experimental technique . The simplest technique is to drive off the hydrogen from a film of known volume by heating it in an enclosure of known volume . The amount of hydrogen given off can then be calculated from the pressure rise in the system [20,51]. This kind of experiment also has the virtue of giving information about the morphology of the film. It is found that films with poor structures , i.e. large internal surfaces, expel hydrogen at lower temperatures than those with very good structures which have to be heated to close to the crystallization temperature [51].

B. Device Research

In Sect. IIA we gave small sampling of fundamentally oriented research and we now turn our attention to more device oriented activities. By this we mean activities whose purpose it is to measure those materials properties which are most relevant to good device performance. In order to know which properties to measure one must clearly understand the principle of the device in question. Todate three main applications for a-Si:H have been targeted: The first and probably best known one is for solar cells or, more generally, for photovoltaic applications. The second one is for electrophotography ("Xerography") . The third one is for large area field effect transistors. The principles underlying all these applications are well known and can be found in any number of books on semiconductor devices (11,52). Here we merely wish to stress that the first two applications involve the creation of electron-hole pairs by light and the drift of these excess carriers across the film or across portions of it. For solar or photovoltaic cells the drift of both types of carriers is involved while only one of the carriers has to move over an appreciable distance in the case of Xerographic

applications. In either case a high mobility of the carriers is
desirable. In a field effect transistors a high mobility of at least
one of the carriers is also desired although it does not undergo drift
in the sense applicable to the other two devices. We shall return to the
relation between device performance and mobility of carriers later on.
First of all, however, we wish to make a few remarks about the use of an
amorphous material in such devices.

The interest in the use of a-Si:H for various applications is partly
due to cost considerations and partly due to the feasibility to produce
large area films. Cost is of particular importance in the case of solar
cells since cells made from crystalline Si are very expensive (it should
be borne in mind, however, that low cost becomes unimportant if the
efficiency of a solar cell is very low because of the fixed costs involved
in the installation of the cells on a large area). The case with which
large areas can be covered with amorphous films is an important asset for
all three applications. In the case of solar cells it is clearly
desirable to produce large area cells so as to reduce the number of
electrical connections in a network of such cells. In the case of Xerographic
photocopier films of uniform quality have to be deposited on a metal drum.
Finally, in the case of the field effect transistor the chief attraction of
using a-Si:H is the ability to produce flat panel displays by coupling such
transistors to liquid crystal cells (53). Although this review article is
not primarily concerned with actual commercial applications of amorphous
silicon it should be mentioned that such applications are already a
reality. While solar cells are not yet used commercially on roof tops to
provide electricity for homes they are used to power pocket calculators
using ordinary room light. Hydrogenated amorphous silicon is also being
used in photocopiers (54). Commercial applications of amorphous FET's,
on the other hand, have not been reported.

In all three device applications mentioned so far the use of amorphous
silicon rather than crystalline silicon is based on cost or convenience,
rather than on inherently superior material properties. By this we mean
that with regard to such parameters as the mobility of carriers, lifetime
of excess carriers, dopability of the material and so forth, crystalline
silicon is certainly preferable to even the best amorphous silicon. As
a result, to give just one example, the efficiency of a solar cell made
from crystalline Si should be, and is indeed, greater than that made from
amorphous silicon. (The efficiency of a solar cell is defined by the ratio
of electrical power produced divided by the incident solar energy flux

under certain standard conditions(55)). Theoretical estimates for
the maximum efficiency attainable are as high as 30% for a crystalline
cell (56) and 15% for an amorphous cell (57). In practice, the state
of the art yields efficiencies of 20% (58) and 10% (57 respectively.
Crystalline materials have superior qualities over their amorphous
counterparts because of their transport recombination properties.
As already pointed out, in a solar cell electrons and holes are created
by photons and they are separated by the internal electric filed of
the p-i-n structure(57) . The electrons which arrive in the n-region and
the holes which arrive in the p-region give rise to a current in the
external circuit. The mobility and lifetime of these excess carriers must
therefore be such as to enable them to drift across the intrinsic i-region
without recombining. A long lifetime and a high mobility of carriers is
also desirable in order to enhance the efficiency of a cell for another
reason. It is not necessary to create electron-hole pairs within the
i-region in order to produce a current in the external circuit. Pairs
which are created within a diffusion length in the doped regions to
either side of the i-region also contribute to the current. A long diffusion
length L will therefore improve the efficiency of a solar cell. L is
proportional to the square root of the mobility and the recombination
lifetime(43) and therefore both a high mobility and a long lifetime are
desirable. Clearly, crystalline silicon will have larger L's than
amorphous silicon. The diffusion length will be larger because, even in
the absence of trapping, the mobility of carriers in the extended states
is orders of magnitude greater than the $10 \ cm^2/V$ sec value mentioned
before. The lifetime may not necessarily be longer but at least
it can be controlled accurately by onctroling the concentration of
recombination centers. In the case of amorphous silicon the best that
can be done to increase the diffusion length of excess carriers is to
reduce the density of bandtail states which are responsible for trapping
and to reduce the density of midgap states which reduce the recombination
lifetime.

    In the case of a Xerographic application one is interested in the
motion of only one of the carriers created by photons, and for this
reason the recombination lifetime is not relevant. It is, however,
desired that the carrier gets across the film under the influence of
the electric field without getting trapped by a deep trap just as in
a time-of-flight experiment . Again, then, the same remark which applies

to material suitable for good solar cells applies to material suitable
for good solar cells applies to material suitable for Xerographic
purposes: one wishes to reduce the number of bandtail state in order to
increase the mobility and the number of midgap states.

Finally, one also wishes to reduce the number of gap states in materials used for FET's (53). Here the reason is that one wishes to change
the conductance of a channel between two p-n junctions by applying a gate
voltage across an oxide layer sandwiched between the channel and a gate
electrode. In order for a change in voltage to have a strong effect on
the conductance of the channel it is clearly desirable to reduce the
number of gap states since these will absorb the additional carriers induced
by the voltage change.

Given the importance of reducing the density of states within the
gap it is not surprising that much of the "device oriented" research is
concentrated on the study of these gap states. In the previous sub-section
we discussed various optical methods used for this purpose. Here we wish
to mention techniques which are more immediately directed towards measuring
the diffusion length of excess electrons and holes. The most commonly used
technique for measuring the diffusion length L is the so-called surface
photovoltage method (59). Here the material is sandwiched between an
ohmic and a rectifying contact, usually an electrolyte. The rectifying
contact produces an electric field due to band-bending and hence, when
light creates electron-hole pairs near the rectifying junction, they are
injected into the bulk and produce a voltage between the two contact in
the open circuit configuration. By measuring the voltage as a function
of the wavelength of the incident light or, equivalently as a function
of its absorption depth one obtains the diffusion length of the carriers.
Recently(60), a simpler technique which is independent of contact
properties has been developed. The technique employs a photograting of
carriers produced by the interference of two coherent laser beams. The
photocurrent in the sample is then compared to the photocurrent for the
case when the two beams are incoherent. One can show that by studying
the ratio of the two currents as a function of the grating period one
can obtain the diffusion length of carriers. The physical effect
underlying the method is that diffusion of the carriers from regions of
high concentration to those of low concentrations tends to reduce the
sharpness of the photograting.

So far we have stressed the point that good device material should be as free of gap states as is possible in a disordered solid. Referring to Fig. 3 one desires that the density of states near E_c and E_v drop off rapidly towards the gap center and that there should be a negligible concentration of states near midgap. At present there is no theoretical guidance for the experimentalist as to what preparation techniques will achieve this goal. Glow discharge decomposition of silane remains the best preparation technique so far but it remains to be seen whether other techniques will prove to be useful or even superior.

Producing material with a low density of gap states is not the only materials problem encountered with a-Si:H. Two other major problems are connected with the stability of the material under illumination and with doping. It is found that illumination of a-Si:H for prolonged period of time at an illumination level somewhat greater but comparable to that found for strong sunlight changes the dark-and photoconductivity of the material (61). These changes are brought about by re-arrangements of local atomic configurations due to a absorption of a photon. Conceptually, it is easy to visualize a situation where a photon is absorbed and creates a localized excited electronic state whose energy is reduced by some kind of relaxation of atoms. After the electron-hole pair has recombined this new atomic configuration is frozen in since it will ordinarily require an activation energy in order for the atoms to get back to their original configuration. Indeed, it is found that heating the films restores its original dark-and photoconductivity. Such instabilities are clearly undersirable and considerable research is therefore devoted to understanding the mechanism involved (62), with the hope of ultimately reducing the effect.

As for doping problems one must recall that doped films are needed both for solar cells which involve p-i-n junctions (57) and for FET's (53) which involve two p-n junctions . The commonly used dopants are phosphorus for n-material and boron for p-material, both added to the silane stream in the form of hydrides(63). One finds, first of all, that the doping efficiency is very low, typically about 1%. This means that most of the P's or B's enter the lattice in a three-fold coordination, i.e. they are surrounded by three Si atoms. Whether such neutral clusters have any deleterious effects on device performance is not clear. The few four-fold coordinated dopants do move the Fermi level towards the respective mobility edges but they create dangling bonds at the same time (63   ). The reason for this undersired side effect is that the energy involved in creating a four-fold coordinated dopant configuration is lowered if, simultaneously, a DB is formed.

Given the potential for industrial applications for a-Si:H and given the intriguing problems with respect to its physics the widespread interest in this material is not surprising. There is good reason to believe that it will continue for many more years and that interesting new phenomena will yet emerge from research on this material.

REFERENCE

1. Semiconductors and Semimetals 21; Volume Editor J.I. Pankove Academic Press 1984.

2. Physics of Hydrogenated Amorphous Silicon, Topics in Applied Physics; J.D. Joannopolous and G. Lucovsky, editors. Springer Verlag, Berlin 1984.

3. Proceedings of the XVIIth International Conference on the Physics of Semiconductors - San Francisco, Cal. 1984; Springer Verlag, Berlin 1985.

4. For proceedings of the latest conference see: Proceedings of the Xth International Conference on the Physics of Amorphous and Liquid Semiconductors - Rome 1985; Elsevier 1985.

5. P.W. Anderson, Phys. Rev. (1985) 109, 1492.

6. K. Seeger, Semiconductor Physics; Springer Verlag New York 1985.

7. N.F. Mott, Adv. in Physics (1967) 16, 49.

8. L. Guttman, Ref. 1, Vol. A, p225.

9. N.F. Mott, Phil. Mag. B. (1985) 51, 19.

10. N.F. Mott and E.A. Davis, Electronic Processes in Non-Crystalline Materials; Oxford University 1979, p219.

11. Xerography and Related Processes, J.H. Dessauer and H.E. Clark, The Focal Press, New York, 1965.

12. C.H. Seager and R.K. Quinn, J. Non-Cryst. Solids (1975) 17, 386.

13. N.F. Mott, J. Non Cryst. Solids-, (1968) 1,1.

14. R.A. Street and N.F. Mott, Phys. Rev. Letters (1975) 35, 1293.

15. W.E. Spear and P.G. Le Comber-Solid State Comm. (1975) 17, 1193.

16. D.E. Carlson and C.R. Wronski, Appl. Physl. Leet. (1976) 28, 671.

17. R.C. Chittick, J.H. Alexander and H.F. Sterling, J. Electrochemical Society (1969) 116, 77.

18. P.E. Vannier, Solar Cells (1983) 9, 85.

19. M.H. Brodsky, R.S. Title, K. Weiser and D. Pettit, Phys. Rev. B (1985) 1, 5311.

20. H. Fritzsche, M. Tamelian, C.C. Tsai and P.J. Graczi, J. Appl. Phys. (1979) 50, 3366.

21. T.D. Moustakas, Reference I, Part A, p 55.

22. M.H. Brodsky and R.S. Title, Phys. Rev. Letters (1969) 23, 1167.

23. R.A. Street, Appl. Phys. Letters (1981) 47, 1480.

24. M. Jannai, R. Weil and B. Pratt, Phys. Rev. B (1985) 31, 5311.

25. G.D. Cody, T. Tiedje, B. Abeles, B. Brooks and Y. Goldstein, Phys. Rev. Letters (1981) 47, 1480.

26. K. Weiser and B. Shapiro, unpublished.

27. J.C. Phillips, J. Non Crystalline Solids (1985) 34, 5311.

28. J.B. Gunn, Solid State Comm. (1963) 1, 88.

29. W. Beyer and H. Overhof, Phil. Mag. (1980) 43, 433.

30. L. Friedman, J. Non Cryst. Solids (1971) 6, 329.

31. W.E. Spear, J. Non Cryst. Solids (1969) 1, 197.

32. T. Tiedje, J.M. Cebulka, D.L. Morel and B. Abeles, Phys. Rev. Letters (1981) 46, 1425.

33. A.C. Hourd and W. E. Spear, Phil. Mag. B(1985) 51, L13.

34. G.D. Cody in Ref. 1, Part B, p11.

35. F. Urbach, Phys. Rev. (1953) 92, 1324.

36. C.M. Soukoulis, M.H. Cohen and E.N. Economou, Phys. Rev. Letters (1984) 53, 616.

37. E. Zeldov and K. Weiser, Solid State Comm. (1985) 56, 867.

38. W.B. Jackson and N.M. Amer, AIP Conf. Proc. (1981) 73, 263.

39. D. Adler, Phys. Rev. Letters (1978) 41, 1755.

40. N.M. Amer and W.B. Jackson, Ref. 1, Part B, p 83.

41. J.D. Cohen, D.V. Lang and J.P. Harbison, Phys. Rev. Letters (1980) 45, 197.

42. L. Ley, Ref. 2, Vol. II, p61.

43. R.A. Smith, Semiconductors; Cambridge University Press, 2nd ed. 1978, Ch. 9.

44. K. Weiser, R. Fischer and M.H. Brodsky, Proceedings of the Xth International Conference on the Physics of Semiconductors, Cambridge, Mass. 1970 (US Atomic Energy Comm. 1971), p667.

45. T.Tiedje and A. Rose, Sol. State Comm. (1980) 37, 49.

46. J. Orenstein and M.A. Kastner, Solid State Comm. (1981) 40, 85.

47. B.A. Wilson, A.M. Sergent, K.W. Wecht, A.J. Williams, T.P. Kerwin, C.M. Taylor and J.P. Harbison, Phys. Rev. B (1984) 30, 3320.

48. R.A. Street, Ref. 1, Part B, p197.

49. G. Lucovsky, in Ref. 2, Vol. II, p301.

50. W. Beyer, Tetrahedrally Bonded Amorphous Semiconductors , edited by D. Adler and H. Fritzsche, Plenum Press 1975, p129.

51. W. Beyer and H. Wagner, J. Non-Cryst. Solids (1983) 59/60, 161.

52. S.M. Sze, Physics of Semiconductor Devices, 2nd ed. Wiley 1981.

53. W.E. Spear and P.G. LeComber, Ref. 2, Vol. I, p63.

54. Commercial advertisements of Cannon, Inc.

55. K.L. Chopra and S.R. Das, Thin Film Solar Cells, Plenum Press 1983, Appendix A.

56. T. Tiedje, E. Yablonovitch, G.D. Cody and B.G. Brooks, IEEE Trans. Electron Devices ED31, 711 (1984).

57. D.E. Carlson, Ref. 2, Vol, I, p203.

58. A.W. Blakers and M.A. Green-Appl. Phys. Letters (1986) 48,215.

59. A.R. Moore, Appl. Phys. Letters (1981) 38,998.

60. D. Ritter, E. Zeldov and K. Weiser, unpublished.

61. D.L. Staebler and C.R. Wronski, Appl. Phys. Letters (1977) 31,292.

62. M. Stutzmann, W.B. Jackson and C.C. Tsai, Phys. Rev. B (1985) 32,23.

63. R.A. Street, Ref. 3, p. 845.

STRUCTURE AND OPTICAL PROPERTIES OF AMORPHOUS SEMICONDUCTORS

Richard Zallen

Department of Physics
Virginia Tech
Blacksburg, Virginia 24061

INTRODUCTION

Amorphous semiconductors are of growing importance in applications which typically involve the use of large-area thin films. A prominent example is the promise of hydrogenated amorphous silicon as a solar-cell photoreceptor (1). Other examples are the use of amorphous chalcogenides in xerography and their potential use in high density optical memories. Also, as described later in this paper, amorphous semiconductors turn up during the technological processing of crystalline semiconductors in the microelectronics industry.

A central challenge in efforts to control the properties of these new materials is the issue of their atomic-scale structure. While for crystalline solids complete information about the atomic-scale structure can be provided (after a great deal of hard work) by the diffractionists, only partial information about the atomic arrangement can be provided for amorphous solids (2). Tools other than the conventional diffraction probes have thus been welcome in the attack on the atomic-scale structures of these materials, and optical techniques (Raman scattering, infrared absorption, ultraviolet reflectivity) have proven to be especially valuable.

Figure 1 displays two-dimensional versions of a pair of continuous-random-network models for covalently-bonded amorphous solids. These sketches convey some of the structural features which are known (2). There is no long-range order, but a high degree of short-range order exists in the form of well-defined bond lengths and coordination numbers. There is a spread of bond angles.

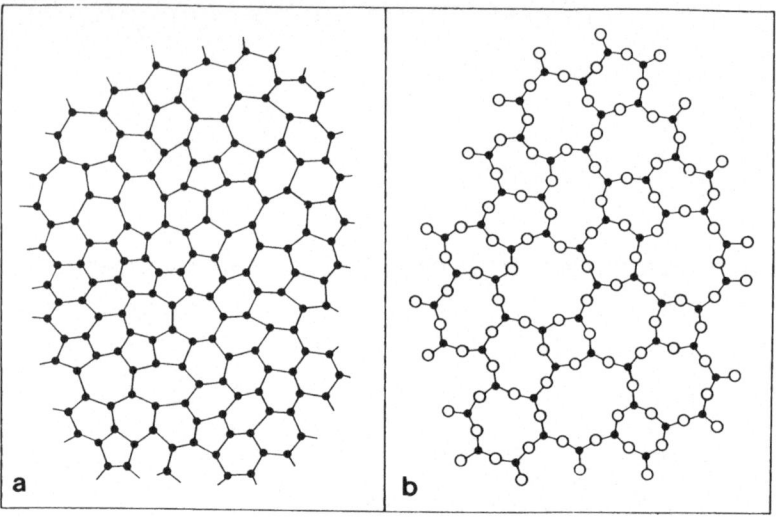

Fig. 1. Two-dimensional representations of continuous random networks, illustrating some aspects of structure in covalently-bonded amorphous solids (after Refs. 2 and 3).

Figure 1(b) reproduces the famous Zachariasen sketch for a glass containing two-coordinated bridging atoms (3), while Figure 1(a) shows a three-coordinated elemental amorphous solid (2). The former serves as an analog of conventional oxide and chalcogenide glasses (with their bridging column-six atoms), while the latter serves as an analog of amorphous semiconductors such as amorphous silicon (a-Si). Of course for tetrahedrally-bonded amorphous semiconductors such as a-Si, the local coordination is fourfold; and the covalent continuous-random network is three-dimensional.

OPTICAL PROBES OF AMORPHICITY

The usefulness of optical techniques, as probes for distinguishing between crystalline and amorphous forms of atomic-scale structure, is illustrated in Figure 2. The upper half of this figure compares the infrared-absorption and Raman-scattering spectra of amorphous silicon with the infrared-absorption and Raman-scattering spectra of crystalline silicon (4). The spectral region displayed is the "first-order" regime of lattice-vibrational fundamentals, i.e. a single lattice-excitation phonon is

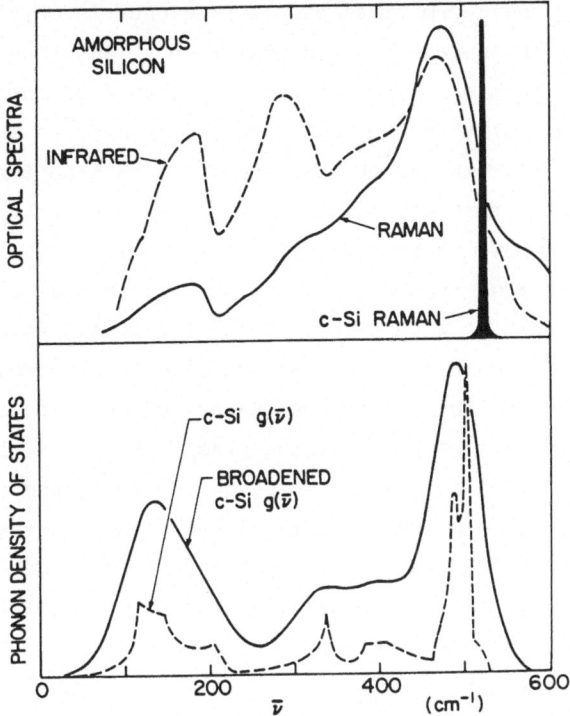

Fig. 2.   Optical spectra of amorphous silicon and crystalline silicon in
the fundamental vibrational-excitation regime (after Ref. 4).

involved (created in the case of the absorption of an infrared photon,
created or destroyed in a Stokes or anti-Stokes Raman-scattering event).

The first-order Raman spectrum of c-Si is very simple. It consists of a
single sharp line at the upper end of the spectrum. This line corresponds
to the transverse-optical (TO) phonon of the crystal at the Brillouin-zone
center (k=0). The first order infrared spectrum of crystalline silicon is
even simpler: it is empty, totally blank! The simplicity of the crystal
spectra has its origin in the "k=0 selection rule." Since the phonon wave-
vector k is a good quantum number for vibrational excitations in a
crystalline solid, and since photons have k values which are essentially
zero on the scale of the Brillouin zone, only k=0 phonons can participate in
these optical processes.

There is only one $\underline{k}=0$ optical phonon frequency in crystalline silicon (corresponding to three degenerate normal modes), and thus only one line appears in the first-order Raman spectrum. Because of the non-ionicity and the unit-cell simplicity (both conditions are needed, as shown in Ref. 5) of this crystal, the $\underline{k}=0$ optical phonons are infrared-inactive, so that zero lines appear in the first order spectrum. Since the number of normal modes in a macroscopic sample is of order 10^{23}, it is clear that in a crystalline solid, only a tiny minority (at most) of the vibrational modes have the privilege of interacting with light. This elitist situation is summarized in the first row of Table 1. For c-Si, as described above, the number of lines which constitute the first-order Raman and infrared spectra are, respectively, 1 and 0. For crystalline GaAs (c-GaAs), a closely-related tetrahedrally-bonded semiconductor to be discussed in the following section, the corresponding numbers are 2 and 1. In the case of c-GaAs, the $\underline{k}=0$ longitudinal-optical (LO) phonon is split off in frequency from the TO phonons by the effect of its macroscopic longitudinal electric field. Both the TO and LO lines are seen in Raman scattering, and the TO line is also seen in infrared absorption.

For the amorphous forms of these semiconductors, the spectroscopic situation is completely different, as illustrated for a-Si in the upper part of Figure 2. Both the Raman and the infrared spectra exhibit broad-band continua, not line spectra. These continua span the full range of frequencies from zero up to the highest-frequency vibrations present. All of the vibrations participate: the spectra mirror, in an overall way, the

Table 1. Connection between structure and vibrational spectra (Si, GaAs)

atomic-scale structure	spectra characteristic of atom-atom vibrations (Raman, infrared)	reason	moral
crystalline	0 or 1 or 2 sharp lines	$\underline{k}=0$ selection rule for phonon creation	crystals are elitist
amorphous	broad continuum which spans the full vibrational density-of-states	$\underline{k}$ is not a good quantum number ("disorder-induced selection-rule break-down")	glasses are egalitarian

full vibrational density-of-states. Although both exhibit three broad, overlapping bands, the Raman and infrared spectra differ from each other in shape because each is modulated by a different energy-dependent matrix-element (or oscillator-strength) function.

Amorphous solids, therefore, in contrast to the elitism of their crystalline counterparts, exhibit spectroscopic egalitarianism. Democracy reigns, all phonons have the right to interact with light. The reason for this crystalline/amorphous spectroscopic dichotomy, summarized in Table 1, is the absence for the glass of any k-conservation requirement. Wave-vector k is not a valid quantum number in the absence of translational periodicity, so the k=0 selection rule is inoperative for the glass. (Another way to look at it: each vibrational mode is characterized -- in an amorphous solid -- by a spread of k vectors, including k=0. So each mode has a k=0 component which allows it to couple to light.) The elitist/egalitarian "moral" in the last column of Table 1 is a convenient analogy or mnemonic device for this dichotomy.

The lower part of Figure 2 indicates a first attempt to account for the spectra of amorphous silicon. The idea here is to start with the phonon density-of-states of crystalline silicon, broaden it by convoluting with a Gaussian, and then assume that disorder-induced selection-rule breakdown allows each mode to contribute equally. Three overlapping broad bands do result, in rough correspondence with the observed spectra. (Note that the low-frequency band is "derived" from acoustic modes of the crystal; acoustic modes are normally hidden in optical experiments.) This approach must, however, be viewed with great caution because the amorphous form contains new structural features (such as five-atom and seven-atom covalently-bonded rings) which are not present in the crystalline form (2). Note the presence of a variety of ring sizes in Figure 1(a); the crystalline counterpart contains only hexagons.

Now let us turn our attention to higher photon energies, moving from vibrational excitations to electronic excitations. The crystalline/-amorphous spectroscopic contrast is again very clear, though it is less dramatic than the vibrational case spelled out in Table 1 and Figure 2. We illustrate the spectral distinction in the electronic case with the ultraviolet absorption spectra shown in Figure 3 for crystalline and amorphous GaAs. Here the spectrum of c-GaAs is the sharply-peaked dashed curve and the spectrum of a-GaAs is the smoothly-varying solid curve. The quantity plotted against photon energy is the imaginary (absorptive) part of

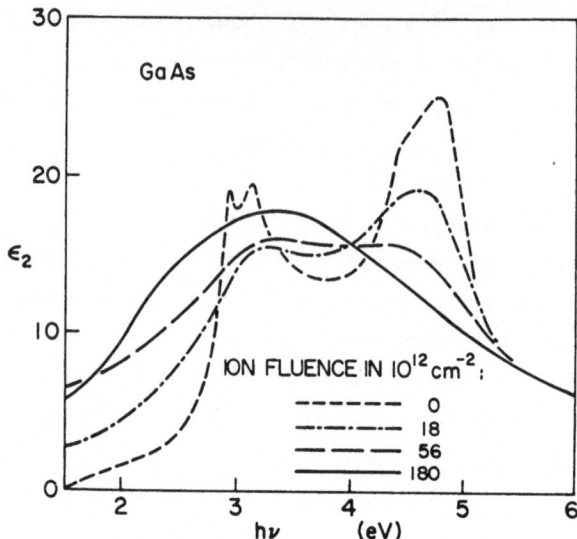

Fig. 3. Optical spectra of amorphous GaAs (solid line) and of crystalline
GaAs (dashed line) in the electronic-excitation regime. These
curves record a bombardment-induced crystalline-to-amorphous
transformation obtained with energetic As⁺ ions (after Ref. 6).

the dielectric function, measured by Aspnes, et al. (6) in ellipsometric
studies of ion-implanted GaAs.

Both c-GaAs and a-GaAs possess continuous absorption spectra in the
ultraviolet which correspond to electronic transitions from the bonding
valence band to the antibonding conduction band. The difference is in the
distribution of oscillator strength within these broad continua. The
distribution is bland and featureless for a-GaAs, but exhibits sharp peaks
and corners for c-GaAs. The sharp structure seen for the crystalline
semiconductor is a consequence of the E($\underline{k}$) band structure and the $\Delta k=0$
selection rule. The crystal interband spectrum is dominated by $\underline{k}$-conserving
"direct" transitions. Thus the sharp features are a consequence of long-
range order, so they disappear in the transformation to the amorphous form.

The disorder-induced selection-rule breakdown for electronic excitations
happens to have special technological significance in the case of amorphous
silicon. Both crystalline and amorphous silicon have roughly similar
optical bandgaps, about 1 eV. But crystal silicon absorbs light only very
weakly in the 1-3 eV range of photon energy (where sunlight lies). The

reason is that c-Si is an indirect-gap semiconductor, so that while the excited states are energetically accessible with photons in this range, the transitions involved do not conserve $\underline{k}$ and are thus forbidden in first order. But amorphous silicon, unburdened by discriminatory $\underline{k}$-conservation selection rules, absorbs light strongly in the 1-3 eV region that dominates sunlight at the earth's surface. This is one reason, as discussed in (1), that a-Si:H is a promising material with respect to solar energy.

CASE STUDY: ION-IMPLANTED GALLIUM ARSENIDE

Electronic properties of the technologically-important crystalline semiconductors, silicon and gallium arsenide, are controlled by small concentrations of intentionally-introduced dopant impurities. Since microelectronic device structures are typically confined to a thin layer near the surface of the semiconductor chips, one effective method of introducing the necessary impurities is by ion implantation, in which a beam of fast-moving ions bombards the semiconductor crystal and penetrates a short distance into it. This method works well. The problem is that the ion beam, in the process of inserting the desired guest atoms, also tears apart the crystalline regularity of the semiconductor host lattice. This structural "damage" is electronically harmful and must be "healed" by a subsequent thermal annealing step in which the disordered (possibly amorphous) damage layer recrystallizes. Technologically, there is strong interest in attempting to understand and control the implantation-induced structural damage, as well as the annealing step which heals that damage. Scientifically, the possibility of using implantation to "tune" the degree of disorder (note the intermediate spectra in Figure 3) has opened up a new research tool for studying such basic phenomena as the influence of amorphicity on solid-state excitations.

In this section I will describe some recent results on the use of optical techniques to study structural changes which accompany ion implantation into crystalline gallium arsenide. Although crystalline silicon remains dominant in the semiconductor industry, GaAs is growing in importance because its high electron mobility is useful for high-speed applications and its direct bandgap is useful for light-emitting applications.

Figure 4 displays a sequence of Raman-scattering results (7) for c-GaAs implanted with 120-keV $^{28}Si^+$ ions at fluences ranging up to 3×10^{15} ions/cm^2.

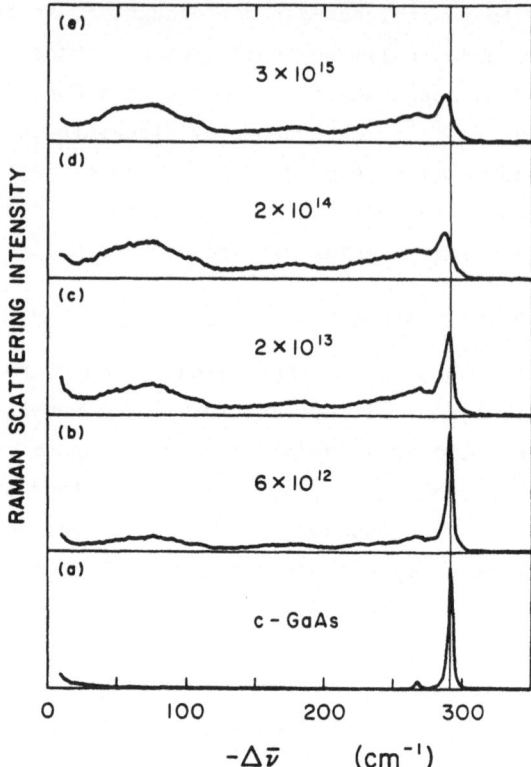

Fig. 4. The effect of ion fluence on the Raman spectrum of silicon-
implanted (120-keV Si⁺ ions) gallium arsenide. The lowest
curve is for the pristine crystal; the other curves are labelled
by the fluénce value in ions/cm².

(Silicon is used as a donor impurity in n-type GaAs, in which it
preferentially substitutes for gallium.) The crystal wafers had (100)
orientation, and the implantation was carried out with the ion beam at 7°
from the (100) normal in order to avoid complications from channeling
effects. The lowest panel [Figure 4(a)] is for the pristine crystal; the
other panels are for samples subjected (from bottom to top in order of
increasing dose) to ion bombardment, with the fluence given in units of
ions/cm².

The Raman spectra of Figure 4 were obtained using a laser photon energy
above the bandgap of the semiconductor (also true of Figure 2). Since this
light is strongly absorbed within the solid, the experiment is carried out

in a close-to-backscattering configuration. For the 2.41-eV laser photons used to obtain the results of Figure 4, the optical penetration depth is 1100 Å in crystalline GaAs and 300 Å in amorphous GaAs. Since the damage-layer thickness is about 2000 Å, these optical experiments are effective in probing just the shallow layer of interest.

The clean spectrum shown in Figure 4(a) for the unimplanted crystal (c-GaAs) exhibits the sharp k=0 LO phonon line at 292 cm^{-1}. Also seen, very weakly, is the k=0 TO line at 268 cm-^{1} which is supposed to be absent for back-scattering along (100). The evolution, in Figures 4(b)-4(e), of the Raman spectrum as a function of increasing implantation dose, shown in Figures 4(a)-4(e), reveals the progressive development of a broad three-band continuum which extends from 0 to 300 cm^{-1}. We interpret this as the Raman signature of amorphous GaAs. Work done in our lab on a-GaAs films prepared by sputtering (8) and by flash-evaporation (9) reveal spectra very similar to the broad component of Figure 4. Thus the amorphous form of GaAs which is generated by extensive ion bombardment of the crystal is very similar to the amorphous form prepared by conventional techniques.

As the amorphous bands grow in Figure 4 to dominate the spectrum with increasing Si$^+$ fluence, the crystal LO line weakens, asymmetrically broadens, and shifts downward in frequency. The downshift and asymmetric broadening agree with similar observations reported by Tiong, et al. (10) for As$^+$-implanted GaAs. These results indicate that the crystalline portion of the damage layer is becoming microcrystalline with increasing dose. Thus the picture that emerges of the implanted region is one of an amorphous matrix in which are embedded microcrystals surviving as remnants of the original bulk crystal.

Figure 5 displays a sequence of Raman-scattering results obtained with SiF$_3$$^+$ ions (also accelerated to 120 keV) instead of Si$^+$ ions. At the same kinetic energy, heavier ions carry more momentum than lighter ones, so that more damage is expected with SiF$_3$ than with Si. In Figure 5 we see that for a fluence of 10^{14} cm^{-2} of SiF$_3$$^+$, the GaAs damage layer is totally amorphous. The crystal LO line is gone, and only the continuum characteristic of a-GaAs remains.

One might think that once the damage layer becomes totally amorphous, at

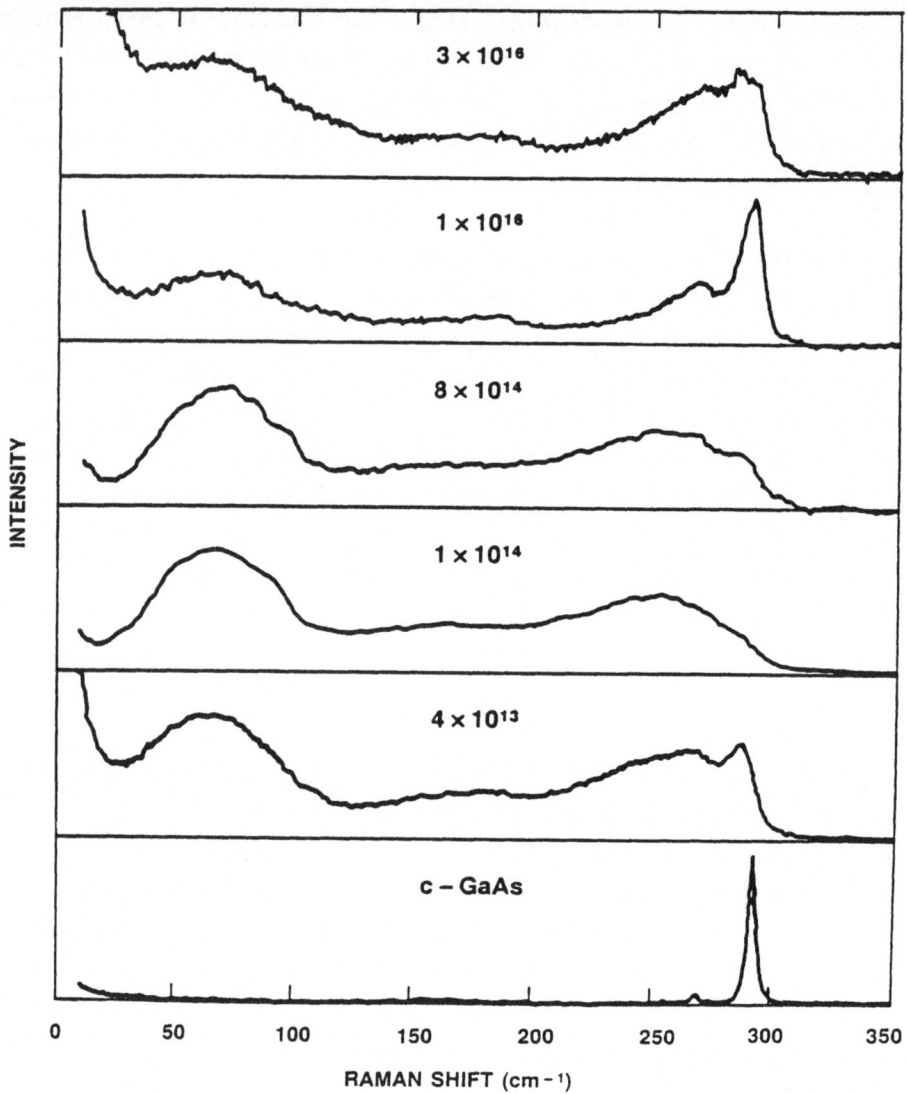

Fig. 5. The effect of SiF$_3^+$ ion bombardment on the Raman spectrum of GaAs.

a fluence of about 10^{14} cm^{-2}, then that is the end of the story. But the upper panels of Figure 5 show this is not so. At 10^{16} cm^{-2} the crystal LO line has reappeared, broadened and shifted (reflecting microcrystallinity) but nevertheless quite strong. The reappearance of a crystalline fraction is <u>not</u> understood, in spite of the fact that a name has been assigned in the literature to this type of phenomenon; "self-annealing." It is true that a

thermal effect is involved to some extent, associated with a temperature
rise which accompanies a long implant (the substrate was not cooled in these
experiments). However, this is unlikely to be the whole story. The top
panel of Figure 5 shows that the effect is cyclical; at even higher fluence
the material is again "going amorphous."

Just as the optical response of a semiconductor changes when the
material's atomic-scale structure changes from crystalline to amorphous, so
the response to an incident particle beam should also change. Thus it may
not be so surprising that, while the effect of ion bombardment on the
crystal is to induce disorder, the effect of ion bombardment on the glass
may be to induce order. Photocrystallization (i.e. the analogous process
using a photon beam instead of an ion beam) is known to occur in other
amorphous semiconductors, notably certain chalcogenides.

SUMMARY

Amorphous semiconductors are interesting and important materials in a
variety of contexts. Optical experiments are effective in providing a
nondestructive probe of these solids. Vibrational spectra in particular
(Figure 2, Table 1) clearly distinguish between the crystalline and
amorphous forms of a given material.

As an illustration of this approach, we have presented recent results
which show how Raman scattering can be used to study structural changes
which occur during the implantation processing of gallium arsenide.
Initially crystalline, the near-surface layer converts, under certain
circumstances, to a totally-amorphous region (Figure 5). A key feature of
this work is the use of strongly-absorbed light to match the optical
penetration depth to the ion penetration depth, so that the incident light
probes just exactly the shallow region of interest. In an extension of this
work, experiments carried out in our laboratory with varying optical
penetration reveal that this technique can yield a depth profile of
structural damage.

ACKNOWLEDGEMENTS

The results on implantation-disordered gallium arsenide, presented here
in Figures 4 and 5, are excerpts from a collaborative experimental
investigation carried out with Mr. M. Holtz, Dr. A. E. Geissberger and Dr.
R. A. Sadler.

REFERENCES

1. See the article by Kurt Weiser in this volume.
2. Zallen, R., "The Physics of Amorphous Solids," John Wiley and Sons: New York, 1983.
3. Zachariasen, W. H., J. Am. Chem. Soc. (1932) $\underline{54}$, 3841.
4. Brodsky, M. H.; Cardona, M., J. Non-Crystalline Solids (1978) $\underline{31}$, 81.
5. Zallen, R., Phys. Rev. (1968) $\underline{173}$, 824.
6. Aspnes, D. E.; Kelso, S. M.; Olson, C. G.; Lynch, D. W., Phys. Rev. Lett. (1982) $\underline{48}$, 1863.
7. Holtz, M.; Zallen, R.; Geissberger, A. E.; Sadler, R. A., J. Appl. Phys. (1986) $\underline{59}$, 1946.
8. Sputtered a-GaAs thin films were from the laboratory of W. Paul, Harvard.
9. Flash-evaporated a-GaAs thin films were from the laboratory of M. L. Theye, Paris.
10. Tiong, K. K.; Amirtharaj, P. M.; Pollak, F. H.; Aspnes, D. E., Appl. Phys. Lett. (1984) $\underline{44}$, 122.

SURFACE MODIFICATION BY RAPID SOLIDIFICATION

LASER AND ELECTRON BEAM PROCESSING

B. H. Kear
Exxon Research and Engineering Company
Corporate Research Science Laboratories
Annandale, NJ   08801

P. R. Strutt
Department of Materials Science
University of Connecticut
Storrs, CT   06268

INTRODUCTION

The use of lasers or electron beams for rapid solidification surface modification of materials has become established metallurgical practice. Typically, processing is carried out by rapidly traversing (scanning) a high power density beam over the material surface, so as to induce melting of a thin surface layer. A high rate of energy delivery facilitates surface melting at very high melting efficiency, such that most of the absorbed energy is used for melting with only a small fraction going into heating of the substrate material. This ability to maintain a cold substrate while melting a thin surface layer of material develops steep thermal gradients, which results in rapid solidification (quenching) of the molten layer, after the cessation of energy input. Calculations show that cooling rates of 10^4-10^8 °K/sec are readily attainable in appropriately thin sections.

The importance of rapid melt-quenching is based on the fact that the unsurpassed homogeneity of the liquid can be preserved (or nearly preserved) in the solid, which can be utilized in that form or after subsequent thermomechanical treatment. A wide variety of microstructures have been developed by such laser, or electron beam 'glazing' techniques, including amorphous metallic solids, extended solid solutions, metastable phases, and ultrafine dendritic and eutectic structures, epitaxially related to the substrate. Laser, or electron beam glazing in conjunction with surface alloying, post-glaze heat treatment, and/or deformation has great flexibility as a means to generate a broad range of novel surface microstructures and properties. Furthermore, although currently being employed primarily for the surface treatment of materials, it also has the potential for processing bulk materials by the sequential build-up of one glazed layer upon another.

Only selected aspects of work carried out in this field will be discussed in this article. Amongst the specific areas of fundamental interest that will be considered are (1) influence of critical solidification parameters, e.g. temperature gradient, solidification rate and cooling rate, on the resulting microstructure, (2) dependence of melt zone geometry on processing variables, (3) effect of fluid flow on solidification microstructure, and (4) differences in response of materials to laser and electron beam surface modification treatments. For more complete information on these and related topics the reader is referred to several recent books dealing with the broad aspects of rapid solidification processing, including rapid solidification surface modification (1-6).

SURFACE MODIFICATION TREATMENTS

The nature of the interaction of a laser, or electrom beam with a material surface is controlled primarily by two variables--the absorbed power density and the available interaction time. Rapid solidification surface modification (glazing) requires high power density of the order of that utilized for deep penetration welding but with substantially shorter interaction times, Figure 1. Under these conditions, energy inputs range from 1-100 J/cm^2, and the heating effect is concentrated in a very thin region at the material surface. This gives rise to extremely high thermal gradients, which promote rapid solidification of the melt.

Laser Glazing

A continuous multikilowatt CO_2 laser is normally employed for laser glaze processing (7). A convenient experimental arrangement is depicted schematically in Figure 2. A nominal 7.5 cm diameter laser beam is directed towards and focused upon the workpiece surface by reflective optics. In a typical test, a 46 cm focal length mirror is used to provide an effective minimum spot diameter of 0.05 cm at the workpiece surface. At 3.0 kW, these optics provide a maximum incident power density of approximately 1.5×10^6 W/cm^2. Such a high power density is essential for localizing the energy input at the material surface, and promotes effective coupling of the laser energy with the material, despite the initial high reflectivity of metallic surfaces to the 10.6 μm wavelength of carbon dioxide laser radiation. The 3 kW level is a convenient one, in that it promotes effective beam coupling, but does not create significant plasma generation problems. As indicated in Figure 2, plasma suppression is accomplished by means of an inert gas shield, which further prevents atmospheric contamination of the melt. Cooling due to the inert gas is negligible, in comparison to the quenching effect of the unheated substrate material.

A range of melt depths may be achieved by varying the translational speed of the workpiece under the focused beam.

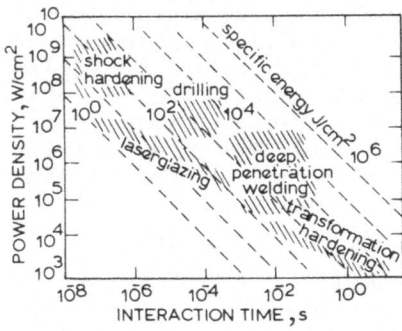

Figure 1--Operational Regimes for Laser Materials Processing Techniques.

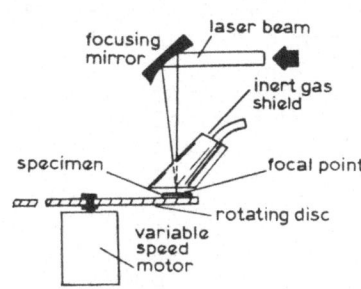

Figure 2--Schematic of Laser Glazing Apparatus.

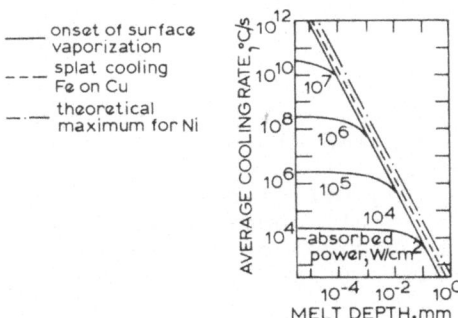

Figure 3--Effect of Melt Depth and Power Density on Average Cooling Rate.

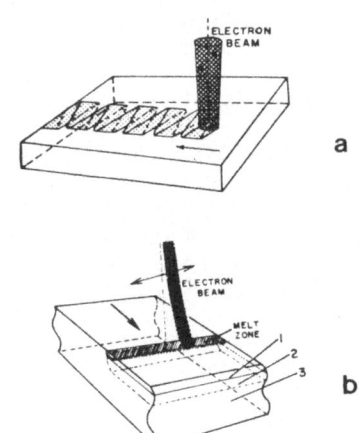

Figure 4--Schematics of Electron Beam Glazing Modes; (a) Point Source Melting, (b) Line Source Melting.

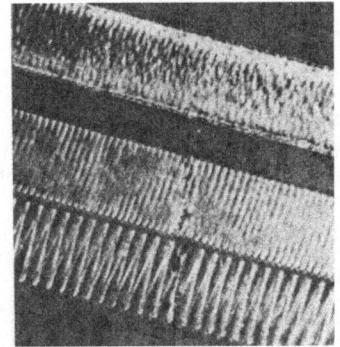

Figure 5--Point Source Melting, Showing Oscillatory Traces, With Varying Degrees of Overlap.

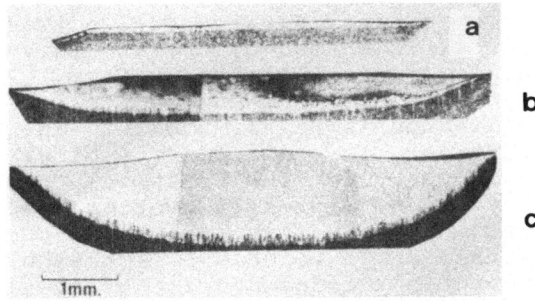

Figure 6--Effect of Beam Traverse Velocity on Profile of Electron Beam Glazed Layer; (a) 4.35 cm.s^{-1}, (b) 1.75 cm.s^{-1}, (c) 0.44 cm.s^{-1}.

Linear speeds of from 1.5 to greater than 60 m/min are attained by using a variable speed rotating disc with specimens located at a fixed radius on the disc surface. Such a speed range yields energies per unit length of melt ranging from 1200 down to 30 J/cm. Higher energy inputs are employed as initial homogenizing passes, and lower energy inputs for subsequent rapid solidification processing. The quench rate is ultimately dependent on melt layer thickness, with cooling rates of 10^4-10^8 K/sec readily attainable in appropriately thin sections.

In Figure 3, the effect of melt depth on cooling rate is shown for selected power densities. The maximum cooling rate that can be attained at each power density occurs for the limiting case in which the melt depth approaches zero. This maximum cooling rate increases with the square of the power density. For a given melt depth, the increase in cooling rate with absorbed specific power is a consequence of the increased temperature gradients; doubling the power density approximately doubles the cooling rate. A maximum melt depth occurs for a given power density, after which surface vaporization begins, and the lowest cooling rates are realized under these conditions.

Electron Beam Glazing

The interaction between an electron beam and a material surface, is controlled by the penetration depth of the beam, which depends primarily on the accelerating voltage and the density of the material. The coupling efficiency of an electron beam is about 60 pct. for steel, compared with no more than 10 pct. for a laser beam, because of the more effective beam absorption in the case of the electron beam. For comparable absorbed power densities, however, similar regimes of processing exist for both laser and electron beam glazing methods, Figure 1. Similarly, the dependence of cooling rate on absorbed power density and melt depth follows the same trends, Figure 3.

An unique feature of the electron beam technique is that a variety of melting patterns can be produced by programmed electromagnetic beam deflection (8). In a typical experimental arrangement, Figure 4, the focused beam is rapidly scanned back and forth to develop the characteristics of moving point, or line source melting. In moving point source melting, Figure 4(a), a strip of rapidly solidified material is formed by traversing an oscillating electron beam over a moving substrate surface. As indicated, the linear motion of the specimen translation stage per cycle of beam oscillation is nearly equal to the beam diameter. Under these conditions, the entire strip of glazed material is formed with minimal overlap of successively melted traces. In general, however, any degree of overlap may be obtained simply by selecting appropriate values of oscillating frequency and specimen stage velocity. Actual oscillatory traces formed with three varying degrees of overlap are shown in Figure 5. In moving line source melting, Figure 4(b), a linear heat source is developed by rapidly oscillating the beam across the entire width of the melt zone, combining this with a relatively low translation motion of the substrate. The requirement for a rapidly oscillating beam to act as a linear melting source is that in any one cycle the material at one extremity remains molten until the beam returns on completion of the cycle.

Using heat transfer analysis, the requisite parameters for producing linear heat source melting can be calculated. The analysis shows that the number of cycles N required to maintain a molten state on the substrate surface along the line of the oscillating beam is given by Equation 1. Here d is the beam diameter, w the width of the oscillation, and ξ the ratio of the power absorbed by the material to the beam power.

$$\sum_1^N n^{1/2} - \sum_1^N (n - d/2w)^{1/2} = \xi \cdot (d/2w)^{1/2} \tag{1}$$

After numerical solution of equation 1 to obtain N, the translation velocity of the specimen is determined from equation 2

$$V = (fd)/N \tag{2}$$

In the present simplification of the complete analysis the assumption has been made that the melting point of the material is approximately half the vaporization temperature; this applies when considering steels.

Once values of frequency, beam diameter, and width have been selected to determine V, the beam power required can be calculated from equation (3)

$$P = \frac{\pi \kappa T_m}{a} (d/2)^{3/2} (wf)^{1/2} \tag{3}$$

The parameter κ is the thermal conductivity and a is equal to $(4\pi/\alpha)^{1/2}$, where α is the thermal diffusivity.

As an example, consider the beam power and velocity V required to produce a 1.25 cm wide track, with d equal to 0.035 cm. In this particular case (w/d) is 35.71, and calculation or graphical solution shows V/fd to be 0.0119. With f=1,000 hertz, the value of V is 0.42 cm.s^{-1}. Solution of equation 3 shows that P is 554 W.

Actual examples of wide zones formed in M7 high speed steel are shown in Figure 6, where the width is 5-7 mm and the frequency 1000 hertz; in each case the beam diameter and power is 0.035 cm and 563 watts, respectively. The shallow zone, Figure 6(a), is formed with a translation velocity V of 4.35 cm.s^{-1}, and is seen to have a reasonably constant depth with a mean value of 62 μm. The effect of reducing the translation velocity V to 1.75 and 0.44 cm.s^{-1} is to significantly increase the melt depth to 510 μm and 1250 μm, respectively. The deep zones in Figures 6(b) and (c), nevertheless, are uniform and symmetrical in shape.

Studies have been carried out to determine the effect of beam power (P), translation velocity (V) and melt zone width (w) on the depth of melting (d_m). Experimentally determined values of d_m obey the relation given by equation 4. This

relation is a simplified form of one theoretically predicted
using heat flow analysis.

$$d_m = 5.94 \times 10^{-5} \ (P/Vw) \tag{4}$$

It is interesting to note that this relation is obeyed over
the particularly wide range of V used in the study, namely
from 0.5 to 50 cm.s^{-1}.

Since surface smoothness of as-glazed surfaces is a
factor of considerable importance, the effect of process
parameters on surface topography has been a topic of study.
In the view of an as-glazed surface in Figure 7, it can be
seen that a high degree of surface smoothness is attainable
using the line source melting technique. Smooth glazed
surfaces of widths of at least 35 mm are readily attainable
using commercially available electron beam sources.

Surface Alloying

For surface alloying, two distinct approaches have been
evaluated: (1) pre-placement of alloying material on the
workpiece surface prior to melting (9), and (2) continuous
delivery of alloying material (wire, ribbon, or powder) to the
interaction or melt zone (10). For high cooling rate
processing, the deposited material must have a thickness $\lesssim 100$
μm. Much thicker deposits of rapidly solidified material can
be achieved by superposition of one deposited layer upon
another in an incremental manner, Figure 8. In this process,
good interlayer bonding and epitaxial growth from layer to
layer can be obtained under proper operating conditions,
Figure 9. A critical parameter is the location of the powder
impingement point with respect to the laser melt zone. Since
the mandrel is rotating, feed stock impingement must occur
slightly ahead of the laser beam for stable, steady state
deposition.

Thick hardfacing deposits laid down by conventional
plasma deposition techniques have also been resurfaced using
laser and electron beam glazing treatments.

THEORETICAL CONSIDERATIONS

Heat Flow Models

An important aspect of rapid solidification surface
modification is the use of appropriate heat flow models to
determine the effect of process variables on (1) melt zone
geometry, and (2) solidification parameters. Suitable models
include the moving point source model of Cline and Anthony
(11), which involves an analytical solution of the thermal
diffusion equation using Green's functions, or the finite
difference method of Mehrabian et al. (12), which involves
lengthy numerical procedures. A convenient and simpler
approach, proposed by Greenwald et al. (13), is to model the
glazing process on the heating and cooling of a semi-infinite
plate produced by exposure and removal of its surface to a
uniform radiant energy flux. In summarizing and further
developing this approach it is convenient to introduce the

234

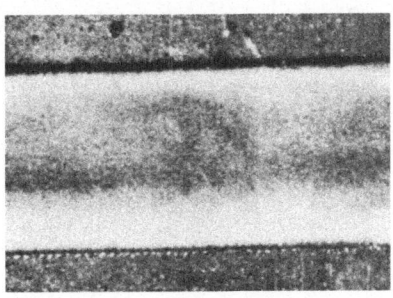

Figure 7--Characteristically Smooth Electron Beam Glazed Layer.

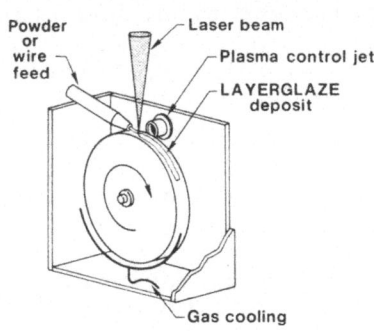

Figure 8--Schematic of the Incremental Glazing Process.

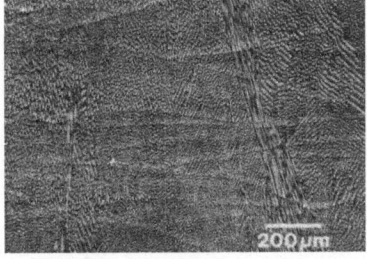

Figure 9--(a) Layer Glaze Test Sample, Produced From Pre-alloyed Powder Feed, (b) Microstructure of Test Sample, Showing Epitaxial Growth Between Successive Deposited Layers.

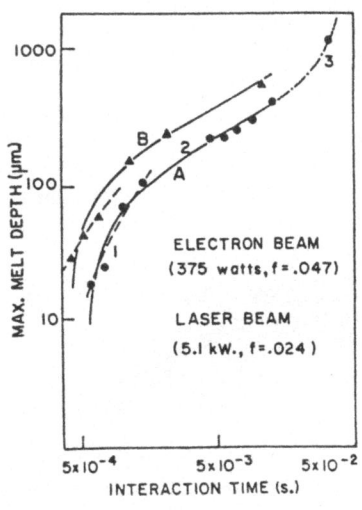

Figure 12. Maximum Melt Depth as a Function of Interaction Time in Glazed M2 Steel.

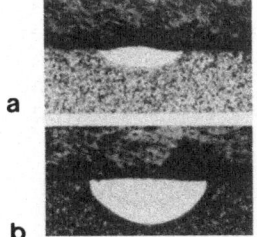

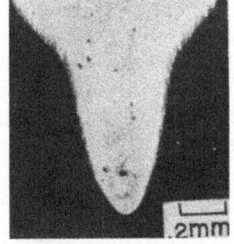

Figure 11--Micrographs of Melt Zone Geometries, Showing Influence of Beam Interaction Time; (a) 1.3×10^{-3} s, (b) 6×10^{-3} s, and (c) 3.1×10^{-2} s.

parameters a, b, and k and the function

$$\psi(x,t) = b \left[a\sqrt{t} \cdot \exp - (kx/\sqrt{t})^2 - x \, \text{erfc} \, (kx/\sqrt{t}) \right] \quad (5)$$

$$a = (4\alpha/\pi)^{\frac{1}{2}}, \qquad b = q_0/K, \qquad k = (4\alpha)^{-\frac{1}{2}}$$

where q_0, α and K are the absorbed power density, thermal diffusivity and thermal conductivity, respectively.

During heating, the temperature at a point x below the surface at time t is given by

$$T(x,t) = T_0 + \psi(x,t) \quad (6)$$

where T_0 is the ambient temperature. During cooling

$$T(x,t) = T_0 + \psi(x,t) + \psi(x,\gamma) \quad (7)$$

where $\gamma = (t-\tau)$, τ is the interaction time, defined for convenience as (beam diameter)/(beam velocity).

The nature of the temperature rise and decay, as predicted by equations (6) and (7), is shown in Figure 10. The time for initial melting (τ^*) and final solidification (τ^1) at a point on the surface, are given by equations (8) and (9).

$$(\tau^*)^{\frac{1}{2}} = \frac{T_m - T_0}{ab} \quad (8)$$

$$\tau^1 = \frac{(\tau + \tau^*)^2}{4\tau^*} \quad (9)$$

From these expressions the value of L, the trailing liquid zone length, may be determined, since L is simply equal to $V(\tau' - \tau^*)$, V is the beam velocity.

Melt Zone Geometry

In order to compare experimental and predicted values of melt zone parameters, surface melting experiments have been carried out using both the CO_2 laser and electron beam. In both cases the beam diameter was 0.5 mm and the beam velocity range was 1.5-100 cm s^{-1}. The material selected for the investigation was M2 high speed steel, since previous laser surface melting studies had been carried out on this alloy (14). Because of the highly efficient beam-specimen coupling with the electron beam the incident beam power was adjusted to 375W, so as to obtain a similar melt depth to that produced by the 3 to 5 kW laser beam. In the calculations the values taken for the thermal diffusivity (α) and the thermal conductivity (κ) in M2 high speed steel were taken as 0.22 cm^{2}s^{-1} and 0.76 J cm^{-1} s K^{-1}; the difference $T_m - T_0$ was taken as 1425°K.

The cross-sectional views of electron beam melted zones in Figure 11 clearly show the strong dependence of melt zone geometry on beam-material interaction time. In Figure 12, where maximum melt depth is plotted vs. interaction time, the discrete points represent the experimental values for the electron beam (curve A) and laser (curve B). As already noted, the incident power required for laser melting is significantly greater than for electron beam melting; curves A and B in Figure 12, correspond to an incident power of 375W and 5.1kW, respectively.

In considering the three distinct regions in curve A, the melt depth at short interaction times (region 1) is a fraction of the width, as shown in Figure 11(a), where τ (the interaction time) is not more than 1.3×10^{-3} s. In this region the heat flow is essentially inwards (normal to the surface), as in the semi-infinite model of Greenwald et al. (13). At higher values of τ (region 2) the melt width increases slowly with τ although the depth becomes about one-third of the width when τ is 6×10^{-3} s, i.e. near the termination of region 2. With this melt zone geometry, Figure 11(b), the heat flow is now essentially radially outwards. When τ increases further (greater than 1.2×10^{-2} s), surface vaporization becomes appreciable and in Figure 11(c) ($\tau = 3.1 \times 10^{-2}$ s), the beam has clearly penetrated, as in the "deep penetration mode", to produce a maximum melt depth of about 1200 μm.

The two dotted lines in Figure 12 are fitted to the experimental data at the shorter interaction times (5×10^{-4} to 1.35×10^{-3} s) from Greenwald's analysis (13), using absorbed power densities of 100,000 and 140,000 W.cm^{-2} for the electron beam and laser. In computing maximum melt depth vs interaction time the latter is taken as (beam diameter)/(beam velocity). Values of maximum melt depth z_{max}, are given in Table I, where there is good agreement between experimental and calculated values. There is also good agreement between observed and calculated values of the molten zone length L; values for the cooling rate are discussed in the following section. It is important to point out that the range of beam velocities that may be used without actual beam penetration is fairly restricted. The actual range of usable beam velocites is readily calculated from equation 6 and it may be shown that the ratio V_v/V_m of maximum to minimum velocity is given by:

$$\frac{V_v}{V_m} = \frac{(T_m - T_o)^2}{(T_v - T_o)^2} \tag{10}$$

where T_v and T_m are the vaporization and melting point temperatures and V_v and V_m are the corresponding beam velocities. For the present alloy $V_v/V_m \approx 3$ and since V_m is in excess of 100 cm s^{-1}, the minimum beam velocity before the surface begins to vaporize is ≈ 30 cm s^{-1}. On the basis of the comparison between predicted and experimental values of maximum melt depth z_{max}, the energy coupling between the beam and the material is about 55% efficient for the electron beam and about 5.5% for the laser. Because of the uncertainty in the values of the thermal parameters under actual melting conditions the absolute values of efficiency are, at best, approximate. This, however, does not detract from the fact that the surface melting efficiency with an electron beam is about a factor of 10 greater than that with a laser.

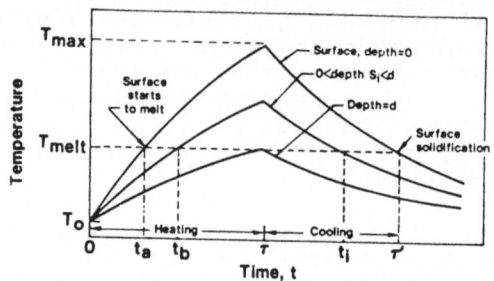

Figure 10--Variation of Temperature with Time at Selected Melt Depths.

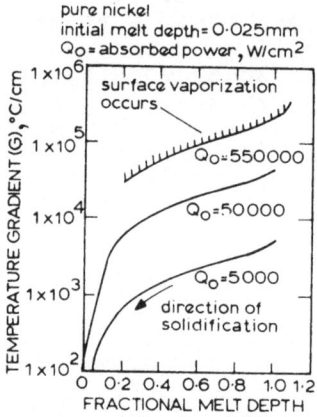

Figure 13--Transient Behavior of Temperature Gradient G.

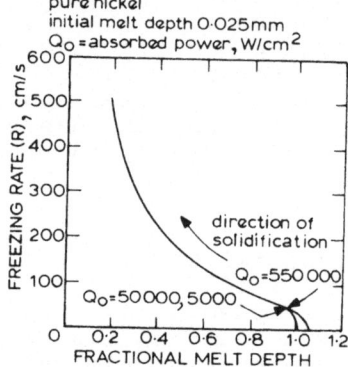

Figure 14--Transient Behavior of Solidification Rate, R.

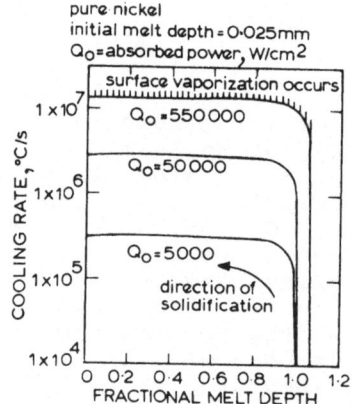

Figure 15--Transient Behavior of Cooling Rate, $\dot{T}$.

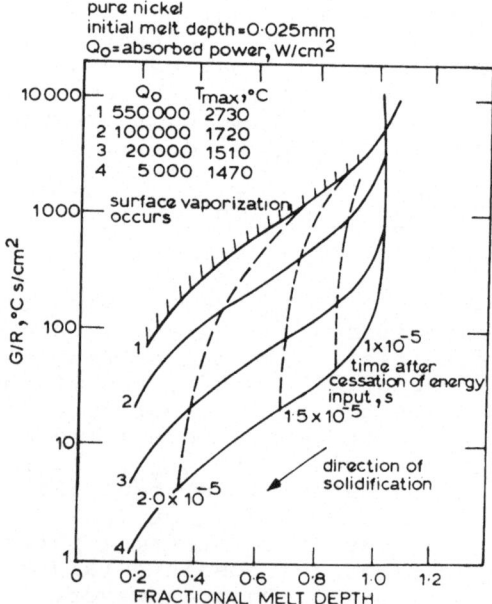

Figure 16--Transient Behavior of Gradient/Rate Ratio, G/R, at Solidification Interface.

Another heat flow model, namely that of Cline and Anthony (11), has been used to compute maximum melt depth values; these are shown by the full curves in Figure 12. Curve A (sections 1 and 2) is for the electron beam and curve B is for the laser; the upper portion is for the region where beam penetration occurs and the model is no longer applicable. On the basis of fitting curves A and B to the experimental points the estimated efficiencies for the laser and electron beam are 9% and 63%, which are in reasonable agreement with those obtained using Greenwald's model (13).

Solidification Parameters

The temperature gradient $G(x,t)$, and cooling rate $T(x,t)$ during solidification are important in determining the nature and scale of the microstructure. These are readily obtained from the theoretical models simply by obtaining the partial derivatives $\partial T/\partial x$ and $\partial T/\partial t$. The solidification rate R can be determined from the equation $\dot{T} = (G.R)$.

These parameters, as determined at the melt interface as it moves from its maximum depth to the surface, are shown in Figures 13 and 14 for temperature gradient and solidification rate, respectively. Figure 13 shows that the value of G is particularly high when solidification commences and that it drops off sharply as the interface approaches the surface; as seen in the figure the value of G depends upon the power density. In contrast, the solidification rate R, as determined by Greenwald (13), is relatively insensitive to power density, Figure 14. This, however, is a result of using a simplified heat-transfer model. By including the effect of latent heat of fusion, Mehrabian (12) has shown that the solidification rate is drastically reduced over the last 50% of the melt zone. Furthermore, it is sensitive to the power density.

The cooling rate $\dot{T} = (G.R)$, as determined by the Greenwald model (13), Figure 15, rises sharply after the onset of solidification and rapidly attains a constant value given by equation 11, see Strutt (14).

$$\dot{T} = \frac{2ab}{(\tau^2 - \tau^{*2})} \tau^{*3/2} \qquad (11)$$

The values of $\dot{T}$ calculated from this equation are shown in Table I and compared with empirical values determined from measurements of the mean solidification cell diameter λ. The scale of the solidification microstructure is determined by the mean diffusional path length, which itself is controlled by the cooling rate $\dot{T}$. There are well established empirical relationships such as that of Matyja et al. (15), who have shown that $\dot{T} = (4.57 \times 10^{-8})\lambda^{-3}$. Values of the cooling rate determined from this relation by measuring the mean cell diameter λ at different beam velocities fit equation 12 with a correlation coefficient of 0.991.

$$\dot{T} = (2.07 \times 10^4) \, V^{1.2} \qquad (12)$$

The comparison of empirical and calculated values of $\dot{T}$ in Table 1 is for the region where the model of Greenwald (13) is reasonably valid, i.e. region 1, Figure 12. The agreement is fairly good since on average the empirical and calculated values of $\dot{T}$ differ by only a factor of 2.

The mode of solidification, i.e. cellular, cellular dendritic, etc., is controlled by the ratio G/R. The calculated transient behaviour of G/R at the melt interface is shown in Figure 16. Characteristically the curves begin at infinity at the onset of solidification and fall off rapidly as the solidification approaches completion. Power density exerts a strong influence on G/R, for a given melt depth, as well as on the maximum surface temperature. Interestingly, the time required to solidify a given depth of melt is nearly independent of absorbed power, as indicated by the dashed isochronal lines, measured from the cessation of energy input.

Fluid Flow Effects

Amongst the many factors controlling the surface topography and microstructure of rapidly solidified layers is fluid flow in the melt zone. Of considerable interest are the surface ripple markings characteristic of glazed layers, and their sensitivity to solidification, or processing parameters.

Figure 17 shows the progressive change in the appearance of ripple markings on electron beam glazed M2 steel. Two distinct ripple periodicities are evident. The mean spacing of the large amplitude ripples is from 100 to 300 μm whereas that between the fine ripples (clearly evident in Figure 17(b)) is about 10 μm. Anthony and Cline (16) and Copley et al. (17) have studied this phenomenon, and it has been proposed (17) that the coarse ripples are created by a periodic overfow of liquid at the trailing edge of the molten pool. The fine ripples presumably result from the freezing in of surface waves, and their absence at higher velocities may well arise from attenuation in the shallow molten pool. An unusual feature noted at beam velocities of 4.0 and 8.5 cm s^{-1} is the highly serrated nature of the fine solidification ripples, with the segments aligned along specific directions, Figure 17(b). This intriguing effect suggests that the solidification surface is stabilized by becoming faceted (on a macroscopic scale) on specific crystallographic planes.

Some fluid mechanics considerations that give rise to ripples of high amplitude and low frequency, and also the central ridge/side trough topography, are summarized in Figure 18. This diagram combines and extends concepts and observations previously discussed. A critical feature is the surface depression formed beneath the incident beam; contributory causes are (i) the progressive increase in the surface tension between the center and edge of the molten pool (16), and (ii) the 'keyhole' due to the actual beam penetration into the material. Motion of the material, relative to the beam, results in formation of a liquid bulge ahead of the beam, Figure 18(a), which causes liquid flow beneath the beam. This results in liquid overflow at the trailing edge (17), Figure 18(b). The formation rate of liquid in the bulge ahead of the beam is determined by the

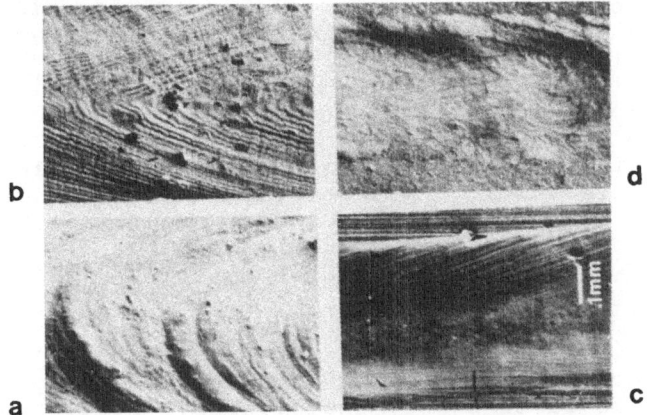

Figure 17--Scanning Electron Micrographs of Surface
Ripple Structures in M2 Steel for Beam Velocities
(a) 1.7 cm s-1, (b) 4.25 cm s-1, (c) 57.5 cm s-1,
and (d) 97.5 cm s-1.

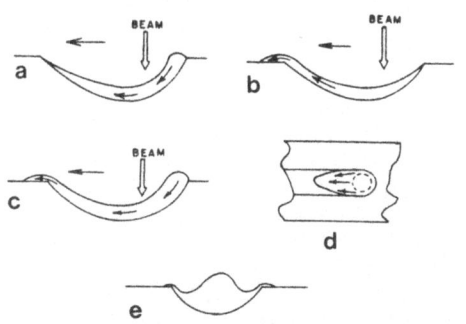

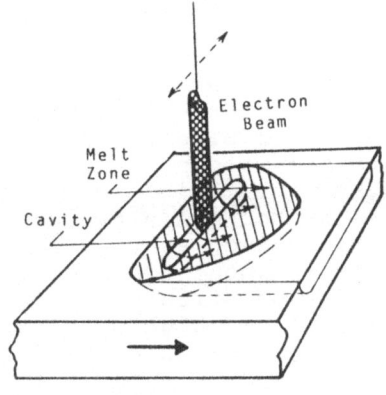

Figure 18--Schematic Diagrams Showing the Formation of the Depression
Beneath the Incident Beam and Liquid Flow to Form (a) and (b) Surface Ripples, (c)-(e) Central Ridge
and Side Trough Morphology.

Figure 19--Formation of Cavity Under
the Electron Beam in Line Source
Melting Mode.

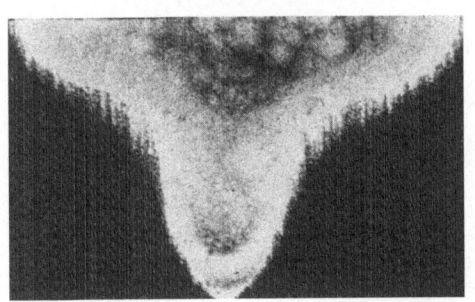

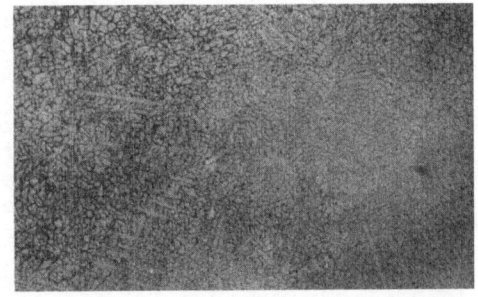

Figure 20--Optical Micrographs Showing Cellular Segregation Patterns
(a) in Point Source Melting Mode When the Beam Penetrates, (b) and (c)
in Longitudinal Sections of Line Source Melting Mode.

velocity and beam power density. For instance, with a low power incident beam, it takes a short time before a sufficient liquid bulge forms for liquid flow to the trailing edge to occur. Overflow at the trailing edge is thus periodic and results in ripple formation. However, with a high power incident beam the excess liquid ahead of the beam forms at a sufficient rate that there is continuous liquid flow beneath the beam, Figure 18(c). This results in the formation of a central ridge with accompanying side troughs (18), Figure 18 (e). In the plane view, Figure 18(d), the arrows show the major flow directions of molten material within and around the beam depression.

In practical applications, surface smoothness may be of overriding importance, in which case the type of profile in Figure 18(e) would be highly undesirable. Fortunately, as shown in Figure 17(d), the effect can be reduced until it is almost negligible by appropriate selection of processing parameters. In other applications, however, where surface smoothness is not crucial, then the extensive fluid flow and relatively high superheat characteristic of the depression-mode of glazing can be exploited to facilitate the dissolution of refractory phases.

Another more complex flow pattern has been revealed in studies of electron beam glazed steel containing a high volume fraction of carbide particles (19). This mode occurs whenever there is significant penetration of the beam into the melt, see Figure 19. Evidence for the fluid vorticity involved is provided by the cellular segregation patterns observed in micrographs, such as Figure 20. The scale of this cellular structure is about two orders of magnitude greater than the solidification structure. A recent detailed study of the phenomenon in M7 high speed steel (19) has shown carbide particle segregation created by centrifugal motion within vortices. In developing a theoretical model, the analysis considers the following aspects:

(i)   The overall driving force for the behavior is the large temperature differential between the plasma cavity and the solid-liquid interface. This creates the primary circulatory flow pattern, or primary cell pattern, depicted in Figure 21. For simplicity the analysis considers flow within a rectangular melt region of length X and depth Y. The onset of instability in the primary convective flow mode is determined by the Rayleigh number (Ra), which is the product of the Grashof (Gr) and Prandtl (Pr) numbers. Usually the critical Rayleigh number for instability is ~ 2×10^5 (20). To determine the dimensionless parameters the temperatures next to the plasma cavity and solid-liquid interface are assumed to be 3000°C and 1500°C, respectively. The coefficient of volume expansion (β) and kinematic viscosity (υ) are taken as those for molten iron; namely, 10^{-3} °K^{-1} and 5×10^{-3} $cm^2.s^{-1}$, respectively. With these values Gr and Pr are 3×10^6 and 0.3; consequently the Rayleigh number is 9×10^5, which is clearly sufficient for turbulent flow.

242

(ii)   To determine the amount of vorticity in the molten
       pool, the maximum flow velocity of the liquid is
       found by an analysis in reference (21). By first
       knowing the value of a parameter η, (see equation
       13) it is possible to determine a function $F(\eta)$ by
       graphical solution (see equation 14):

$$\eta = Y/X^4 \sqrt{Gr/4} \qquad\qquad (13)$$

$$F(\eta) = uX(2\upsilon\sqrt{Gr})^{-1} \qquad\qquad (14)$$

When $F(\eta)$ is determined the maximum fluid velocity
u is known; this is 7.0 cm.s^{-1} when the melt zone
perimeter $2(X+Y)$ is assumed to be 0.6 cm; the
corresponding value of the vorticity in the pool
is 4.2 cm^2.s^{-1}.

(iii) Instability of the primary flow cell just
       considered, see Figure 21, gives rise to secondary
       cells, which are represented by eddies in the
       schematic diagram. Centrifugal motion of carbide
       particles in these results in the segregation
       cells seen in micrographs, such as Figure 20.
       Such micrographs show the small vortex radius (r)
       is $\sim 10^{-3}$ cm.; consequently the vorticity should
       be > 0.3 cm^2.s^{-1}. Viscous damping causes the
       small vortices to decay in time τ, which is given
       by equation 15, where R is the outer cut-off
       radius.

$$\tau = (2r^2/\upsilon)/\ln(R/r) \qquad\qquad (15)$$

Estimates of r and R from micrographs (i.e. Figure
20) show τ is ~ 5 ms. This, as shown by detailed
analysis, is the mean time of transit of a carbide
particle from the center to periphery of a vortex
under the action of the centrifugal force.

(iv)   The transit time, just discussed, is obtained by
       numerical solution of the equation of motion of
       the carbide particle. Particle displacement vs.
       time curves in Figure 22 are for a vorticity of
       0.1 cm^2.s^{-1}. This figure shows that a 1 µm radius
       particle is displaced 10 µm from the vortex center
       in 2 ms; beyond this time the carbide particles
       slow down considerably. Thus, the central region
       becomes rapidly denuded of carbide particles and a
       high concentration is accumulated at a radius
       somewhat > 10 µm. A critical factor in the
       process is that both the particle transit time and
       vortex lifetime are significantly less than the
       average time the material is molten. This time is
       L/V, where L is the length of molten pool and V
       the specimen velocity; for $L \sim 0.2$ cm and $V \sim 4$
       cm.s^{-1}, the time is ~ 50 ms.

The salient feature of the vortex model (19) is the rapid
aggregation of carbides into roughly circular regions,
followed by a significantly longer time for complete
dissolution in the melt. This results in a cellular

243

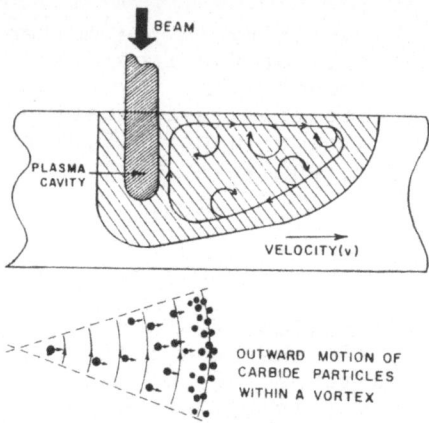

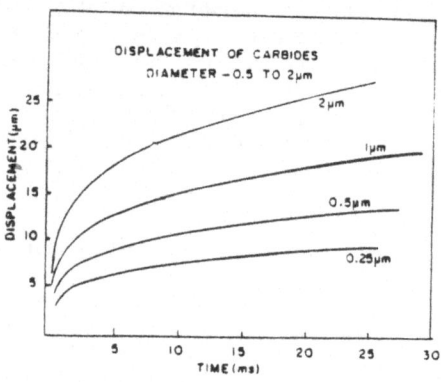

Figure 21--Schematic of the Melt Pool Geometry and the Fluid Flow Driven by Temperature Difference Between the Cavity and Solidification Front.

Figure 22--Plot of the Displacement of Carbide Particles in a Vortex as a Function of Time and the Size of Carbide Particles.

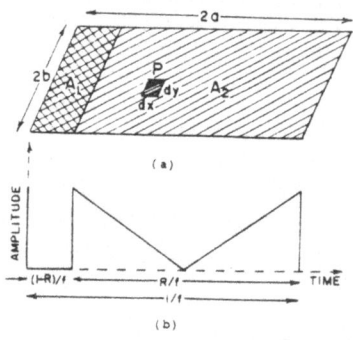

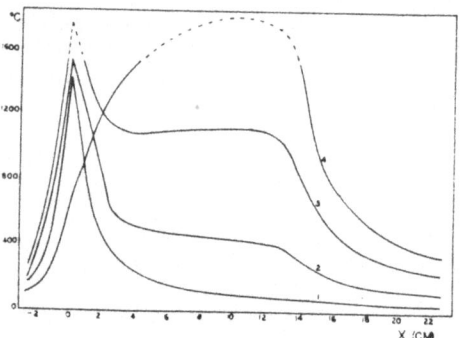

Figure 24--Steady State Thermal Distribution by Computer Controlled Scanning Electron Beam; (a) Distribution of Power Density on Specimen Surface, (b) Computer Generated Pattern Supplied to the Deflection Coils Defining Amplitude and Interaction Time of the Beam.

Figure 25--Temperature Profile in the Direction of Beam Deflection 600 μm Below the Surface. Value of R is 0.0, 0.5, 0.75, and 1.0 for Curves 1, 2, 3, and 4 respectively.

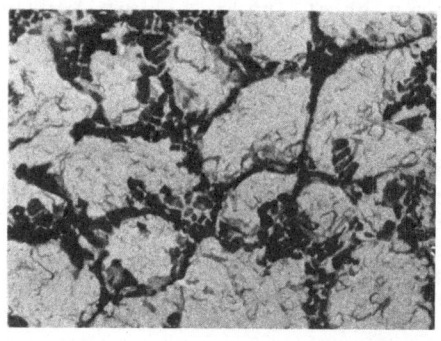

Figure 23--Transmission Electron Micrographs of Carbide Extraction Replicas Showing High Density of Carbides Within the Diffuse Walls of the Solidication Segregation Cells, see Figure 20.

segregation distribution of carbide forming elements. It is within the diffuse walls of these cells that new carbides form, as shown in Figure 23. Interestingly, these carbides, unlike the original ones, are particularly rich in W and Mo and only contain a small amount of Fe and Cr (19).

Solid-State Cooling Effects

The potentialities of laser and electron beam glazing of materials, such as high speed steels, are only actualized after post-glazing thermal treatment. During this process, controlled carbide distributions are obtained by solid-state precipitation from the highly solutionized rapidly solidified material. This, however, when accomplished by conventional furnace heat-treatment, involves fairly long processing times (2×10^3 to 8×10^3 s). More recently, it has been found that solid-state treatment times can be drastically reduced (22) by using an electron beam to momentarily reheat rapidly solidified material. This is because the controlability of an electron beam in surface heating enables brief excursions to be made to a high temperature in a precise and reproducible manner. Consequently, it is possible significantly to increase the kinetics of phase reactions by using high temperatures for short durations. An important and natural extension of the processing is the possibility of performing rapid solidification and post-glaze heat treatment in an integrated process using programmed beam deflection.

The generation of a steady-state thermal distribution $T(x,y)$ by a scanning electron beam on a moving surface is depicted in Figure 24, where v is the substrate velocity. This is formed by programming the beam to trace out a selected pattern (or successively dwell at a selected array of positions) and then to repeat the process at a high repetition rate (preferably in the kilohertz range). Any desired pattern may be obtained by exciting the X and Y coils of the electromagnetic beam deflection yoke in a preselected manner.

As an illustration of the technique, a simple power density distribution is considered; this has a "hot leading-edge" for line-source melting, followed by a low power-density region for solid-state post-glaze heating, see Figure 24(a). The pattern is generated by exciting the X-axis beam deflection coils with an oscillatory current whose waveform is shown in Figure 24(b), and exciting the Y-axis beam deflection coils with a triangular waveform. In Figure 24(b), the time for the complete cycle is 1/f, where f is the oscillatory current frequency. Defining the ratio R of the dwell-time for the triangular pattern to the constant portion (see Figure 24(b)), the respective dwell-times in these regions are Rf and (1-R)f.

At a sufficiently high repetition rate it may be assumed that the beam power for surface melting is $(1-R)\cdot P$ and that for heat treating is RP; P is the actual beam power. The power densities q_1 and q_2 are for melting and post heating are given by equations 16 and 17:

$$q_1 = (1-R)P/wd \tag{16}$$

$$q_2 = RP/wL \tag{17}$$

Consequently

$$q_2 = q_1(d/L)R/(1-R) \tag{18}$$

From equation 18, for a given q_1, any desired value of q_2 may be obtained by selecting the values of R and L. Figure 25 shows computed curves of the temperature profile (along the X-axis) for a ~ 600 μm depth below the surface of a steel specimen; the specimen velocity is 2 mm s^{-1}. Each curve depicted corresponds to a value of R ranging from zero (curve 1) to unity (curve 4); as previously discussed, R is the ratio of the deflection time to the total time in each deflection cycle. An interesting and important feature is the plateau for R values in the range 0.5 to 0.75. For rapid solid-state transformations the height and duration of this thermal arrest will determine the microstructural changes that occur. A secondary feature is the fairly rapid drop in temperature following the momentary thermal arrest; this is quite appreciable for R = 0.75. The temperature level of the thermal arrest is of prime importance in steels since it determines whether transformation occurs in the austenitizing or tempering region.

Detailed microscopic observations on M7 high-speed steel (22) show that a transient thermal arrest during solid-state cooling induces an ultrafine intra-cellular dispersion, Figure 26(a), accompanied by a martensitic structure, Figure 26(d). At high magnification, Figure 26(c), the dark-etching carbides are revealed to be of unusual morphological form. As well as the clearly defined twinned martensitic platelet at A, Figure 26(c), there is a fine-scale martensitic structure. This is particularly evident in the heavily etched SEM micrograph in Figure 26(d). Here, in spite of the heavy etching, the martensitic platelets appear in "bas-relief". The existence of a refined martensitic structure of mean platelet length less than the solidification cell diameter is noteworthy. This shows that the carbide particle precipitation during the thermal arrest in the asustenitizing range provides the necessary distribution of nucleation sites and obstacles to ensure fine-scale martensite formation upon subsequent quenching.

These considerations show how programmed manipulation of the cooling curve in the process cycle can be used to modify rapid solidification microstructures. Clearly, by such an approach it becomes possible to perform precipitation reactions at significantly higher temperatures than are normally possible. An additional advantage is the potential for developing innovative procedures, such as producing a desired carbide distribution prior to forming martensite.

MICROSTRUCTURAL ASPECTS

The microstructural consequences of rapid solidification

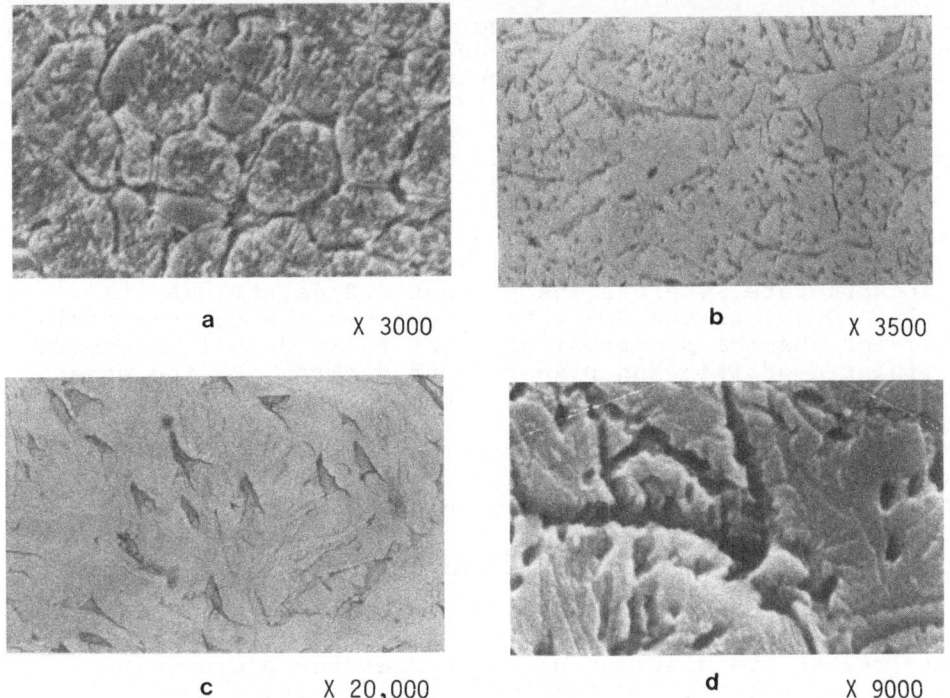

Figure 26--Solid State Transformations by Simultaneous Melting and Post Heating Under Electron Beam; (a) Uniform Phase Distribution Within Cells, (b) Intracellular Carbide Distribution Revealed by TEM of Two-Stage Replica; (c) Higher Magnification of (b) Showing Refinement of Martensitic Plates, (d) SEM Micrograph of Etched Out Carbides and Martensitic Plates.

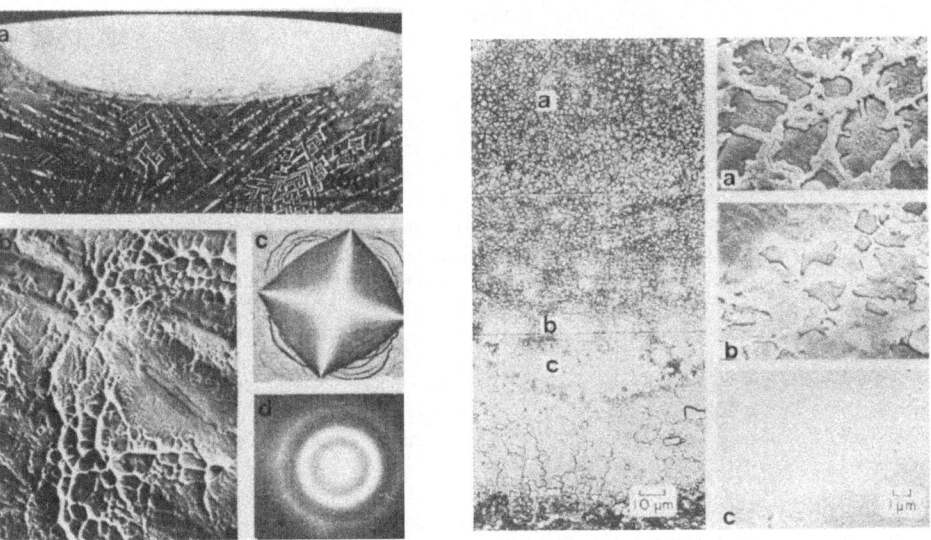

Figure 27--Amorphous Structure in Glazed Pd-4.2 Cu-5.1 Si Alloy; (a) Cross-Section of Glazed Region, (b) Fracture Surface, (d) Hardness Indentation, and (d) Electron Diffraction Pattern.

Figure 30--Microstructural Changes Within the Laser Glazed Region of Alloy M2. (a), (b) and (c) are Higher Magnification SEM Micrographs of the Regions Indicated in the Adjacent Optical Micrograph.

laser and electron beam glazing have been examined in a wide
variety of materials. The range of microstructures observed
include refined dendritic structures, ultrafine eutectics and
dispersed phases, extended solid solution phases, metastable
phases, and amorphous metallic solids. Systematic
modifications of the microstructures within the glazed zone
have also been reported, which are due to variations in the
critical solidification parameters, G,R and T. These
solidification parameters are of particular interest because
the ratio G/R controls the character of the solidification
microstructure, whereas the product G.R determines its
scale. Thus, as the G/R ratio increases the solidification
behavior changes progressively from fully dendritic growth to
cellular dendritic and planar front growth. On the other
hand, increasing the gradient-rate product, or cooling rate,
gives rise to shorter diffusion paths and finer structures.
Under extremely high solidification rates, in some alloy
systems, planar front growth can be achieved more or less
independent of the imposed thermal gradient (23).

Low Melting Point Eutectic Alloys

Under surface melting conditions, the crystalline
substrate necessarily is in intimate contact with the melt.
This raises the question as to whether the presence of
favorable nucleation sites for crystallization would prevent
the undercooling necessary to achieve an amorphous
structure. Test experiments with the low melting point
eutectic alloy (Pd-4.2 Cu-5.1Si) have demonstrated that a thin
layer of amorphous material can be developed on a crystalline
substrate. The melted layer (~0.18 mm maximum in thickness)
was obtained with an incident power density of 6×10^6 W/cm^2
and an interaction time of 3×10^{-5} sec. As shown in Figure
27(a), within the melted region there are no discernible
microstructural features, in contrast to the eutectic
structure in the substrate. One manifestation of the non-
crystalline amorphous nature of the glazed material is the
symmetrical pattern of curved shear bands formed around the
hardness indentations, Figure 27(c). Another manifestation is
the vein-like character of the fracture surface, Figure
27(b). Confirmatory evidence was obtained by TEM; the thin
foils had a speckled appearance and the electron diffraction
pattern showed diffuse rings, Figure 27(d). Similar tests
have been performed on other low melting point eutectic
alloys, and with comparable results. A good example is
provided by the work on the technically interesting alloy
$Fe_{40}Ni_{40}P_{14}B_6$, which exhibits exceptional mechanical and
corrosion resistance properties (24). Because of the much
higher critical cooling rates required (~10^6 °K/sec compared
with 10^2 °K/sec for Pd-Cu-Si) the amorphous layer was very thin
and could be detected only by TEM.

Systematic studies on rapidly solidified Pd-Cu-Si alloys
have shown that the microstructure is controlled essentially
by the interface velocity, or solidification rate (25). For
example, in off-eutectic alloys, a structure composed of
dendrites and interdendritic eutectic at low interface
velocity (~0.25 mm/sec) becomes a fine eutectic-like structure
at velocities of ~1 mm/sec. At even higher velocities
approaching 2.5 mm/sec, an abrupt transition occurs to the

amorphous state. Because of the need for diffusional sorting of the components and the creation of solid/solid interfaces, the kinetics for eutectic solidification are relatively slow. Thus, the transition to the amorphous state at a critical interface velocity can be attributed to the difficulty of maintaining coupled growth, i.e. eutectic solidification, at high interface velocities. In this alloy system, crystallization into a metastable single phase solid with the same composition as the liquid is not possible, because of the special nature of the phase diagram.

Precipitation Hardened Nickel-Base Superalloys

In glazing studies on the nickel base superalloys a relatively deep penetration homogenization pass is usually applied to the surface prior to one or more superimposed rapid solidification laser glazing passes. Without such an homogenizing treatment the glazed layers exhibit incomplete dissolution of MC carbide particles and other refractory constituents in the initial microstructure.

Figure 28 shows the typical response to laser glazing of a superalloy crystal with the <100> orientation parallel to the direction of the incident beam. As indicated, glazing produces epitaxial growth in the resolidified layers, accompanied by a marked refinement in the scale of the microstructure. Within a given re-cast layer, the scale of the dendritic structure remains reasonably constant, although it undergoes obvious discontinuous changes with varying melt depth. This is to be expected, since Figure 15 shows that the cooling rate, which controls the scale of the microstructure, is relatively constant, after an initial transient stage. Close examination of Figure 28 shows a very thin layer of plane-front solidified material in contact with the melt/substrate interface. With increasing distance from this interface, the structure becomes cellular (one-dimensional dendritic), cellular dendritic, and finally fully dendritic. Such changes are clearly in accord with the calculated high initial G/R ratio and its rapid fall-off as solidification proceeds towards the free surface, Figure 16. The second pass in Figure 28 shows that cellular growth extends to the free surface, Figure 29, which is indicative of a persistently high G/R ratio in this high cooling rate regime of solidification.

Figure 28 shows that the region encompassing the second pass, including a heat-affected zone (HAZ) in the underlying first pass, does not etch-up as well as the rest of the material. The reason for this is that the cooling rate is fast enough to prevent precipitation of γ' particles only in the second pass and its associated HAZ. Apparently, in the HAZ, the elimination of the γ' particles is a consequence of much faster kinetics for the solution of the γ' phase than for its precipitation.

Martensitically Strengthened Steels

A representative alloy of this class of materials is M2 steel, which has been studied after both laser and electron beam glazing (26,27). In the conventionally heat treated condition, the microstructure consists of a mixed carbide

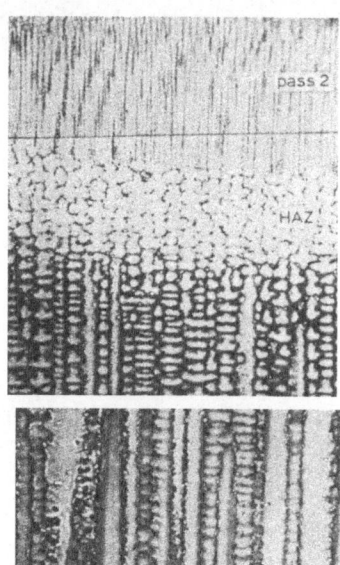

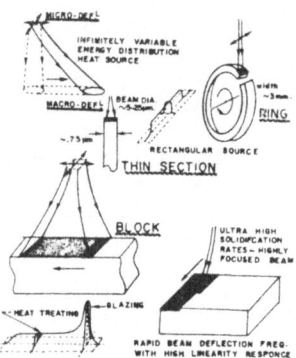

Figure 28--Refined Dendritic Structure in Glazed B1900 Superalloy Crystal (Double Pass); the (100) Surface is Normal to the Direction of the Incident Beam. Note Evidence for Epitaxial Growth Between Substrate and Pass 1, and Between Superimposed Passes 1 and 2.

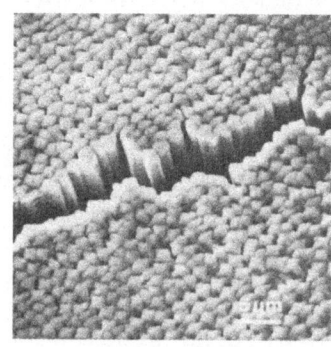

Figure 29--External Surface of Pass 2, Figure 28, Showing Cellular or Cellular-Dendritic Structure, Due to Inadequate Melt Feeding During the Last Stage of Solidification.

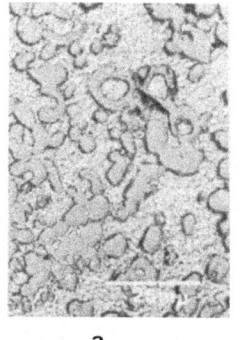

a   b

Figure 31--Microstructure of Ferro-TiC Alloy; (a) As Sintered, (b) After Electron Beam Glazing, Showing Refined Dendritic Carbides.

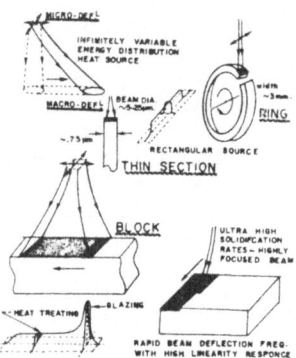

Figure 33--Examples of Electron Beam Glazing Treatments.

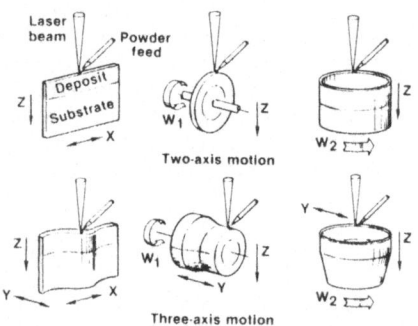

Figure 32--Examples of Layer Glaze Processing for Near-Net Shape.

phase dispersion in a martensitic matrix, with hardness values typically ~800 VHN. After glazing, the microstructure depends upon position within the melt zone, Figure 30, with hardness values ranging from 650-1100 VHN from top to bottom of the zone. Austenitizing this structure by annealing at 1230°C for 300 mins, followed by quenching into liquid nitrogen, gives a rather well-defined substructure, which is decorated with extremely fine carbide particles. The corresponding hardness values are ~1100 VHN, and show little variation throughout the melt zone. Near the top of the melt zone, i.e. in the last material to solidify, a cellular δ-ferrite structure is formed, Figure 30(a). The actual cells are surrounded by boundaries of γ-austenite, together with fine carbides formed by the peritectic reaction δ + liquid → γ + carbide. A region lower in the melt zone, Figure 30(b), shows less δ-ferrite so that the peritectic reaction has proceeded to a further stage, whereas near the bottom of the melt zone, Figure 30(c), there is no evidence for δ-ferrite, so that the peritectic reaction must have gone to completion.

The progressive morphological changes within the melt zone, i.e. from the top surface to the maximum melt depth, reflect variations in solidification parameters during glazing, and also concomitant alloying element redistribution effects. A particularly striking feature is the effect of the high value of G/R at the bottom of the melt zone in determining the growth of dendrites along radial heat flow directions, Figure 30(c). It is more difficult, however, to find a plausible explanation for the progressive decrease in degree of completion of the peritectic reaction from bottom to top of the melt zone. The cooling rate $\dot{T}$ does not seem to be a factor, since this is invariant with position, except in the vicinity of the maximum melt depth. In contrast, the solidification rate R is initially zero and then progressively increases as the liquid-solid interface moves toward the external surface. As the interface accelerates, therefore, progressively less time is available for alloying elements to redistribute themselves by liquid-phase diffusion. This increasing restriction of alloying element redistribution, in turn, limits the extent to which the peritectic reaction can proceed. Thus, only in the lower melt zone, where solidification occurs slowly, does the peritectic go to completion.

Carbide Dispersion Strengthened Alloys

A characteristic feature of many carbide strengthened alloys is that they contain a high volume fraction of carbide phase embedded in an alloy matrix. An example is the pseudobinary commercial alloy, known as Ferro-TiC. Conventionally processed material consists of large titanium carbide particles (~25 vol. pct.) in a tool steel matrix, Figure 31(a). After glazing a highly refined dendritic carbide microstructure is formed, Figure 31(b). An important consequence of this rather dramatic change in carbide particle morphology is that it gave a 3 to 5 times increase in lifetime in metal-to-metal wear tests using fully hardened M42 tool steel as a counterface material (28).

Electron beam glazing has also been used to reprocess hardfacing deposits of Stellite 6 alloy, and TiC and TiB$_2$ particle strengthened Stellite 6 (29). Prior to electron beam reprocessing the hardfacing layers were formed either by laser cladding or by plasma deposition. In applying the glazing technique to such dispersion strengthened materials it is necessary to pre-heat the surface to avoid quench cracking. The pre-heating may be accomplished by sweeping a wide diameter defocussed beam across the surface prior to point source or line source melting. By appropriate selection of the process parameters it is possible to form very smooth surfaces.

SURFACE MODIFICATION TECHNOLOGY AND APPLICATIONS

Laser and electron beam glazing treatments have been employed to modify the surface structures and properties of very thin edges of samples using a single pass of a sharply focussed beam. On the other hand, to obtain continuous surface coverage of glazed material it has been necessary to generate a multiplicity of overlapping passes by scanning the focussed beam over the workpiece surface, or by indexing the workpiece with respect to a fixed beam. A laser beam may be scanned by making use of special coupled arrangements of mirrors, whereas an electron beam may be scanned by electromagnetic means. For laser glazing, a numerically controlled work station with at least two axes of motion is generally preferred, Figure 32, whereas for electron beam glazing programmed electromagnetic beam deflection has proven to be more versatile.

A selection of possible approaches for electron beam glazing are indicated diagrammatically in Figure 33. As shown in the lower right hand corner, a well-focussed beam may be deflected rapidly so that the energy on an incident surface is concentrated into a microscopic area sweeping over an entire surface. The energy distribution within the microscopic area and the scanning pattern are selected by a micro-computer. Figure 33 also shows how a short line source with a rectangular energy distribution may be used for producing a rapidly solidified hardened layer along the edge of a thin strip of material. The thin section material may be in the form of a ring, such as a typical piston ring. The potentiality for combining cycles of rapid solidification and post solidification heat treatment, in steels for instance, is economically important. Using programmed rapid beam deflection the surface of a block may be suitably processed with the energy distribution shown in the lower left hand corner.

Both laser and electron beam glazing treatments have been used to achieve beneficial modifications in the surface properties of materials. In sensitized 304 stainless steel (16), laser glazing has the effect of resolutionizing harmful carbide phases at the grain boundaries and restores the resistance to stress corrosion cracking. In 614 aluminum bronze (30), laser glazing homogenizes the surface, which increases its resistance to corrosion in chloride solutions. In M2 high speed steel (26,27), heat treatment of laser or electron beam glazed surfaces generates a uniformly fine distribution of hard carbide particles in an austentic/martensitic matrix, which improves its cutting performance, e.g. in applications such as saw blades, drill bits, and end mills. In a pseudobinary Fe-TiC alloy (28), electron beam glazing and tempering produce a threefold increase in the wear life in test performed on a fully hardened M42 steel counterface material. Laser glazing has also been applied to eutectic-type alloys that are ready glass formers. Thus, amorphous surface layers have been developed on crystalline substrates in Pd-4.2Cu5.1Si, and in the technically more interesting $Fe_{40}Ni_{40}P_{14}B_6$ alloy, which exhibits exceptional mechanical properties and corrosion resistance (24). The high hardness and corrosion resistance of metallic glasses containing P and Cr, together with their ability to accept and maintain a sharp cutting edge, suggests such as surgeon's scalpels and even long-life razor blades.

Laser glazing in conjunction with surface compositional modification is also an area of obvious high potential. Methods of processing typically involve preplacement of alloying material (powder, electrodeposit, etc.) on the workpiece surface prior to glazing, or particle injection during glazing. Carbide particle injection into alloy substrates has been used to develop wear resistance surfaces (31). Much thicker deposits have also been laid down by the continuous delivery of prealloyed powder to the interaction, or melt zone. Surface alloying by this means is being developed for a wide range of applications, including hardfacing of valve seats, turbine blade tips, bearing surfaces, and gas-paths seals. Experimental work has also been conducted on the fabrication of bulk rapidly solidified structures by incremental solidification processing (10). Simple axisymmetric shapes, such as a demonstration turbine disc, have already been fabricated by this process. Typically the deposited material exhibits a pronounced columnar grained dendritic structure, with grains extending through many successive layers of material. The inherently strong tendency for epitaxial growth between layers ensures good mechanical strength at the interfaces between layers, even when the composition is deliberately changed, e.g. by changing the composition of the powder feed. Applications for this process are currently limited by the requirement that the deposited material possess good weld-cracking resistance, and by the need to improve the shape-defining capabilities of the process. As indicated in Figure 32, the fabrication of more complex shapes requires the use of a numerically controlled work station, which is capable of simultaneous motion about two or three axes.

FUTURE PERSPECTIVE

Although laser and electron beam glazing techniques have been shown to be capable of providing modifications in surface microstructures and properties in a variety of materials, much remains to be done to optimize these processes for specific applications. Future work is likely to be concerned primarily with this problem, now that the major effects of rapid

solidification and subsequent solid state cooling are
reasonably well understood. Amongst the many aspects of the
problem that need to be explored are (i) influence on surface
properties of post glaze heat treatments and/or mechanical
deformation, e.g. shot peening, (ii) effect of substrate pre-
heating on cracking propensity in susceptible materials, and
(iii) the use of reactive gases for more convenient alloying
of glazed layers.

In laser glazing, multikilowatt CO_2 gas lasers have
proved to be versatile tools for processing most metallic
materials. Unexplored opportunities exist for utilizing other
laser systems, both in the pulsed and continuous modes of
operation. Furthermore, even though considerable progress has
been made in surface alloying using preplacement or continuous
feed techniques, the full potential of ambient environment
laser processing has yet to be exploited. For example, the
use of gas phase precursors as a means of surface alloying
needs to be investigated in detail. In electron beam glazing,
improved electromagnetic beam deflection methods need to be
developed for combining pre-heat, glazing, and post-glaze heat
treatment in a single operation. The rapidity of
metallurgical transformations induced by post-glaze heat
treatment should reduce processing time, and thereby
manufacturing cost.

Applications of these surface modification technologies
are limited to relatively simple shapes, with line-of-sight
accessibility to all surfaces to be treated. However, it
appears that there is no limit to the extent of scale-up of
the processing. Thus, laser surface modification of large-
scale structures seems technically feasible.

Experience with the incremental glazing process has also
pointed the way to a potential means for controlling
structures and properties in thick sections, and even bulk
parts, more closely than has previously been possible. The
technique has been termed 'directed energy processing' (32).
Processing starts with material being added by the laser or
electron beam melting and deposition of feedstock. This step
essentially creates the part. Other materials processing
operations can be sequentially accomplished on each
incremental layer. These include inspection, mechanical
deformation and/or heat treatment. Like deposition, which

TABLE I

Values of molten zone length L, cooling rate T, and maximum
melt depth z_{max} for various beam velocities

$V(cm\ s^{-1})$	L (μm)		$\dot{T}(°Cs^{-1}) \times 10^6$		z_{max} (μm)	
	Obs.	Calc.	Emp.	Calc.	Obs.	Calc.
38.0	10.0	9.9	1.6	0.5	108	108
57.5	8.5	6.9	2.7	1.1	67	60
76.0	5.0	5.3	3.7	2.1	23	34
97.5	3.0	4.1	5.0	3.8	18	17

derives advantages from rapid solidification of a thin section, each of these subsequent operations also derives advantages from being applied to a small volume element. Incremental inspection can be performed without destructively testing the part. Mechanical processing can be applied to small increments without resulting in significant distortion, and thermal treatments, or even remelting can be carried out by interrupting the material feed.

References

1. "Rapid Solidification Processing: Principles and Technologies II," Eds. R. Mehrabian, B. H. Kear, and M. Cohen, Claitor's Publishing Division, Baton Rouge, LA, 1980 (Proc. 2nd Int. Conf. at Reston, VA, March 1980).

2. "Rapidly Solidified Amorphous and Crystalline Alloys," Eds. B. H. Kear and B. C. Giessen, Elsevier North Holland, New York, 1982 (Proc. Mats. Res. Soc. Meeting at Boston, MA, November 1981, Symposium F).

3. "Rapidly Quenched Metals IV," Eds. T. Masumoto and K. Suzuki, The Japan Institute of Metals, 1982 (Proc. 4th Int. Conf. at Sendai, Japan, August 1981).

4. "Rapid Solidification of Metals and Alloys," H. Jones, Monograph #8, The Institution of Metallurgists, London (1982).

5. "Laser Materials Processing," Ed. M. Bass, North Holland Publishing Company (1983).

6. "Rapidly Quenched Metals V," Eds. S. Steeb and H. Warlimont, North Holland Publishing Company, Amsterdam, 1984 (Proc. 5th Int. Conf. at Wurzburg, Germany, September 1984).

7. B. H. Kear, E. M. Breinan, and L. E. Greenwald, "Solidification and Casting of Metals," The Metals Society, London, 1979, p. 501 (Proc. Int. Conf. on Solidification, Sheffield, England, July 1977).

8. P. R. Strutt, ONR Report #N00014-78-C-0580 (1982).

9. D. S. Gnanamuthu and E. V. Locke, E. V. U.S. Patent # 4015100 (1976).

10. E. M. Breinan, D. B. Snow, C. O. Brown, and B. H. Kear, as reference 1, p. 440.

11. H. E. Cline and T. R. Anthony, J. Appl. Phys. Vol. 48, p. 895, 3888 (1977).

12. R. Mehrabian, S. Kou, and A. Munitz, Laser-Solid Interactions and Laser Processing, Am. Inst. of Physics, NY, p. 129 (1978).

13. L. E. Greenwald, E. M. Breinan, and B. H. Kear, Laser-Solid Interactions and Laser Processing, Am. Inst. of Physics, NY, p. 189 (1978).

14. P. R. Strutt, Mat. Sci. and Engr., Vol. 44, p. 239, (1980).

15. H. Matyja, B. C. Giessen, and N. J. Grant, J. Inst. Met., Vol. 9, p. 30 (1968).

16. T. R. Anthony and H. E. Cline, J. Appl. Phys. 49, p. 1248 (1978).

17. S. M. Copley, D. Beck, O. Esquivel and M. Bass, Laser-Solid Interactions and Laser Processing, Am. Inst. Phys., NY, p. 161 (1979).

18. P. Moore, C. Kim, and L. W. Weinman, Laser-Solid Interactions and Laser Processing, Am. Inst. Phys., p. 221 (1979).

19. A. Tauqir, P. G. Klemens, and P. R. Strutt, Proc. Materials Research Society Meeting, Boston, (November 1985).

20. M. E. Newell and F. W. Schmidt, J. Heat Transfer, $\underline{92}$, p. 159 (1970).

21. S. Ostrach, Report #1111, 39th Annual Report, National Advisory Committee for Aeronautics (1953).

22. P. R. Strutt, Annual Report, ONR Contract N00014-78-C-0580 (1985).

23. S. R. Coriell and R. F. Sekerka, Proc. Second Int. Conf. on Rapid Solidification Processing, Claitor's Publishing Division, Baton Rouge, LA, p. 35 (1980).

24. R. Becker, G. Sepold, and P. L. Ryder (private communication).

25. W. J. Boettinger, F. S. Biancaniello, G. M. Kalonji, and J. W. Cahn, Proc. Second Int. Conf. on Rapid Solidification Processing, Claitor's Publishing Division, Baton Rouge, LA, p. 50 (1980).

26. P. R. Strutt, M. Tuli, H. Nowotony, and B. H. Kear, Mats. Sci. & Eng. $\underline{36}$, p. 217 (1978).

27. Y. W. Kim, P. R. Strutt, and H. Nowotony, H. Met. Trans. $\underline{10A}$, p. 881 (1979).

28. P. R. Strutt, B. G. Lewis, S. F. Wayne, and B. H. Kear, "Specialty Steels and Hard Materials," Eds. N. R. Comins and J. B. Clark, Pergamon Press, p. 389 (1982).

29. M. Kurup, A. Tauqir, P. R. Strutt, and B. H. Kear, Proc. Mats. Res. Soc. Meeting, Symposium F (1983).

30. C. W. Draper, et al. Corrosion $\underline{36}$, p. 405 (1980).

31. J. D. Ayers, T. R. Tucker, and R. J. Schaefer, in ref. 3, p. 212.

32. E. M. Breinan and B. H. Kear, as reference 5, p. 237.

MATERIALS-BY-DESIGN: PROSPECTS AND PROMISE

James J. Eberhardt

U.S. Department of Energy
Energy Conversion and Utilization Technologies
 (ECUT) Program
Washington, DC 20585

INTRODUCTION

The following is a somewhat different perspective from that of the other papers presented in this symposium on materials design. This is the view of a theoretician with an acute appreciation of experimental data and the new instrumental techniques of materials science.

The concept of "a priori" designing materials and predicting their properties by tailoring the microscopic structure through the use of computational models was described in a recent paper.[1] In addition, the feasibility of such an approach was comprehensively covered by a recent assessment[2] of all the theoretical and experimental tools currently available or becoming available to materials science. These include not only such exotic tools as EXAFS which makes use of the intense radiation of synchrotron light sources,[3] but chemical instrumentation such as FTIR[4] which did not exist fifteen years ago. Such advanced instrumental techniques are now widely available[5] to chemists and materials scientists both at universities and in private industry.

This paper will concentrate on the theoretical aspects of materials design but with the instrumental as well as preparative advances clearly in mind. These latter would be the subjects of separate papers in themselves.

STRUCTURE, BONDING AND FUNCTIONALITY

To the uninitiated, it would seem that chemists and solid state physicists are obsessed with molecular structure, atomic arrangements and the properties these structures confer to the molecule or material of interest. Even so, it is easy for scientists, because they are so close to the problem, to lose sight of the fact that it is combinations of just the 92 naturally occurring elements of the periodic table which result in the tremendous diversity of materials which we see about us. It is the rules of bonding that control these structures and the resulting properties of materials and even of biological systems such as ourselves. Perhaps it is best to step back and view the total picture to get the appropriate perspective; a simple example of structure-property relationships will make this point clearer.

Carbon[6] is perhaps a good case to consider. It exhibits very interesting structure-property relationships. Figure 1 shows a comparison of the molecular or crystalline

GRAPHITE

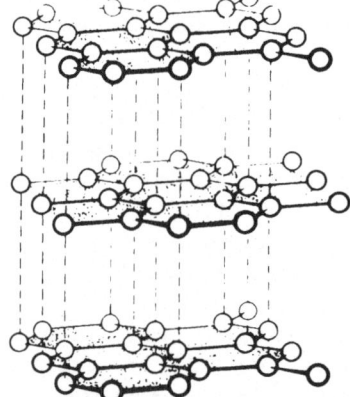

DIAMOND

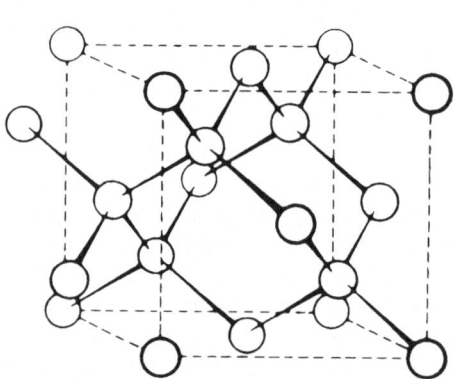

C–C BOND LENGTH (IN PLANE) 0.142 nm

C–C BOND LENGTH (BETWEEN LAYERS) 0.335 nm

HEXAGONAL UNIT CELL: a = 0.2456 nm
                     b = 0.6696 nm

C–C BOND LENGTH 0.154 nm

CUBIC UNIT CELL: a = 0.3567 nm

PHYSICAL PROPERTIES

	GRAPHITE	DIAMOND
HARDNESS	VERY LOW	HIGHEST KNOWN
COLOR	BLACK	COLORLESS

Fig. 1. Comparison of graphite--diamond properties and structure.

structures of graphite and diamond. As indicated, the
bonding results in differences in structure which in turn
result in totally different properties. Graphite is soft,
whereas diamond is the hardest substance known. Graphite
is black, and diamond is colorless and clear. This
dramatically illustrates how bonding and resulting
structure affect material properties. This bonding and
subsequent structural effects on properties are worth
keeping in mind as the subject of materials design is
approached.

DEVIATIONS FROM IDEAL STRUCTURES

In real materials, in addition to the atomic
arrangements, defect structures play a (perhaps dominant)
role.[7,15] Molecular structures are well known to chemists,
and defects do not occur in simple molecules. However,
material scientists deal with solids where defects
influence the properties and performance of the material.
Figure 2 is a transmission electron micrograph of one
type of defect, dislocations in nickel aluminide, Ni_3Al.

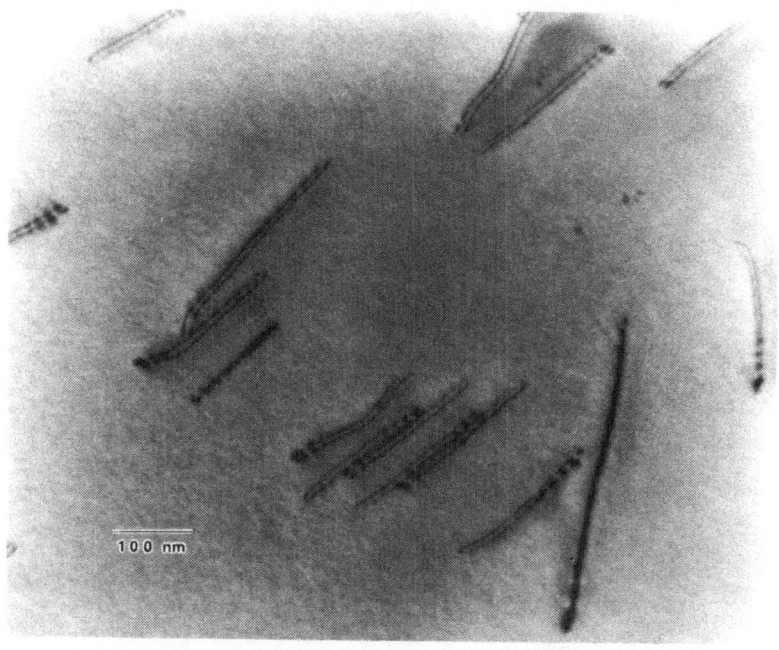

Fig. 2. Transmission electron micrograph of dislocation
pairs in B-doped Ni_3Al.

This material is causing a revolution in metallurgy;[8] why this is so will be covered in detail later.

Figure 3 shows another transmission electron micrograph of Ni$_3$Al with the dislocations piling up at the grain boundaries. The grain boundary is seen as the straight line in the lower right-hand corner. This micrograph illustrates the complexity which defect structures may exhibit. If real materials are to be designed a priori, such complex defect structures may have to be modeled.

MATERIALS BY DESIGN--DEFINITION

What is meant by the term "materials by design"? Many definitions are possible. In the context of the feasibility study[2] carried out by Allied-Signal and Texas A&M, the following tentative definition is put forth:

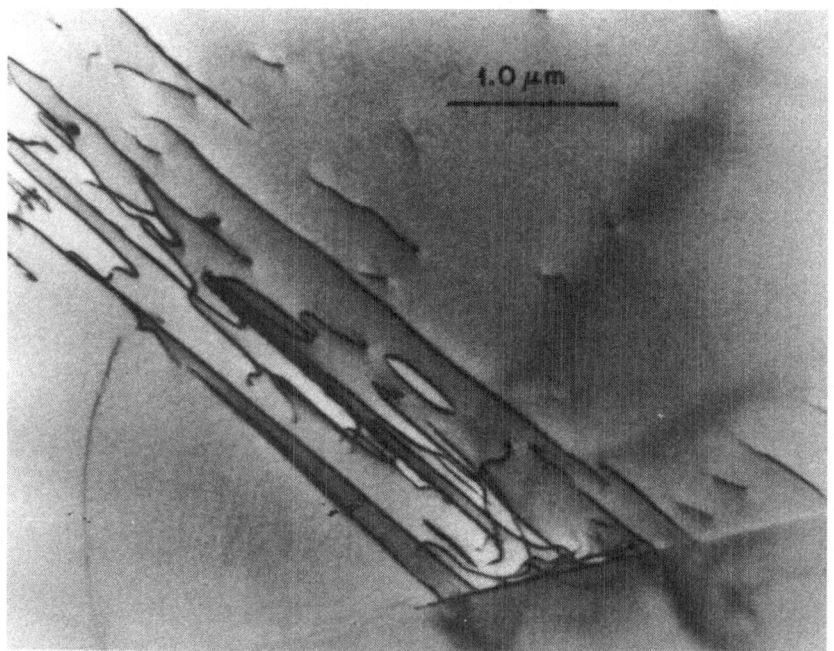

THE GRAIN BOUNDARY IS THE FAINT GREY STRAIGHT LINE
IN LOWER RIGHT CORNER

Fig. 3. Transmission electron micrograph of dislocations piled up against a grain boundary in B-doped Ni$_3$Al.

"The use of scientific principles to predict
materials properties and performance
characteristics."

This definition, as can be seen, is not the same as
the one given by Professor Cocke in a previous paper and
illustrates the different perspectives different people
will have at this early stage of the concept. In
addition, it would be preferable to have some reference
to structural features in the definition; the advantage
of the above definition is that it is completely general
and permits the design of materials by analogy to known
materials. The emphasis of this report, however, is on
the use of the principles of bonding and structure to
design totally new materials "a priori". Accordingly then,
the more restrictive definition of materials by design
will be given as:

"The use of computational structural models to
tailor *a priori* the microscopic structure and
properties of a material to be optimal for a
specific application."

HIERARCHY OF MODELS FOR MATERIALS DESIGN

Figure 4 shows a conceptual hierarchy of models[1] needed
to effect such an "a priori" design of materials. The
models span the spectrum of needed processes from the
systems performance level to the electronic level.

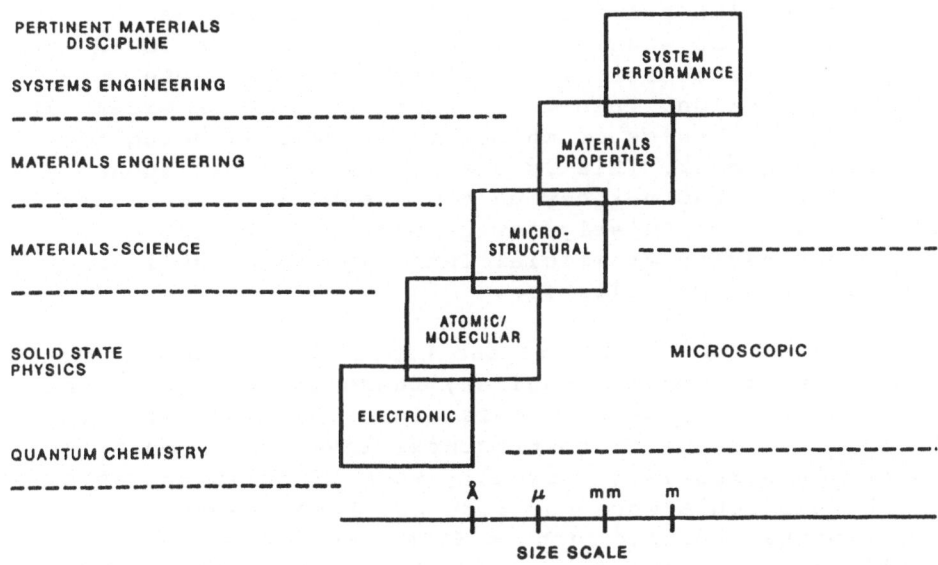

Fig. 4. Materials-By-Design hierarchy.

On the ordinate are displayed the pertinent materials disciplines that are involved at each level of this hierarchy of conceptual and working models. This covers the range from the most basic quantum chemistry and theoretical solid state physics through materials science to the quite pragmatic systems engineering. Each of these disciplines focuses on a given aspect which is a necessary component of materials design. The abscissa indicates the relative order of magnitude of the phenomena being modeled. The models in the lower left-hand corner of the hierarchy focus on atomic and electronic (microscopic) interactions, whereas models at the upper right-hand corner involve macroscopic objects. For a complete materials design process the models must overlap. Such a completely overlapping system of linked models might appropriately be labeled an "expert system" since it would embody the knowledge of many different experts in materials science and engineering. Such an expert system could be useful as a thought structuring, predictive tool to design into materials the properties needed to achieve a specified system performance. It can be seen that the concept of materials by design is based ultimately on electronic and atomic interactions which describe the type of bonding needed to achieve a structure associated with the needed properties.

This process is already done intuitively but not in any systematic fashion and certainly not in an automated manner with a complete set of necessary models. Presently each individual brings his/her own individual knowledge base to the problem of selecting materials for a specific application. The materials generally are not *designed* for the application but rather are *selected* for the application. No single individual's data base could ever contain the necessary information for a complete materials design. This would involve being an expert in theoretical physics or chemistry as well as being a systems engineer. This is clearly beyond most human capabilities but not beyond the capabilities of computers, particularly new generations of super-computers having Artificial Intelligence as well as computational capabilities.

It should be noted that the hierarchy of models can be entered at any point required; there is no requirement that the full range of models always be exercised. It can be entered at the microstructural level where most materials scientists presently work. This is the level at which this conference has focused. Present-day experimental tools,[9] such as STEM and HREM and their application to materials characterization are giving new insights at this level. The scanning tunneling microscope

(STM)[10] is projected to go to the subatomic level and "see" the atomic orbitals and their directions at surfaces. This will bring even more emphasis to the atomic and subatomic level models. This will be reflected in the theoretical world where this new experimental emphasis will spur new or improved concepts from theory. The concepts will then feed back to improve the models.

HISTORIC CONVERGENCE OF THEORY, INSTRUMENTAL TECHNIQUES AND COMPUTATION

About seven years ago it began to become evident to the author that rapid advances were being made in instrumental techniques and that each probe provided complementary information which, when taken as a whole, might lead to data bases sufficient to fully characterize materials according to their bonding and structure (including defect structure). At the same time theoretical methods were rapidly advancing. Quantum chemistry as well as solid state physics was moving rapidly. Finally, the power of computers was rapidly increasing. Thus, heretofore impossible computations were becoming feasible and, perhaps more relevant to materials design, more economical. It is now approaching the point where "computer experiments" may be in some instances more economical than "trial and error" laboratory experimentation. The convergence of these three activities makes the process of materials design a real possibility within our generation.

Three years ago the DOE Energy Conversion and Utilization Technologies Program (DOE-ECUT) advertised a request for proposals to address the status and possibility of "materials by design." Universal Oil Products, then a division of the Signal Companies, won the contract. Texas A&M was commissioned by UOP to hold several workshops to gather material for the assessment. At about the same time the DOE Basic Energy Sciences (BES), Division of Materials Science was also carrying out an assessment[11] focused on the more basic aspects of materials structure/property relationships in metals and ceramics. The ECUT assessment focused on applications of electronic and atomic-level computations to certain technical areas of interest for energy conservation, the more efficient conversion and use of energy. The BES assessment focused mainly on atomic level computations. These two assessments are quite complementary. The rapid advances being made in experiment, theory and computation call for an effort in materials by design. The advances in computational capabilities are particularly relevant.

IMPLICATIONS OF ADVANCES IN COMPUTATIONAL SPEED

There is about to be an incredible quantum leap in computational power. Figure 5 shows the VAX-11/750 with floating point arithmetic as the standard scientific computer of today. This gives a baseline with relative speed and cost per unit of speed of unity.

A comparison can be made with the CRAY-1S, which now is a relatively "old" computer--that is, ten years old. The CRAY-1S is the first of the so-called "supercomputers." Its speed relative to the VAX baseline is about a factor of 100, and its relative cost per unit of speed is about one quarter. However, the next generation CRAY-3 or ETA-10 supercomputer is estimated to have a speed relative to the VAX of about a factor of 10,000 with a relative cost per unit of speed of 1/100 (and may be as low as 1/1000 with a less conservative acquisition and maintenance cost estimate). Clearly, there is tremendous computational power just on the horizon. However, conversion of the applications software to take advantage of the architectural (speed producing) features of these machines is lagging badly and may be the next major barrier to materials by design. Many of the problems of interest for designing materials by simulating features at the electronic and atomic/molecular levels are quite amenable to both vector processing and parallel processing. This reflects the nature of the problem and of the techniques which have been developed to construct solutions to the partial differential equations describing the problem.

TECHNOLOGY AREAS OF INTEREST

The ECUT assessment looked at three areas of applications interest: heterogeneous catalysis, tribology

SYSTEM	RELATIVE SPEED	RELATIVE COST PER UNIT OF SPEED
VAX 11/750+FPA	1	1
CRAY-1S (1976)	~100	0.25
CRAY-3 (1986?)	~10,000 (est.)	0.01 (est.)

Fig. 5. Relative speeds and computational costs of various computer systems.

and materials interfaces. All three have interfacial aspects as the dominant theme. This choice was dictated by the enormous recent advances in experiment and theory applied to interfaces. In addition, surface and interfacial phenomena may require fewer atoms to model as compared to bulk or solution phenomena.

Upon examination of these three technological areas, some commonalities among them become evident. Virtually every process in these three technology areas involves one or more of the following fundamental phenomena:

1. defect stability or movement;

2. physisorption or chemisorption;

3. surface reactivity;

4. interfacial bonding;

5. intermolecular bonding;

6. electron transfer at the interface.

ELECTRONIC/ATOMIC LEVEL COMPUTATIONAL MODELS

The fundamental phenomena listed above can, in principle, all be modeled at the two lowest levels of the hierarchy of models discussed previously.

These phenomena all involve either electronic motion/transfer or atomic/vacancy motion. The disciplines of quantum chemistry and solid state theoretical physics that have been focusing on these aspects have been doing so in isolation from practical problems for many years. They have, however, developed many theoretical computational techniques[1,2] which may be capable of simulating these phenomena in real materials. The quantum chemists have been calculating electronic motions (orbitals) and electronic probability distributions for atoms and molecules. The solid state physicists, on the other hand, have been modeling similar quantities (band structure) in solid materials. Each group has developed its own techniques which for some years were divergent from each other. Recently,[12] the two groups have begun to work toward a convergence of the methods.

Imagine a small cluster of atoms; a molecule, if you will. Now add more atoms and the solid state is approached. The point at which this occurs is currently the subject of much research. However, physical intuition tells us that the distinction between a small cluster of atoms (a molecule) and a large cluster (bulk material) and the concepts of covalent, ionic, van der Waals and metallic bonding in such cluster systems are all artificial to an electron. This is because the forces among all the constituent particles (protons and electrons) are coulombic in nature and, therefore, the energy operator (Hamiltonian) of the wave equation, which constitutes the foundation for the entire hierarchy of models, should apply to any assembly of such particles no matter what kind or how many they are. Thus, the subject of modeling interfaces either as local structure, i.e., active sites, or as long range structure, i.e., the extended solid state, is bringing the fields of solid state physics and quantum chemistry closer together. The increasing use of supercomputers should help this convergence process since the computations can be more exact with fewer shortcuts and assumptions. Hopefully, a common theoretical language will develop between the two groups.

Both the quantum chemical and solid state physics approaches tend to merge at the next higher level in the hierarchy which attempts to simulate atomic motions. The two techniques used by both groups are classified as either molecular dynamics or Monte Carlo simulations. For either technique the same input is required, namely good interatomic potentials. These potentials describe the many-body forces among the atoms bound together by the electronic motions.

APPLICATION OF ELECTRONIC/ATOMIC LEVEL MODELS TO MATERIALS DESIGN

What can the models at the two lowest levels in the hierarchy really do in terms of materials design? These models can provide information or insights complementary to experiment. For example, consider a grain boundary. It is very difficult to characterize a grain boundary experimentally without changing it in the process. These models have the power to predict composition and the nature of the bonding in materials both qualitatively and quantitatively. Thus models can aid in the production of information needed for materials design without having to perform quite difficult, costly or even impossible experiments. Hazardous materials would be another area where modeling would be advantageous.

ECONOMIC BENEFITS TO DESIGNING MATERIALS

In the development of any research program, attempts must be made to quantify the potential benefits. This is particularly difficult for a program such as materials by design where success thus far has been limited but where the potential for future success is very large. In lieu of a better approach, case studies can be carried out by examining the benefits of designing a given material. In this case the benefits of designing an ideal catalyst were chosen. One of the main reasons for choosing a catalytic case was that the chemical industry is fairly well aggregated. It is almost one standard industrial classification code, and it can be easily identified. The availability of economic data also is relatively good in the catalyst case.

Suppose an ideal catalyst could be designed. What would the characteristics of such a catalyst be? An ideal fluid catalytic cracking (fcc) catalyst was chosen for a case study[13] with the UOP RCC process being chosen as a baseline. The characteristics of an ideal catalyst were identified by interviewing companies in the fcc business. The characteristics of an ideal fcc catalyst are:

1. Production of one product--that is highly selective to gasoline--no light gases given off.

2. Just enough coke would be produced to fuel the endothermic cracking reactions.

3. The catalyst would be resilient--would have a long life and resist deactivation.

4. It would also resist contamination by V and Ni, common light gas producers.

Assumed potential benefits were identified for this ideal catalyst. These are:

1. Reduced energy consumption--fcc feed provides energy from the coke by-product.

2. Reduced capital and operating costs--a regenerator would not be needed and ideal selectivity would eliminate separation fractionators.

3. Feedstock flexibility would be improved.

This flexibility in feedstocks would allow processing those crudes that contain more V and Ni.

The bottom line to the study was quite surprising. The energy savings were on the order of 15% of the feedstock and process energy costs. But the main saving was in the annualized capital costs which was on the order of 50%. The total production cost savings were substantial (35%). Several other case studies were performed with similar results. The capital savings were large (15-90%) because secondary processes such as regeneration and separation were eliminated. One only need extrapolate these capital and energy savings to the $10 billion or more capital expenditures made annually by the Chemical Process Industry to see the tremendous economic potential for materials by design. The main point to be gleaned from this analysis is that there is tremendous potential from the materials by design approach for saving energy and for making U.S. industry more competitive in the international marketplace.

A REAL WORLD EXAMPLE OF MATERIALS BY DESIGN

In the materials study conducted by the DOE Basic Energy Sciences program, one of the recommendations was to use modeling to attack an applied problem so as to provide an economic justification for the rather large and long term (10-20 years) government funding of such a program. ECUT, in developing its materials by design program, is carrying out this recommendation. A real problem in materials design which ECUT is attempting to carry out is to design a totally new ductile ordered intermetallic alloy analogous to the nickel aluminide mentioned previously. The background to this problem is as follows.

Ni_3Al is a binary intermetallic alloy exhibiting long range atomic ordering. It would be a very good structural metal for high temperature applications because of its high melting point (in excess of 1400°C) if it were ductile. Unfortunately, it is very brittle in the normal polycrystalline form as are most of the long range ordered alloys and intermetallics. So, solving the ductility problem with Ni_3Al could have wide spread ramifications for metallurgy because of the very large number (more than 3,700 by direct count of the phase diagrams) of other possible binary intermetallic alloys.

Conventional wisdom indicated that these intermetallic ordered alloys, lying somewhere between ceramics and superalloys in high temperature capability, might have interesting properties; but, they were essentially worthless as structural materials because of their brittleness. Many research groups over the last 30 years had tried to make the materials ductile but had failed until about three years ago when C. T. Liu and co-workers at Oak Ridge National Laboratory[8] succeeded in producing an ordered intermetallic alloy having 50% ductility. How was this done, and was it a step toward materials design? The breakthrough came, interestingly, through an interaction of Oak Ridge researchers with Japanese researchers.

It was known that single crystals of Ni_3Al are very ductile; they are cubic in structure and have the right crystalline slip systems. However, polycrystalline nickel aluminide is very brittle. This implied that something was wrong at the grain boundaries. It was known from the Japanese work[14] that boron was a grain boundary strengthener for Ni_3Al. Liu used boron to make polycrystalline Ni_3Al hold together and, therefore, ductile. It worked beautifully. Figure 6 shows a picture of the polycrystalline Ni_3Al without, and with boron, in comparison with stainless steel. This so called "cap" test is a very severe test of ductility; it can be seen that the stainless steel cracks during this test but the boron-modified nickel aluminide does not. One additional feature of nickel aluminide, besides its high melting point, is the fact that the yield strength of the material actually increases as a function of temperature. Figure 7 shows a graph of yield strength of Ni_3Al as a function of temperature along with some conventional materials for comparison.

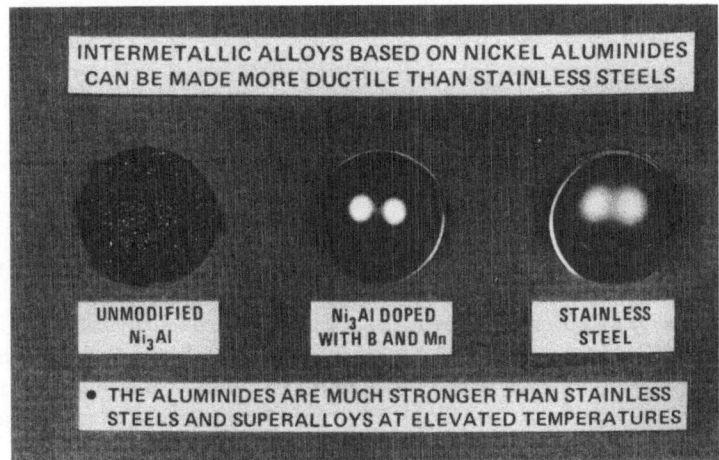

Fig. 6. Effect of boron in improving the ductility of Ni_3Al. Cap test comparison to stainless steel.

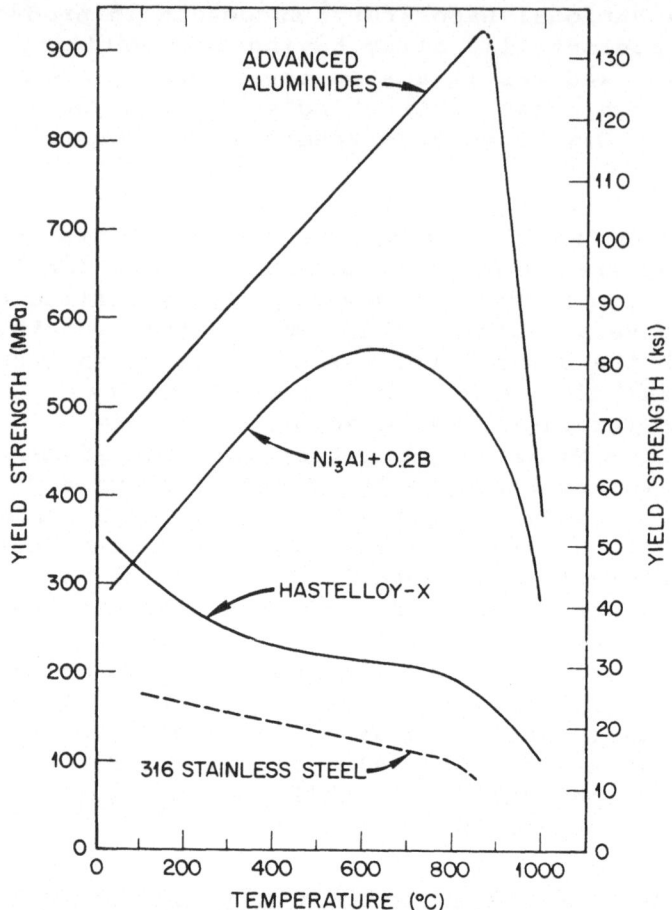

Fig. 7. Variation of yield strength with
temperature for Ni₃Al alloys and
selected high-temperature alloys.

    The material, exhibiting the best high temperature
strength, contains Ni₃Al with 200 ppm of B and 1.5% Hf.
As seen, the yield strength peaks out at 900 MPa
(134,000 psi) at 950°C. This is indeed an interesting
material. It may find application in the new suborbital
aircraft (the "orient express") in components such as the
outer skin. Reentry temperatures are expected to be on
the order of 1000°C, just about the point where this
material is strongest. Oxidation studies have been
performed at 1000°C for several hours and no deleterious
oxidation is observed. Weldability studies have been
performed with good welds being obtained. Creep strength
is also very good. Many companies have expressed an
interest in this material for various high temperature
applications.

The mechanism by which boron functions is unknown but it is hypothesized that it somehow "strengthens" the grain boundaries. It is known from Auger electron spectroscopy that boron preferentially segregates to the grain boundary. In addition to Auger electron spectroscopy, other types of materials characterization tools are being brought to bear on this problem. These include the use of the atom probe, positron annihilation spectroscopy and ion beam channeling. Unfortunately, as previously stated, experimental techniques which study a free grain boundary are very onerous. This is where computational simulations have a decided advantage. Grain boundaries with and without boron are being simulated. The computations to date show, in agreement with experiment, that the boron energetically prefers to be at the grain boundaries. The materials by design concept now enters the picture by searching for materials analogous to Ni_3Al. That is, A_3B type intermetallics with grain boundary embrittlement problems will be simulated and microalloyed computationally before actually attempting to make them in the laboratory. This approach, if it succeeds, will truly be an example of materials by design.

SUMMARY

In summary, the "a priori" design of materials at the atomic level is now a reasonable possibility due to the historic convergence in the development and availability of theoretical, instrumental and computational tools. Several attempts are currently underway to apply these tools to the design of metallic materials. The DOE Energy Conversion and Utilization Technologies Program which has created and pioneered the concept of materials by design is prepared to make the long term programmatic commitment of federal government resources needed to prove the concept. Such a long term commitment is the apotheosis of an appropriate federal role on the research and development of new materials for increasing efficiency of conversion and use of energy and enhancing U.S. international competitiveness.

ACKNOWLEDGMENTS

The author wishes to acknowledge many useful discussions on this subject with Dr. Joseph A. Carpenter, Jr., of the Oak Ridge National Laboratory and Dr. P. Jeffrey Hay of the Los Alamos National Laboratory and Ms. Anne R. Ehrenshaft of Oak Ridge National Laboratory for preparation of the manuscript.

271

REFERENCES

1. Eberhardt, J. J.; Hay, P. J. and Carpenter, Jr., J. A., "Materials by Design--A Hierarchial Approach to the Design of New Materials," in *Proceedings of a Symposium on the Computer Based Microscopic Simulation of Materials Properties and Structures*, Fall 1985 Meeting of the Materials Research Society, Boston, MA, December (1985).

2. Assessment of Theoretical and Experimental Tools for Applied Research and Exploratory Development in Certain Energy Technologies, Energy Conversion and Utilization Technologies--Allied/Signal Engineered Materials Research Center (1986).

3. Parr, A. C., "Synchrotron Radiation: Applications in Chemistry" in *New Directions in Chemical Analysis* (Texas A&M University Press, College Station, 1985) 161-198.

4. Marshall, A. G., "Fourier Transform Methods in Spectroscopy" in *New Directions in Chemical Analysis* (Texas A&M University Press, College Station, 1985) 111-134.

5. Shapiro, B. L., Ed., *New Directions in Chemical Analysis* (Texas A&M Press, College Station, 1985).

6. Harrison, W. A., *Electronic Structure and the Properties of Solids* (W. H. Freeman and Co., San Francisco, 1980)

7. Catlow, C. R. A. and Cormack, A. N., *Chemistry in Britain* (1982) 627.

8. Liu, C. T. and Stiegler, J. O., "Ductile Ordered Intermetallic Alloys," Science 226 (1984) 636.

9. Spence, J. C. H., *Experimental High Resolution Microscopy* (Clarendon Press, Oxford, 1981).

10. Bennig, G. and Rohrer, H., Helv. Phys. Acta 55 (1982) 726.

11. "Theory and Computer Simulation of Materials and Imperfections," Division of Materials Sciences, US Department of Energy, Washington, DC 20545 (Aug. 1984).

12. "Sanibel Symposium on the Interface Between Electronic Structure and Dynamics," April 21-25, 1986, Snowbird, UT.

13. "Catalysis by Design: Maximum Theoretical Energy Efficiency Limits and Maximum Economic Benefits from Ideal Catalysts," J. K. Young, et al., Battelle Pacific Northwest Laboratories (1986).

14. Nippon Kinzoku Takkaishi 43, 1190 (1979), K. Aoke and O. Izumi.

15. Hirth, J. P. and Lothe, J., *Theory of Dislocations*, Second Edition, John Wiley & Sons, New York, 1982, pp. 5-9.

CRYSTALLOGRAPHIC ENGINEERING

R. E. Newnham

Materials Research Laboratory
The Pennsylvania State University
University Park, Pennsylvania, 16802

INTRODUCTION

The age of crystallographic engineering — the design and fabrication of carefully patterned crystals to carry out complex functions — is upon us. In this paper we examine the present scope of the subject and predict some future trends.

The history of crystallography can be divided into three eras (Fig. 1). Classical crystallography — the study of the morphology and physical properties of crystals — culminated in the compendiums by Groth (1) and Voigt (2). Shortly after the discovery of x-ray diffraction, many crystallographers took up the study of the atomic structure of crystals, beginning with the simple arrangements found in salts and metals, and moving on to the large

```
┌─────────────────────────────────┐
│    Classical Crystallography    │
│          19th Century           │
│       Crystal Morphology        │
│         Crystal Physics         │
└─────────────────────────────────┘

┌─────────────────────────────────┐
│        Structure Analysis       │
│          20th Century           │
│   Mineral and Metal Structure   │
│       Biological Molecules      │
└─────────────────────────────────┘

┌─────────────────────────────────┐
│  Crystallographic Engineering   │
│          21st Century           │
│       Solid State Systems       │
│     Biomolecular Engineering    │
└─────────────────────────────────┘
```

Fig. 1.   Three eras in the history
of crystallography.

complex structures of vitamins and proteins. The 1985 Nobel Prize in chemistry honors two crystallographers who developed the mathematical techniques required to solve crystal structures (3). Nobel laureates Jerome Karle and Herbert Hauptman are the latest in a long line of distinguished crystallographers.

But, although the Age of Structure Analysis remains in full swing, many engineers and scientists are now engaged in the study of miniaturized solid state systems functioning inside crystals. Crystallographic engineering builds on all we have learned of structure-property relations in the past, and attempts to optimize the performance.

This progression from classification (Classical Crystallography) to scientific study (Structure Analysis) to application (Crystallographic Engineering) is a natural progression observed in many sciences. The study of electricity, for example progressed from the amateur observations of Benjamin Franklin, to the theories of James Clerk Maxwell, to the sophisticated circuit design of contemporary electrical engineers. The trend toward engineering applications is the mark of a mature and confident science.

Crystallographic Engineering

A list of present-day solid state systems is presented in Table 1. Systems such as these are engineered from and into crystals of silicon, lithium niobate, yttrium iron garnet, gallium arsenide, niobium, alumina, and lead zirconate titanate. Some of the systems are in an advanced stage of development, while others are relatively primitive. A survey of recent developments in crystallographic engineering is presented in this review, together with some predictions regarding future developments.

ACOUSTIC ARRAYS

We begin with a relatively simple sensor illustrating the connection with structure-property relationships. A hydrophone is a transducer designed to detect weak pressure waves in water. Most hydrophones utilize the piezoelectric effect in single crystals or poled ceramics such as $Pb(Zr,Ti)O_3 (=PZT)$. A sensitive transducer requires a large electrical response to a small pressure change. This implies large hydrostatic charge coefficients ($d_h = 2d_{31} + d_{33}$) and large voltage coefficients ($g_h = d_h/K\varepsilon_o$). A starting point for the materials engineer is the hydrophone figure of merit $d_h g_h$.

The problem with most piezoelectrics is that $2d_{31}$ and d_{33} partially cancel and the dielectric constant K is too big. But this is not true in all piezoelectrics. Longitudinal coefficient d_{33} is far larger than the transverse coefficient d_{31} in chain structures like ferroelectric SbSI (Fig. 2). Under pressure, the chains bear all the stress parallel to the chain, but relatively little of the stress components perpendicular to the chain. Hence SbSI has a much larger sensitivity to hydrostatic stresses than do the pseudocubic perovskites.

And now the materials engineering begins. We build the chain structure idea into a composite transducer. Stiff PZT fibers are embedded in compliant polymers to form a macroscopic version of the SbSI structure. An additional advantage of the PZT-polymer composite (Fig. 3) is its low dielectric constant, leading to improvements in the figure of merit as large as two orders of magnitude (4).

Before going on to other examples, let us review what we are trying to do. We are attempting to maximize the figure of merit of the device by

Table 1. Contemporary Solid State Systems, Materials and Applications

System	Material	Application
Semiconductor Microelectronics	Si	Electronic Circuitry
Integrated Optics	$LiNbO_3$	Optical Communication
Bubble Domains	$Gd_3Fe_5O_{12}$	Magnetic Memory
Microwave Semiconductor	GaAs	Signal Processing
Josephson Devices	Nb	Superconductor Memories
Multilayer Ceramics	Al_2O_3	IC Packaging
Ultrasonic Transducers	$Pb(Zr,Ti)O_3$	Biomedical Scanners

engineering the optimum connectivity pattern. For diphasic systems there are ten different connectivity patterns descriptive of series and parallel connections in three dimensions (4). The PZT-polymer composite shown in Fig. 3 has a 1-3 connectivity pattern with one-dimensional PZT rods embedded in a three-dimensional polymer matrix. Internal mechanical stress patterns and electric flux patterns are controlled by the connectivity geometry.

PZT rods can be poled continuously to any legnth using conducting brush electrodes (5). Using pre-poled rods, a number of multiply-poled acoustic arrays have been constructed from ceramics and polymers. MUPPET is an acronym for multiply-poled piezoelectric transducer. Fig. 4 shows a 3-3 MUPPET polarization detector. By positioning electrodes on all three faces, the transducer is capable of sensing the longitudinal and shear components of an acoustic wave. Note the resemblance of the composite to the perovskite structure, but without the troublesome cross-connections at the junctions. Each set of fibers is electrically independent.

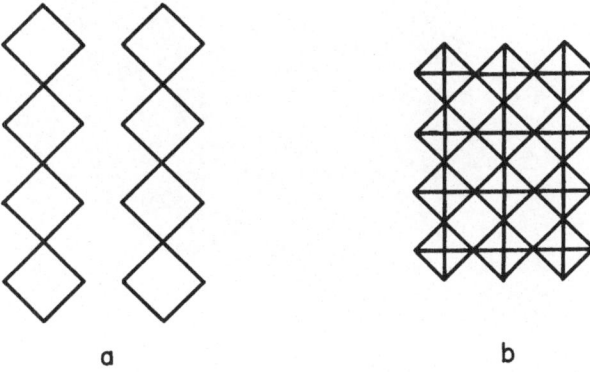

a   b

Fig. 2. Structure-property relationship in piezoelectric crystals: (a) chain structure of SbSI leads to a large hydrostatic d_h coefficient, while the network structure of perovskite (b) promotes isotropic response and a small d_h.

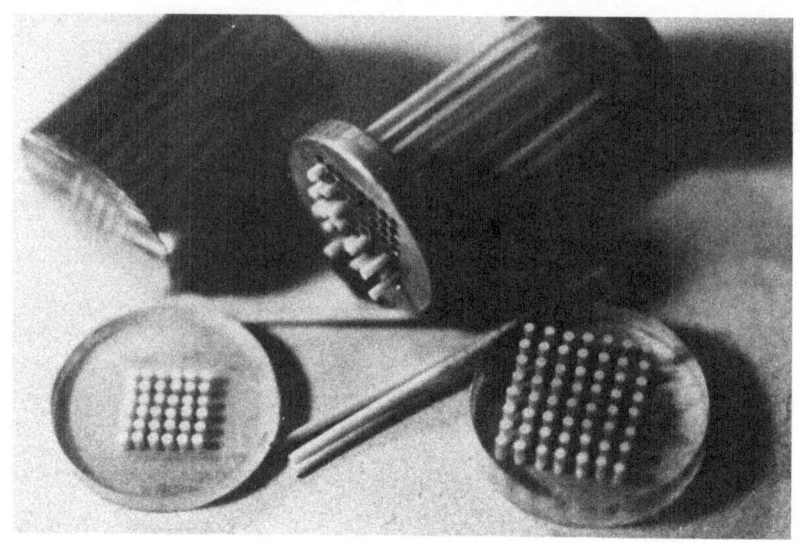

Fig. 3. Composite hydrophone transducer made from piezoelectric PZT fibers embedded in epoxy. The 1-3
structure resembles the structure of SbSI.

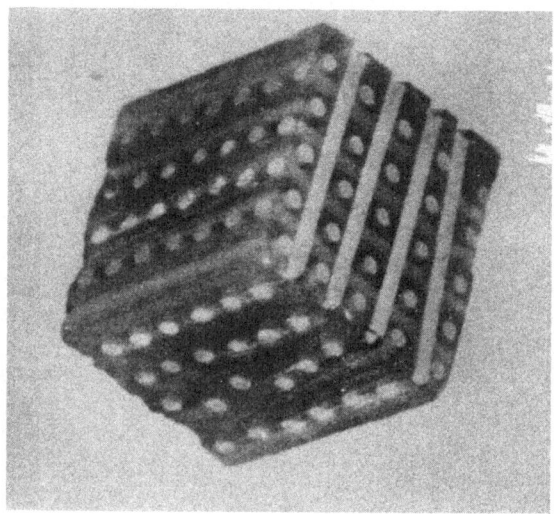

Fig. 4. A 3-3 MUPPET polarization analyzer
made from prepoled PZT fibers embedded in an epoxy matrix (5).

The crude hybrid acoustic arrays just discussed hardly fit the definition of crystallographic engineering with its carefully patterned circuitry inside single crystals. Domain engineering is one way of incorporating ultrasonic arrays into a crystal (6). By installing twin patterns in piezoelectric crystals, one can emulate many of the effects produced in prepoled composite systems. Surface acoustic wave devices utilizing finger-like ferroelectric domains have been constructed in lithium niobate (7).

Tailored domain patterns can be installed in a number of ways. Laser beams can generate twins in ferroelastic and ferrobielastic solids through localized heating, while non-uniform electric (or magnetic) fields control domain configurations in ferroelectric (or ferromagnetic) crystals. Fig. 5 shows a 50 μm grid of domains in ferroelectric $Pb_5Ge_3O_{11}$ constructed using a fine-mesh Cu screen as one electrode. Optical activity makes the domains visible; left-handed squares are separated by right-handed stripes. Lead germanate is an ambidextrous crystal (8) in which spontaneous polarization reversal is accompanied by a change in handedness.

A simple stripe pattern of growth twins has been developed for optical phase-matching (9). With this composite crystal of lithium niobate, second harmonic light generated by a laser beam can be amplified to intensities thousands of times greater than that of a single crystal (Fig. 6). As light enters the first twin segment, second harmonic light is generated in $LiNbO_3$ because of its acentricity. However, the harmonic is soon diminished in intensity because of dispersion. The fundamental and harmonic beams travel with different velocities causing a phase problem through destructive interference. Twinning, however, reverses the sign of the polar axis and also the sign of the nonlinear optic coefficient. Phase matching is achieved by spacing the twin walls about 10 μm apart, corresponding to the coherence length.

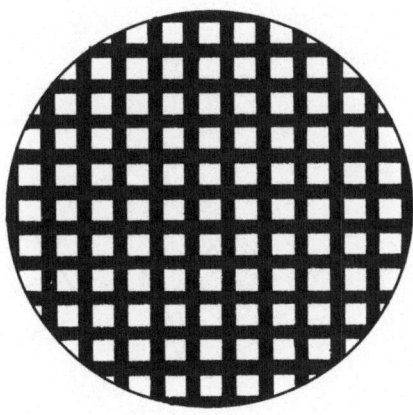

Fig. 5. Tailored domain structure in optically active $Pb_5Ge_3O_{11}$.

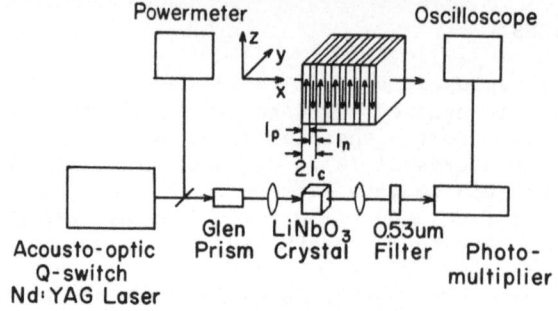

Fig. 6. Growth twins in $LiNbO_3$ used in phase-
matching second harmonic light to the
fundamental.

Bubble Domain Systems

In more complex domain systems, such as bubble domain memories, the
walls are not fixed in position, but move about the process information.
Bubble domains are 180° magnetic domains in ferrimagnet crystals such as
the rare earth iron garnets, orthoferrites with the perovskite structure,
and barium ferrites with the magnetoplumbite structure (10). Because of
magnetic anisotropy, the magnetization is constrained to point perpendicular
to the surface of the crystal. With no external field, up and down 180°
domains are equally abundant, but with bias field one set collapses into
bubbles which can be manipulated about using magnetic circuitry (Fig. 7).
Permalloy electrode patterns imprinted on the surface of the magnetic
crystals allow each location to be addressed magnetically. Up domains
representing "1" in binary code are shifted from location to location.
The absence of such a domain is "0".

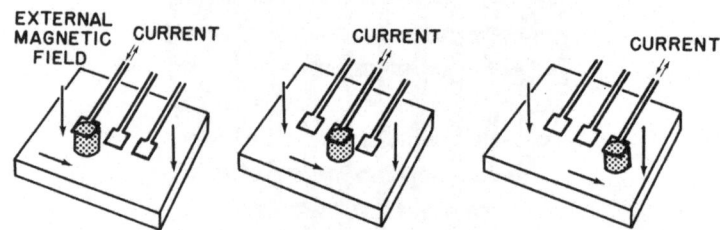

Fig. 7. Magnetic bubbles are 180° cylindrical-shaped
domains which are moved about by the magnetic
fields generated by permalloy circuitry.

State of the art in bubble domain technology is a 100,000 bit memory measuring a few millimeters on edge. Bubble domain memories feature serial access with speeds comparable to semiconductor charge coupled devices, but slower than random access memories. The non-volatility of magnetic memories is an advantage (11).

The examples just discussed illustrate the first law of crystallographic engineering: <u>Circuitry gets smaller and more complex with time and even-tually disappears inside a crystal</u>. Various stages in this process can be recognized: discrete components, hybrid circuits, composite arrays, partial integration, and full scale integration.

INTEGRATED OPTICS

Integrated optical systems are at a very interesting stage at present; hybrid systems are giving way to integrated systems in which a single material carries out many different functions. The loss levels in fiber optics have been reduced to the point where long range communication has become a reality. And now other components must be integrated into the system: prisms, wave guides, switches, modulators, deflectors, external couplers, and fiber couplers. Fig. 8 illustrates some of the components and their function. At present $LiNbO_3$ is the most widely used host crystal, but compound semiconductors, perhaps GaAs or InP, will probably win out in the long run (12).

In a hybrid system, the components in Fig. 8 might be millimeter-size prisms and lenses of ZnS embedded in polystyrene, but the integrated systems of the future will have etched diffraction gratings and laser light sources built into single crystals on a micron scale. The semiconductor lasers will have several different wavelengths by adjusting the band gap through doping. Detectors operate on a similar principle: photodetectors with graded compositions are tuned to sense the wavelengths of interest.

Optical switches, modulators and deflectors often make use of magneto-optic, electro-optic, and acousto-optic effects. The latter two phenomena are present in GaAs and other III-V and II-VI compounds with the acentric zincblende and wurtzite structure. Electro-optic coefficients are smaller than those of ferroelectric crystals like $LiNbO_3$ but the possibility of full integration with light sources, detectors, and associated semiconductor circuitry compensates for this disadvantage.

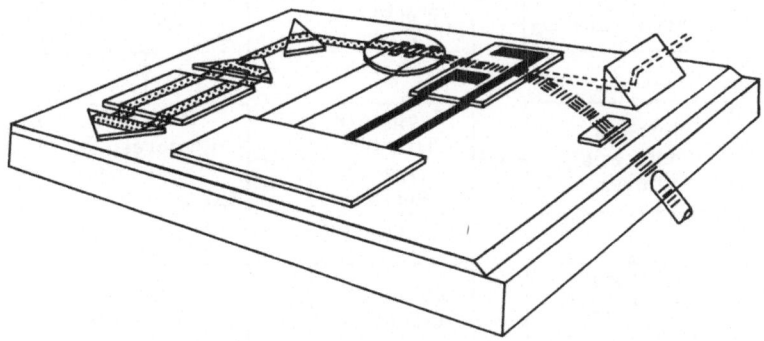

Fig. 8. Optical components in integrated optic systems manipulate light beams for information processing and communication (12).

As illustrated in Fig. 9, integrated optical circuits are made by photolithographic methods similar to those used in the semiconductor industry. Photoresist layers and Ti evaporation followed by diffusion are steps used in incorporating waveguides into $LiNbO_3$ crystals. Titanium dopants modify the refractive indices of the host crystal. Electrodes are screened on later to provide a switching matrix.

SEMICONDUCTOR MICROELECTRONICS

Semiconductor microelectronics is the giant of the Solid State systems field. In this brief account, only a few recent developments will be described.

Monolithic microwave integrated circuits (MMIC) are currently under intense development, mainly for military systems, but direct satellite TV for home receivers seems certain to have an immense consumer market. Components are laid down on a GaAs substrate: transistors, resistors, capacitors, and miniature transmission lines (13). One of the more interesting components is a thin film spiral inductor with air-bridge crossovers to reduce parisitic effects. Representative of the current stage of integration is the direct coupled GaAs monolithic FET used for RF signal generation in the 2GHz range.

Taking an overall look at semiconductor microelectronics, we see that the industry is growing at a very fast rate—much faster than world population. Not only is the number of microelectronic circuits growing, but the complexity of each circuit is growing too. Moore's rule states that in state-of-the-art circuitry, the number of components per circuit doubles every year (14). It has done so for nearly the past two decades, but there are some interesting physical limitations in sight which may slow down the process (Fig. 10).

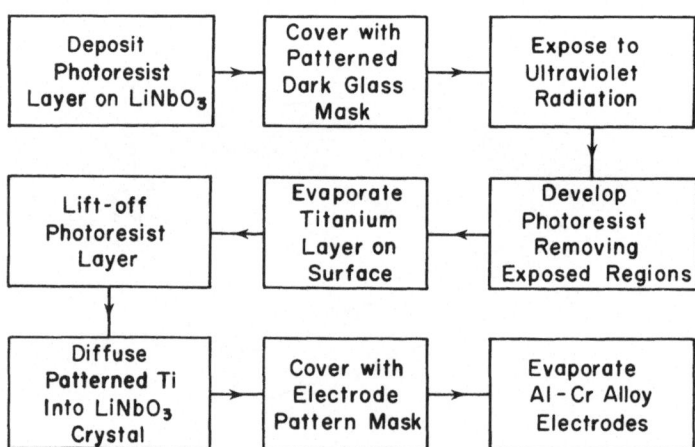

Fig. 9. Processing steps in building in electro-optic reversal switches in $LiNbO_3$ crystals (12).

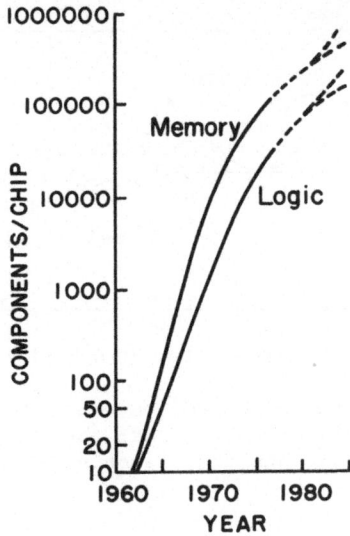

Fig. 10. Growth of component density
during the past two decades.

Another important factor is the price per component. During the past decade the costs have decreased by more than an order of magnitude. In 1973 the price/storage bit in 1K memories was about half a cent, but by 1983 it had dropped to 0.02 cents in a 64K memory. There are two factors involved here: the usual benefit from mass production, but also the value of large scale integration. This is one of the major driving forces for large scale integration.

It should be remembered that the cost of fabricating an IC chip is only a small fraction of its testing and mounting cost. Therefore the more functions that can be done on one chip, the lower the cost. In a typical application, the fractional costs per chip might be wafer and circuit fabrication 5%, testing 30%, packaging and package testing 20%, circuit board costs 30%, panel, cabinet, wiring, and power supply costs 15%. Recent progress in ceramic package integration will be discussed later.

Time is another important consideration. Fig. 11 illustrates two important points: (i) speed costs money, and (ii) two solid state systems sometimes compete head-to-head for the same market. Random access memories — both bipolar (conventional p-n junction devices) and MOS (metal oxide field effect devices) have captured the high speed computer market, while the less expensive magnetic memories have the low-speed high-volume market. The charge coupled device (CCD) is a serial access memory similar in principle to the bubble domain systems. Instead of magnetization packets, it passes along charge packets from electrode to electrode. CCD memories are comparable in size, cost, and speed to the magnetic bubble devices. Their main advantage appears to be compatability with silicon processing techniques, and judging from the number of CCD manufacturers, bubble domains appear to be on the way out.

This raises an interesting question: Will one solid state system eventually gobble up all the others? It certainly appears that silicon technology has vanquished most of the present competition, but there are

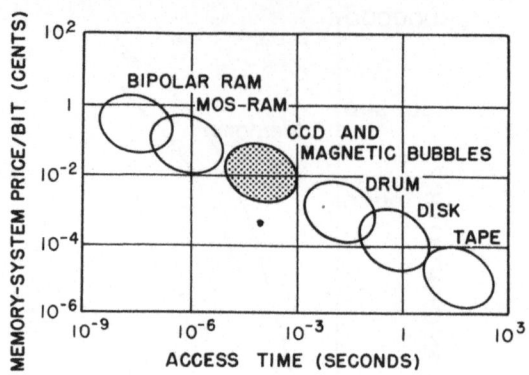

Fig. 11. High speed semiconductor systems are
more expensive than the magnetic tape,
drum, and disk memories.

reasons why this may not happen in the long run. As the size of semiconductor circuit elements drops below 1 μm, a number of new effects and
possible limitations appear. These will be dealt with later. A second
factor is the emergence of new systems such as those based on superconductivity.

Superconductor Microelectronics

Compared to silicon devices, superconductor systems based on the
Josephson Junction are an order of magnitude faster and operate at power
levels down by several orders of magnitude (15). The basic element is the
Josephson Junction formed by a thin insulating layer of 20-50 Å sandwiched
between two metallic superconductors. Because of the extreme thinness of
the insulating layer, electronic tunneling can take place between the
metals. Two types of electrons are involved: normal and superconducting.
The tunneling process can therefore be in two states (normal or superconducting) and a small external magnetic can switch the tunneling state.
This is the physical basis of the Josephson Junction.

When used as a memory cell, two Josephson Junctions are connected
back-to-back in a device called a SQUID, or superconducting quantum
interference device. Superconductor systems have two obvious drawbacks.
First, they must be operated at very low temperatures, and secondly, the
materials must be well mated to withstand repeated thermal cycling.
Neither of these objections appear insurmountable for large scale computers,
but the cryogenic requirement does eliminate many consumer markets.

INTEGRATED CERAMIC PACKAGING

The dramatic increase in circuit density of very large scale integrated
chips presents a number of challenges to packaging engineers. Switching
times in IC chips are approximately one nanosecond, but 10 nanoseconds may
be required to transmit the signal through the package to a neighboring
chip. It is therefore important to shorten the interconnections and incorporate as many components as possible within the package.

Multichip modules (MCM) made from multilayer ceramics (MLC) are used to achieve high interconnection density. The MCM contain about thirty alumina layers with fine-line (0.1 mm) molybdenum conductors, and are made by tape-casting and screen printing. Great care must be exercised in the design of these packages to optimize pulse propagation (16).

New monolithic multicomponent ceramic substrates with integrated capacitors and resistors are now fabricated by a similar process (17). Glass-bonded Al_2O_3 is the basic dielectric, with additional tape-cast layers of $Pb(Fe,Nb,W)O_3$ as capacitors, and screen-printed Ag-Pd interconnects and RuO_2 resistors. All four components are co-fired in air 900°C to give three-dimensional all-ceramic circuitry, with IC chips later mounted on the outer surface. An exploded view of a monolithic multicomponent package is shown in Fig. 12. Other circuit elements (thermistors, varistors, transducers, chemical sensors) are likely to be incorporated into multilayer packages in the future. Sol-gel and thin film processing will lead to further miniaturization.

ULTRAFINE PATTERNS

Many of the solid state systems require special techniques for ultrafine pattern generation, and most involve some rather imaginative crystallographic engineering. Optical lithography can be used down to 1 μm resolution, but further miniaturization will necessarily take advantage of the greater resolving power of x-rays and electron beams.

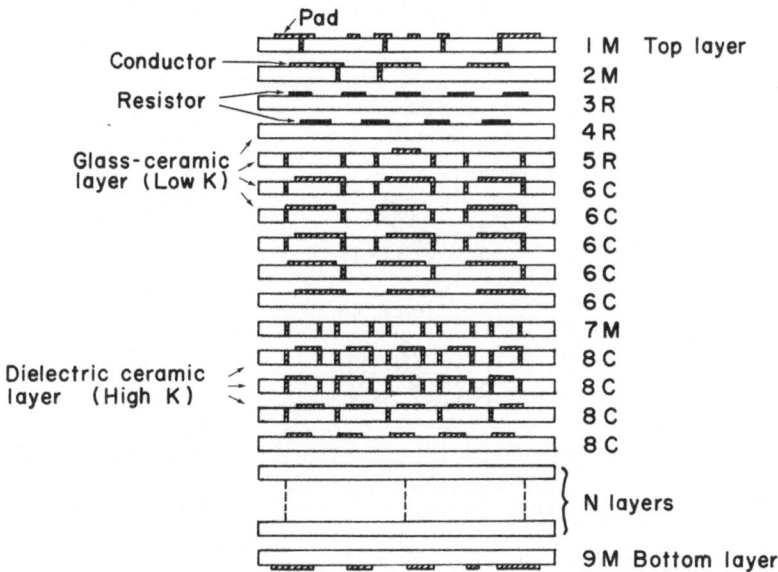

Fig. 12. Expanded vertical section of a monolithic multicomponent ceramic substrate incorporating a three dimensional array of capacitors, resistors, and metal interconnects in a dielectric package.

X-ray lithography has recently broken the 1 μm barrier to about 0.2 μm, still far above the diffraction limit, but scattering is a big problem (18). X-ray photoresists are complex polymer-monomer mixtures which polymerize further under irradiation. Gold masks are engraved with an electron beam, followed by plasma etching.

Another technique for "personalizing" integrated circuits and for repairing mistakes makes use of lasers with ultrafine focal spots. Holes 1 μm in size can be drilled in IC chips, or in spot-welding two metal layers (19).

Another class of ultrafine line methods falls under the heading of "geometric tricks" (20). They have some appeal for crystallographers who also love geometrical tricks. An example is shown in Fig. 13. Very fine metal wires are laid down on glass by a four-step technique: masking, ion-etch, metal evaporation, and ion-etch, making use of shadowing.

Communication Channels

The importance of dimensionality in connectivity patterns was pointed out earlier for piezoelectric composites. Connectivity is no less important in transport phenomena, especially to VLSI designers, to whom the term dimensionality takes on a somewhat different meaning.

Hypercubes are computers made of many processor chips with dedicated communication channels to neighboring processors. The patterns shown in Fig. 14 illustrate 2-D, 3-D, and 4-D machines with two-, three-, and four-neighbors per mode. Custom-designed 6-D hypercubes are now available (21).

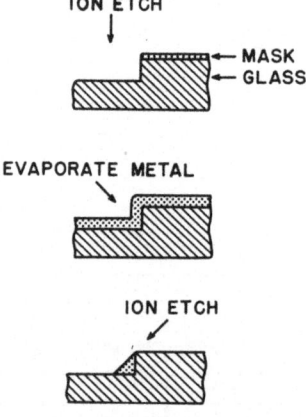

Fig. 13. Step-edge fabrication of thin wires,
(a) a masking layer is laid down on glass, followed by vertical ion etching, (b) metal is evaporated onto the etch surface at an angle, and (c) ion etching is carried out at a reverse angle, thereby removing all the metal except the shadowed portion.

An amusing story (22) told about the eminent scientist J.B.S. Haldane stresses the importance of dimensionality. J.B.S. was a guest at a scientific dinner and was called upon to speak. He said that he had enjoyed his dinner and that he had particularly enjoyed the company of the two charming ladies who had sat on either side of him at the table. J.B.S. then explained that heaven as he imagined it would be N- dimensional, for then it would be possible for him to sit simultaneously side-by-side with N-1 charming ladies.

The same principles apply to the architecture of microcircuits in assembling microprocessors or memories.

FUNDAMENTAL LIMITATIONS AND OPPORTUNITIES

On how small a scale can crystals be engineered, and what are some of the limitations on ultrasmall devices? Thin gold wires about 300 Å in diameter have been formed by ion etching, and multilayer structures with 100 Å layers can be fabricated by molecular beam epitaxy. These layered ultrathin coherent structures (LUCS) have some interesting properties and applications including modified band structure, modified phonon spectra, excitonic superconductivity, new magnetic structures, neutron mirrors, and x-ray monochromators. Some of the most spectacular work has been done on GaAs-GaAlAs layers producing highly modified energy band structure and two-dimensional electron channeling (23).

Regarding the fundamental physics of engineered crystals, theoretical analyses of silicon devices indicate that the changeover from the classical to the quantum regime occurs at about 0.2 μm, the so-called medium small device (24). For smaller sizes many changes occur: breakdown of Ohm's law, hot electron emission, highly anisotropic diffusion, tunneling phenomena, interface and surface effects. The statistical averages underlying kinetic theory no longer apply.

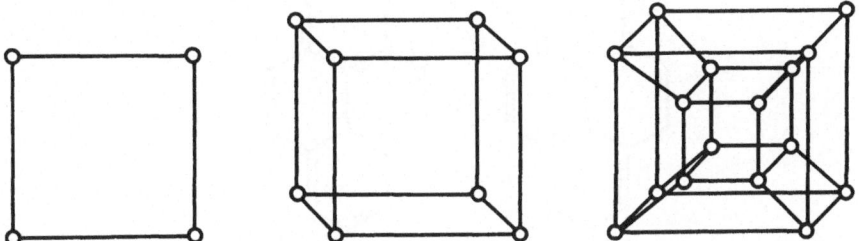

Fig. 14. Hypercube computer connections illustrating 2-, 3-, and 4- dimensional spaces (21).

Component sizes of 100 Å are smaller than many organic molecules, and a lot smaller than DNA and other biological molecules. If a switch could be built in cube measuring 100 Å on edge, a three-dimensional array would contain 10^{18} switches/cm^3, about a billion times greater than a modern solid state computer. Dr. Forrest Carter of the Naval Research Laboratory has organized two workshops on molecular electronic devices to explore future developments in the field (25).

Molecular computers already exist as human brains, and for many investigators, the message is clear: "If nature can do it, why can't we?". Molecular memories might involve bistable molecules (Fig. 15) in which protons store binary code in a double potential well. It has also been proposed that pulses might be transmitted along chainlike molecules such as polyacetylene, which consists of a chain of carbon atoms connected by alternating single and double bonds. An electronic disturbance in this chain would travel along as a soliton. Progress in molecular engineering will draw on a number of ideas from molecular biology. The aim is to make molecular computers with picosecond response times, no wiring costs, and miniscule heat leads.

Future Developments

Looking ahead, crystallographic engineers will be developing complex three-dimensional systems with ever diminishing component sizes. Higher speed and cheaper prices will continue to spur integration, not only in semiconductor chips, but in the associated sensors, packages, and actuators.

In the far future, molecular systems with very high bit densities seem likely to appear. The interactions between human beings and these bimolecular systems will be very complex. It even seems conceivable, as Arthur C. Clarke has suggested (26), that information processing systems may be incorporated into our bodies — or we into theirs!

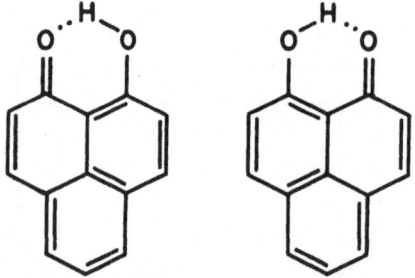

Fig. 15. Bistable molecule of 1, 9 - Hydroxyphenalenone for use in molecular memories.

Personal Philosophy

I would like to close with a few personal thoughts regarding the age of crystallographic engineering. The prospects of future developments both fascinate and frighten me. I am troubled by the fact that much of the research is being undertaken for military systems—weapons that may one day kill us all. I also have a vague fear of the unknown power of these solid state systems—a fear personified by the computer Hal in 2001.

But I am by nature an optimist, and the thrill of scientific and technological achievement overrides these fears. My father and grandfathers were carpenters and builders. I too love to build things with my hands—wooden ship and airplane models, miniature armies of lead soldiers, crystal structure models, morphological models, moving models of phase transitions, composite structures, and polyphasic ceramics. I love to see little things carefully designed, beautifully made, and serving a useful purpose. The craftsmanship delights me, and as Richard Feynman said, "There is Plenty of Room at the Bottom" to do all kinds of exciting crystallographic engineering. It makes me glad to be alive—here and now—to be taking part in this great adventure as we learn what we can from the structures nature has shown us, and then do our best to improve upon the designs.

REFERENCES

(1) Groth, P. "Chemische Krystallographie" W. Engelmann, Leipzig, 1919.
(2) Voight, W. "Lehrbuch der Kristallphysik" B.G. Teubner: Leipzig, 1928.
(3) McLachlan, D.; Glusker, J. P. "Crystallography in North America" Amer. Cryst. Assoc. Publishers, New York, 1983.
(4) Newnham, R. E.; Bowen, L. J.; Klicker, K. A.; Cross, L. E. Mat. in Eng. (1980) II, 93.
(5) Gururaja, T. R.; Christopher, D.; Newnham, R. E.; Schulze, W. A. Ferroelectrics (1983) 47, 193.
(6) Newnham, R. E.; Miller, C. S.; Cross, L. E.; Cline, T. W. Phys. Stat. Solidi (1975) 32, 69.
(7) Farnow, S. A.; Auld, B. A. Appl. Physics Lett. (1974) 25, 681.
(8) Newnham, R. E.; Cross, L. E. Endeavour (1974) XXXIII, 18.
(9) Feng, D. Appl. Phys. Lett. (1980) 37, 607.
(10) Nielsen, J. W. I.E.E.E. Trans. (1976) MAG-12, 327.
(11) Kryder, M. H.; Bortz, A. B. Physics Today (1984) Dec. p. 20.
(12) Tien, P. K.; Giordmaine, J. A. Bell Lab. Rec. (1980) Dec., 371, (1981) Jan., 8, Feb., 38.
(13) Pucel, R. A. I.E.E.E. Trans. (1981) MTT-29, 513.
(14) Noyce, R. N. "Microelectronics" W. H. Freeman & Co. Publishers: San Francisco, 1977.
(15) Gheewala, T. Proc. I.E.E.E. (1982) 70, 26.
(16) Venkatachalam, P. N. I.E.E.E. Trans. (1984) CHMT-6, 480.
(17) Utsumi, K.; Shimada, Y.; Ikeda, T.; Takamizawa, H.; Nagasako, S.; Fujii, S.; Nanamatsu, S. N.E.C. Res. & Dev. (1985) 77, 1.
(18) Lepselter, M. P. I.E.E.E. Spectrum (1981) May, 26.
(19) Logue, J. C. I.B.M. J. Res. & Dev. (1981) 25, 107.
(20) Watts, R. K.; Bruning, J. H. Solid State Tech. (1981) May, 99.
(21) Wallich, P; Zorpette, G. I.E.E.E. Spectrum (1986) Jan., 36.
(22) A. L. Mackay, Private Communication.
(23) Hess, K.; Holonyak, N. Phys. Today (1980) Oct., 40.
(24) Barker, J. R.; Ferry, D. K. Solid State Elec. (1980) 23, 519, 531, 545.
(25) Peterson, I. Science News (1983) 123, 378.
(26) Clarke, A. C. "Profiles of the Future"; Harper & Row Publishers: New York, 1963.

STRUCTURE - PROPERTY RELATIONSHIPS IN METALLIC AND OXIDE GLASSES

Philip H. Gaskell

University of Cambridge
Cavendish Laboratory
Madingly Road
Cambridge CB3 OHE U.K.

ABSTRACT

The link between current knowledge of the structure of glasses and amorphous solids and their physical properties is tenuous and insubstantial.
Where a property is primarily dependent on the immediate coordination
polyhedron centred on a given atom — the "local structure" — then a semi-
quantitative understanding is possible. This is the case for the optical
properties of oxide glasses at energies above about 1eV. If, however, a
physical property reflects longer-range correlations — often termed
"medium-range structure" — then any understanding may be more primitive.
A surprising example of the latter circumstance is represented by density-
composition relationships in metallic and oxide glasses which are understood only in outline. Density-composition relationships are discussed
for the specific cases of transition metal-metalloid alloys and alkali
silicate glasses, together with attempts to model the latter. In
addition, experimental measurements of the optical constants of silica and
simple binary silicate glasses are reviewed. Attempts to interpret these
data are also discussed: from simple "effective oscillator" models to band
structure calculations.

Four years ago, we celebrated the fiftieth anniversary of Zachariasen's speculative proposals (1) for the structure of network glasses.
Zachariasen's paper began with the disarming sentence - "It must be
recognised that we know almost nothing about the structure of glasses".
Fifty four years on we no longer claim such comprehensive ignorance of the
detailed structure of glasses but, nevertheless, there are three areas
which we still find to be very mystifying.

1. In a majority of cases we either know, or have methods to discover,
the <u>local</u> structure of glasses. By this I mean the geometry and symmetry
of the immediate environment of particular atoms in glasses. For example,
in the case of SiO_2, we know that silicon is tetrahedrally-coordinated
and, to a very good approximation, can be considered to be 'equivalent' to
the SiO_4 group in crystalline polymorphs of silica.

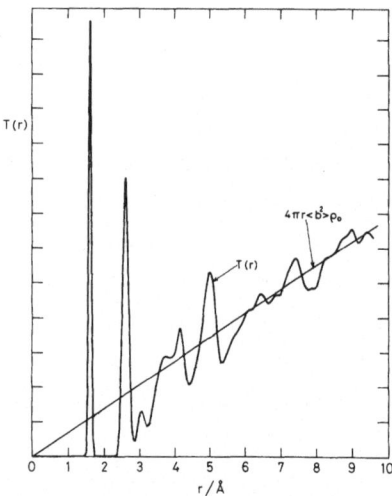

Fig. 1   Reduced radial distribution function, T(r), (barn $\AA^{-2}$) corres-
          ponding to a model for vitreous silica (2) which produces a neutron
          interference function in good agreement with experimental measure-
          ments (3). This diagram gives a realistic impression of the narrow-
          ness of near-neighbour correlations in the glass since the peaks are
          not broadened by the effects of Fourier transformation over truncated
          reciprocal space data.

Figure 1 makes this point: the first two peaks correspond to the Si-O
and O-O distance distributions which are sharp and relate to the almost
undistorted tetrahedra (the Si bond angle distribution has a standard
deviation of about $0.6°$ as indicated by ESR studies (4). Similarly the
oxygen environment is also known — in this case the bond angle is very
broad, but can be estimated from NMR data (5,6) or modelling studies (7).

In the case of the metallic glasses, the most accurate information is obtained from neutron scattering studies, for example on Ni-B or Ni-P alloys with isotope substitution (8,9) which indicate a well-defined 9-atom coordination shell of Ni around the metalloid. NMR data provide supplementary evidence of the symmetry of this group (10).

Notwithstanding the very real progress which has been made in understanding the local structure of glasses, there remains the considerable problem of discovering whether any medium-range order exists and, if so, what it looks like. Thus for the examples mentioned above, while the majority view would certainly be that a continuous random network model is the most appropriate description — interconnected SiO_4 tetrahedral groups joined at apical oxygen atoms with a random distribution of ring sizes - there seems to be no definite proof that this is so (7). Nor is the distribution of ring sizes known with any certainty (7). Similarly, although it seems likely that the structure of a-Ni_4B alloys is not just a random packing of the two elements and while there is some evidence for extended groups of interconnected 9-atom trigonal prismatic units, no-one is convinced that the question has thereby been adequately answered.

In summary, although on the whole, local structure presents few real problems, answers, theories (whatever the current theory) to the questions concerning medium-range structure have the appeal, subtlety and bedrock foundation of "Four legs, good; two legs, bad" in "Animal Farm" (12).
2. All this being so, it is not surprising that the ground rules for glass formation are themselves the subject of considerable uncertainty and speculation — certainly at the moment. Clearly the central question concerns the avoidance of crystallisation during quenching from the melt (13). Answers may be based on the kinetics of the crystallisation process (14,15) with the influence of structure entering implicitly through the viscosity, through relations between the glass temperature and liquidus temperature, or via the compositional equivalence (or not) of the glass and the corresponding crystalline phase and so on. Alternatively, or complementarily, the barrier to crystal nucleation may be fundamentally and explicitly related to a geometrical or topological inequivalence between the structure of the melt and that of the crystal. Thus a local packing arrangements of atoms — tetrahedral packing for example — that is inconsistent with space-filling of ordinary three dimensional, Euclidean space, can be regarded as a preferred structure for the liquid or glass. Transformation to the crystal then requires extensive reorganisation. Such 'local packing' or 'curved space' models have become seductively popular in recent years — thanks to developments by Sadoc and co-workers

(16), and latterly by Nelson et al (17), although the subject has a long history -F. C. Frank (18) is often considered as the parent of these ideas but he suggests that 'countless generations of schoolboys' may have been there before him. These ideas have received an important fillip with the discovery of icosahedral structures in Al-Mn and similar alloys.

Moreover, the idea that glasses may be essentially microcrystalline is not completely dead as might have been imagined some years ago. Phillips (19) has provided (almost single-handed) a spirited revival of these concepts (which in fact predated random network models for oxide glasses).

A summary of this section is difficult but, generally, it seems that transformation of the liquid to the crystal may be either kinetically or structurally frustrated. Any resulting glass may, or may not, contain vestiges of crystallographic (or non-crystallographic) order, depending to some extent on the particular composition.

3. The third major area of uncertainty, which follows logically from the previous two, concerns the relationships which can be forged between the structure of a glass and its macroscopic physical and chemical properties and this will be the subject of this paper. Clearly any attempt to gain structural knowledge requires a fit between simulated and observed micro-scopic properties — such as, the X-ray or neutron structure factors, total or partial pair correlation functions, electronic and vibrational dens-ities of states etc. But the correlations between structure and macro-scopic physical and chemical properties are very largely below the horizon — even now. To take some examples of simple properties — density, optical constants, thermal expansion coefficient, specific heat, viscosity — in no case has an adequate structural theory been advanced which relates micro-scopic and macroscopic quantities. Empirical, composition-based rules exist linking, for instance, the refractive index to composition or wave-length in terms of arbitary, adjustable parameters; 'explanations' in profusion exist for the anomalous thermal properties of silica at low and high temperatures, for viscosity/temperature relations and the glass-liquid transition, yet no current theory quantitatively relates the known structural facts to such properties.

In what follows I do not hope to change the situation — merely report a number of attempts — my own and others — to remedy a totally inadequate situation. It seems sensible to try to walk before running and this paper will concentrate therefore on two simple properties: firstly, the compos-ition-dependence of the density of metallic and oxide glasses and, sec-ondly, optical constants of simple silicate glasses.

At the outset it must be clear that no attempt will be made to try to

understand such properties by an ab initio simulation. In principle, it
may be possible to envisage construction of suitable atomic models — by
molecular dynamics — say, which adequately fit experimental structural
data, followed by a predictive calculation of optical properties involving
atomic wavefunctions (20). In practice this problem is still some way
beyond the horizon representing the limits of cost benefit, computing and
real time and, I suggest, scientific good sense. The alternative is to
present an account which ultimately relies on experimentally derivable
quantities, comparison with crystalline materials of known structure, but
which includes obvious physical and chemical insight — so that, even if
these are wrapped up in empirical constants, the meaning is clear.

COMPOSITION-DEPENDENCE OF THE DENSITY OF METALLIC AND OXIDE GLASSES —
PROSPECTS FOR STRUCTURAL MODELS

Perhaps, because density is such a simple quantity, commonly studied
in a first-year school general science course, it seems almost an affront
to the intelligence of the reader to claim space for such a trivial
problem. Those who have attempted to build models representing amorphous
solids know better: this simple yet precise quantity presents a stern
challenge and several otherwise satisfactory models - even for monoatomic
amorphous solids — have foundered on this rock. Even phenomenological
descriptions of composition-density relations for binary alloys and oxide
glasses present difficulties of interpretation — containing arbitary
empirical constants with no obvious reference to the 'reality' of an
atomic model.

Paradoxically perhaps, in view of their relatively recent discovery,
it is possible to claim greater understanding of the density of metallic
glasses than silicates and the subject is discussed in that order.

Metallic Glasses — specifically transition metal-metalloid alloys

In most glasses, it is commonly observed that the density of the amorphous phase is only slightly less — typically 2-5% less - than that for
the compositionally-equivalent crystal and this rule applies also for
amorphous transition metal-metalloid (T-m) alloys. In a binary T-m alloy
it is tempting to try to express the average atomic volume in terms of
separate contributions from T and m atoms but this is difficult in the
absence of any satisfactory model for the glass. Difficulties stem from

the fact that while the effective atomic radius of an element is obtainable from partial pair correlation functions, it is only rarely possible to measure the latter with sufficient accuracy. Alternatively, it may be reasonable to assume constant values obtained from the 12- coordinated 'Goldschmidt' radius of the elemental metal or the tetrahedral covalent radius of the metalloid. What we cannot do is to assume that the interstitial volume in the (unknown) structure of the amorphous solid is partitioned between the two elements in any obvious way. The total 'effective' volume of an element — expressed in terms of a Voronoi polyhedron centred on that element is clearly determined by the detailed packing arrangements and there is no reason, therefore, to extrapolate rules which apply to packing in the lattices of the two pure elements. A number of authors have attempted to do this leading, in my view, to considerable confusion.

It seems preferable to partition the total volume between the metal atoms alone — a process which is unambiguous — and to consider the variation of this quantity with composition etc. and then try to interpret it in terms of structural models (21).

For an alloy $T_{1-x} m_x$, the mean atomic volume (per metal atom), V_M is

$$V_M = \frac{< A >}{N_A \rho (1-x)} \; 10^{24} \; \text{Å}^3$$

where $< A >$ is the compositionally-averaged atomic weight and N_A is Avogadro's number. The metal atom packing fraction,

$\eta_M = 4\pi \, R^3/(3 \, \bar{V}_M)$, can be defined where R is the Goldschmidt radius of the metal or a compositionally-weighted average if two or more metallic elements are present.

Much of the rather extensive data relating ρ with composition can be represented by a series of linear relations describing the variation of ρ with concentration of the metalloid, x, and of ρ with the relative size of the metalloid expressed through the radius ratio $p = R_m/R$. Thus the variation of η_M with x for a series of T-M alloys is shown in Figure 2. Changes in η_M at constant composition for alloys containing metalloid elements with different radius ratios are shown in Figure 3 and similar data for a number of crystalline T_3m compounds are shown in Figure 4.

These data reveal a number of important features. The gradients in Figure 2 are functions of the radius ratio, the slope being large when p is large. Secondly, the lines appear to converge at a packing fraction $\eta_M^0 \sim 0.70 - 0.72$ — a value which is close to ideal close packing (0.74) and is significantly higher than that for the packing of hard spheres

(0.63). Both of these features were noted by Johnson & Williams (22) in their work on Mo-Ru borides. Thirdly, and perhaps most interestingly, the η_M - p curves for the 25% glass and for the corresponding T_3m crystals are closely similar — as shown in Figure 5.

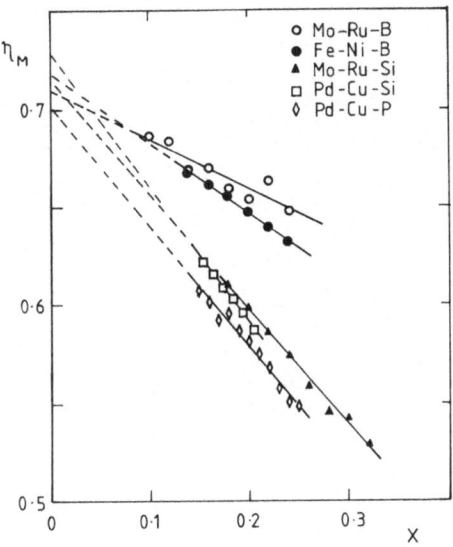

Fig. 2 Metal packing fractions η_M for amorphous borides and silicides.

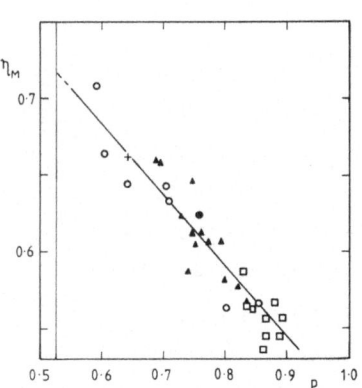

Fig. 4 Dependence of η_M on radius ratio, p, for crystalline T_3m solids

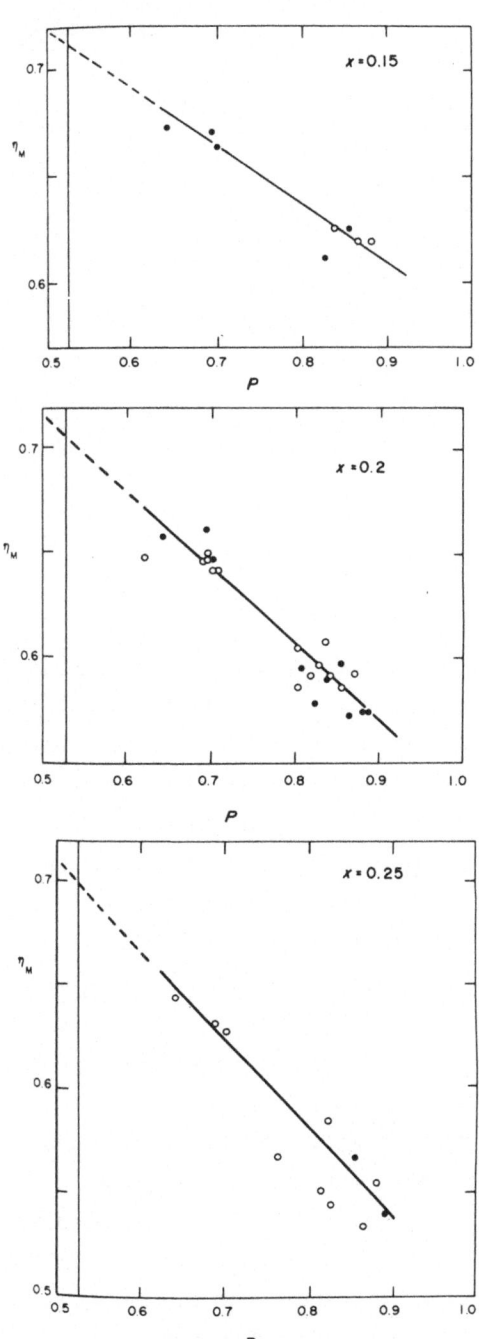

Fig. 3 Metal packing fraction versus radius ratio at several metalloid concentrations, x.

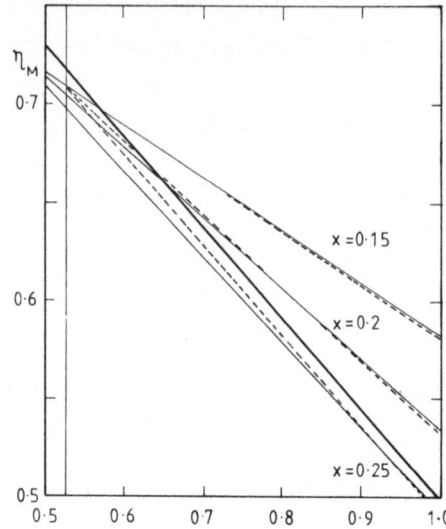

Fig. 5. Metal packing fractions as a function of the radius ratio. Fine
lines represent least-squares fits to data for amorphous alloys (Fig.
3), the bold line refers to crystalline T_3m alloys (Fig. 4). Dotted
lines represent values calculated from eq. 1.

It is then possible to postulate a simple empirical relation which
adequately represents these data by assuming that the metal atoms are
almost close-packed with atomic volume V^O, except in the neighbourhood of
metalloid atoms where the volume per metal atom V' increases due to dilatation of the metal 'cage'. If we assume that the number of dilated regions is proportional to x and that the extent of dilatation is proportional to p — or more particularly to a reduced radius ratio $p' = p - 0.528$
(the significance of which will become clear) — then it is possible to
write simple relations for V_M and η_M . Thus,

$$\eta_M(x) = \eta^O (1 - x\gamma p') \qquad \qquad ...(1)$$

where η^O and γ are empirical constants with values $\eta^O = 0.705$, $\gamma = 2.57$.

Detailed interpretation of such relations is not conclusive. As Johnson and Williams (22) show, an attempt to fit the data with compositionally-weighted atomic volumes calculated from the close-packed volume for
the metalloid leads to a composition-dependent value of the effective
volume of the metalloid (rising to infinity as $x \to 0$). If the effective
packing fraction of the metal is taken to be 0.705, as the experimental
data suggests, then a constant value of the effective metalloid volume is
obtained which is about 25% smaller than the close-packed atomic volume.
This suggests that the metalloid lies within a cage of metal atoms thus

allowing the atomic volume of the metalloid to fall below the value corresponding to the close-packed element.

The question then arises: can we establish the size and 'shape' of the metal cage? The fact that the densities of a large number of alloys can be fitted by a 'universal' relationship <u>suggests</u> that the environment of the metalloid has some constancy — invariant coordination number perhaps — and this is borne out by analyses of partial pair distribution functions. Secondly, the fact that the behaviour of glasses is so similar to crystals, in which the coordination of the metalloid is represented by a distorted trigonal prism, and the fact that η_M tends to η_M^o at a value of $p \sim 0.53$ corresponding to the radius ratio of a regular trigonal prism centred on the metalloid, <u>strongly suggests</u> that the environment is trigonal prismatic.

Clearly, though, these pieces of evidence do not constitute definitive proof but there is now ample direct structural evidence for a well-defined coordination polyhedron and it is possible to suggest the following working model to qualitatively and semi-quantitatively 'explain' the composition-dependence of the density of T-m alloys.

Addition of a metalloid element to a transition metal alloy converts part of the pre-existing local arrangement of metal atoms, which appears to be almost close-packed but is otherwise undefined, as yet, into trigonal prismatic cavities centred on the metalloid; one prism per metalloid atom. For the hypothetical case of a small metalloid, $p < 0.53$, the prismatic cavity is regular and the resultant volume change reflects the conversion from close-packing to trigonal prismatic packing — which can be very efficient ($\eta \rightarrow 0.72$ in $c\text{-}T_3m$ compounds, Figure 4). The mean atomic volume per metal atom thus becomes almost independent of concentration. Small metalloids in large transition metal hosts approach this behaviour. At the opposite extreme, $p > 0.8$, corresponding to a large metalloid in a small host metal, for each m atom added, large distortions are introduced into the trigonal prism, and this 'dilatation volume' becomes relatively large so that the volume per metal atom increases sharply with the concentration of m. Phosphides of the 3d transition metals ($p \sim 0.83$) behave in this way, and it can be shown that the dilatation volume is just equal to the atomic volume of the crystalline metal V^o_{cp}. The result is that the average effective atomic volumes of both constituents are equal to V^o_{cp} — a fact which had been noted by Turnbull [23] and has been subsequently discussed and extended by Bennett and Watson [24]. Both ideas can be shown to be special cases of the more general treatment described here.

Several of the experimental facts described for metallic glasses are also relevant for glassy oxides, particularly the observation that densities of glasses and equivalent crystals agree to a few percent. Experimental data for a range of silicates is shown in Figure 6 (25). Again, it is convenient to consider the volume as partitioned between the oxygen atoms and define a quantity V_O and an oxygen atom packing fraction, η_O.

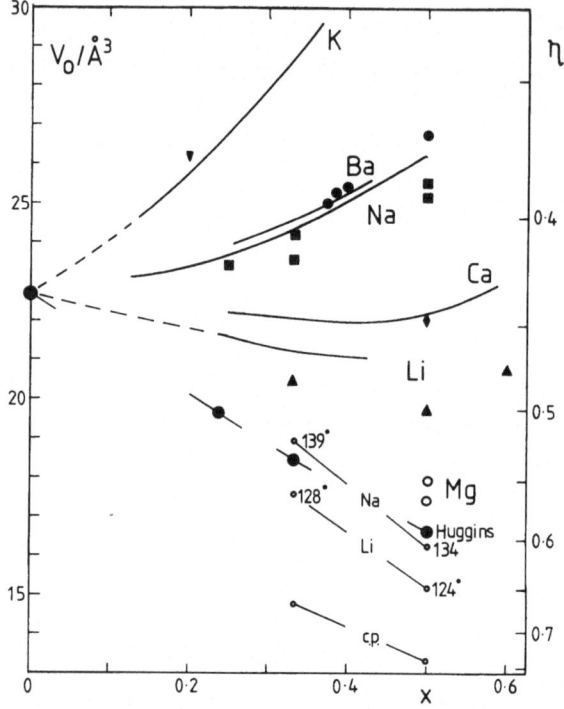

Fig. 6. Values of V_O and η_O for silicate glasses (full lines) and crystals (points) are shown in the upper part of the diagram. In the lower part, the estimated contributions to V_O from the Si-O "framework" are given. Filled circles represent values calculated from Huggins' relation (eq. 2): other values correspond to the calculations described in the text for the partially close-packed Li and Na meta- and disilicate structures. Numbers refer to oxygen bond angles.

Oxides are, however, different from transition metal-metalloid alloys in the following respects. Although the packing fraction generally decreases from the value of 0.43 for SiO_2 as metallic oxides are added to form mixed oxides (designated as $xM_nO \ (1-x)SiO_2$), this does not occur for the small, or highly charged ions such as Li^+, Ca^{++} or Mg^{++}. Secondly, with the exception of $MgSiO_3$, oxygen packing fractions do not approach close-packed values (for any glass-forming composition). Thirdly, and most obviously, the effects on the oxygen packing of two other atoms must be considered.

Because the density, ρ, of commercial oxide glasses is an important parameter (for example in the design of windows or lens systems), a number of empirical relationships have been devised which express the composition-dependence of ρ. Furthermore, ρ enters into relations describing the composition-dependence of other properties, for example, the refractive index, elastic modulus etc. (Somewhat surprisingly, density differences between crystals and silicate melts corresponding to one part in 10^4 are sufficient to cause convection in geothermal processes and this, perhaps represents the ultimate requirement for accuracy (26)).

The best known examples of the empirical relationships are those due to Huggins, and Stevels (27). Huggins expresses V_O in terms of the ratio of Si and M atoms to oxygen, $(1-x)/(2-x)$ and $nx/(2-x)$, respectively. Thus,

$$V_O = b_{Si} + c_{Si} \frac{1-x}{2-x} + c_M \frac{nx}{2-x} \qquad \ldots (2)$$

In this equation, the constants b_{Si}, c_{Si} relate to specified ranges of x: the result is that the contribution of the silicon-oxygen sublattice to η_O consists of a series of linear relations shown in Figure 6.

For an explanation of this behaviour, the structure of equivalent crystalline phases again provides a good guide. The structure of crystalline lithium metasilicate is shown in Figure 7; sodium metasilicate differs only in detail. The lattice consists of corrugated silicon-oxygen chains parallel to the c-axis, with distorted Li-O tetrahedra occupying the interchain regions. The oxygen sublattice is closely related to the Wurtzite (B4) structure as discussed by Hesse (28), McDonald and Cruickshank (29) and approaches this structure for small cations which fit within a regular oxygen tetrahedron. The oxygen sublattice is then almost hexagonally close packed (h.c.p.) with Si and M cations lying in tetrahedral interstices: with each Si-O chain surrounded by six Li-O chains, Figure 8. For the close-packed oxygen sublattice, the volume per oxygen, $V_O = 4\sqrt{2} R_O^3 = 13.31$ Å^3 and $\eta_O = \pi \sqrt{2}/6 = 0.7405$.

The close-packed model for metasilicates is unrealistic because this implies a bond angle at oxygen of 109.5° whereas observed bond angles, ε, are 124° and 134° in Li- and Na- metasilicates respectively. There are also smaller discrepancies — differences in Si-O bond lengths to bridging and non-bridging oxygens, and associated O-Si-O angular variations: all of which will be ignored. If we introduce the actual value of the oxygen bond angle, ε, and retain close-packing otherwise (i.e. normal to the c-axis) then V_O increases from $4\sqrt{2}R_O^3$ to $V_O = (k + 2/\sqrt{3})\, lm\, R_O^3\, /6$ where $k = 2\sqrt{2}\sin\varepsilon$, $l = 3$ and $m = 4\sin\varepsilon/2$. Similarly, $\eta_O = 8\pi/[(k + 2/\sqrt{3})\, lm]$. This value represents the volume per oxygen atom for the silicon-oxygen sublattice in the limit that the cations M lie in undilated tetrahedral interstices and is compared with values obtained from the actual coordin- ates of c-Li_2SiO_3 and Na_2SiO_3 in Figure 9. Agreement is good. A value of $V_O = 16.64$ $Å^3$ for the Si-O contribution is calculated from Huggins' formula <u>for glasses</u> (with $c_M = 0$). This value corresponds to Si—O chains with bond angle, ε, of $137°$ — a value typical of a large number of pyrox- ene minerals and representing a reasonable value for commercial glasses.

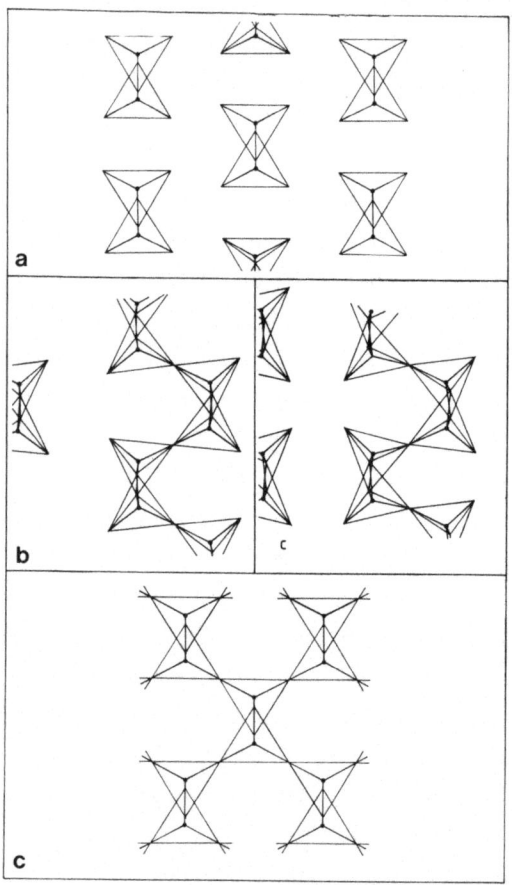

Fig. 7. Schematic representation of the progression from the metasilicate chain structure (a), through disilicate sheets (b,c), to the framework tridymite structure. Each step involves the removal of non-bridging oxygens between metasilicate chains normal to the plane of the diagram a) $LiSiO_3$ (x,y) projection, b),c) similar projections of α-Na_2SiO_5 and Li_2SiO_5. Note increased corrugation of $Si_2O_5^{--}$ sheets in Li_2SiO_5. Si atoms are marked by small filled circles: alkali atoms are not shown but lie appoximately above and below non-bridging oxygens in a)-c).

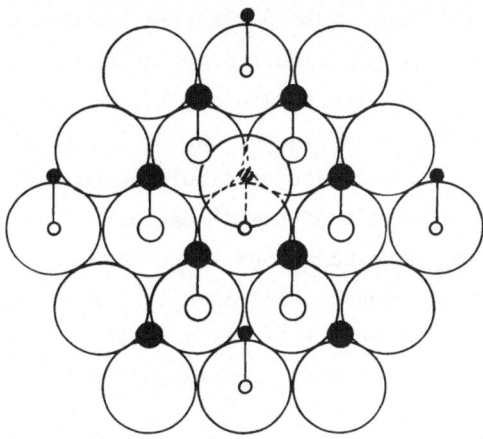

Fig. 8. The hexagonally close-packed oxygen sub-lattice corresponding to a metasilicate structure. Si atoms are shown by small circles and occupy tetrahedral interstices at two levels, forming chains parallel to the hexad axis. Each Si-O chain is surrounded by six parallel Li-O chains - the Li atoms are represented by medium-sized circles.

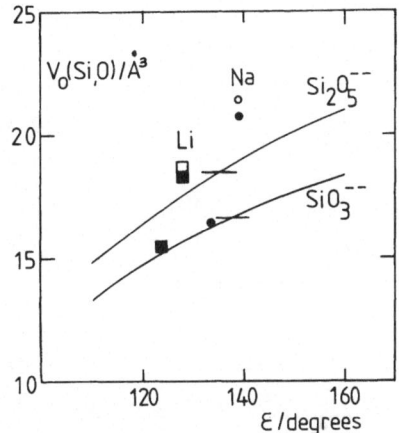

Fig.9. Contributions to V_O from the Si-O "framework" for meta- and di-silicate structures as functions of the oxygen bond angle, ε. Filled points for metasilicates represent crystal structure data and horizontal bars, predictions from Huggins' relationship. Experimental data for disilicates are compared with calculated values (open points) after adding the <u>intra</u> planar expansion.

With the proviso that we can allow the oxygen bond angle to increase above the tetrahedral angle corresponding to close-packing, we can vis-ualise the progression through several structural types of silicates in terms of a progressive condensation of the metasilicate lattice, Figure 7a. Thus the disilicate lattice is generated by elimination of non-bridging oxygens (and associated M cations) to give the corrugated sheet structures shown in Figure 7b,c. Similarly, the disilicate sheet struct-ure may be condensed to give the framework tridymite lattice, Figure 7d.

In the process, the silicon-oxygen sublattice becomes progressively more open and this is the major contribution to the composition-dependence of Huggins' factor for the silicon-oxygen sublattice. For example, it is possible to calculate values of V_O for the Si-O sublattice in disilicates. The calculation is now more difficult because the intra-planar puckering varies with the size of the M^+ cation. This has been calculated from purely geometrical considerations as described below. Then, if we retain the assumption of oxygen close-packing normal to the c-axis, $V_O = 8\ R_O^3\ \sin(\varepsilon/2)\ (2\sqrt{2}\ \sin\varepsilon/2 + 1/\sqrt{3})/\ 3$. Values are shown in Figure 9. Again Huggins' formula gives V_O corresponding to an oxygen bond angle of about 136°.

Table 1. Calculated values of the contributions to V_O compared with experimental data for the crystals

V_O(Si,O)	Intra planar expansion	Inter-planar expansion	Total V_O (calc)	V_O (exp)
Li_2SiO_3	15.2	-	4.2	19.4	19.7
Na_2SiO_3	16.2	-	8.9	25.1	25.6
Li_2SiO_5	17.6	1.1	1.7	20.4	20.4
Na_2SiO_5	19.0	2.2	3.7	24.9	24.2

The final step to reality involves an estimate of the dilatation of the lattice resulting from interchain or interplanar expansion of the close-packed oxygen sublattice due to the fact that cations do not fit within undistorted tetrahedra. The effect is that the cations push apart the close-packed planes of oxygen atoms leading to a volume increase which is clearly some function, $f(R_M)$, of the cationic radius, R_M. Thus since the relative number of cations and oxygen atoms is $nx/(2-x)$, the volume increase can be written as $nx\ f(R_M)/(2-x)$. The quantity $f(R_M)$ has been calculated for chains and sheet structures from the increased separation of notionally close-packed oxygen planes, assuming tetrahedral coordination for the cations. Results are shown in table 1, together with the total calculated value for V_O and experimental data. Agreement to better than 3% is observed which, considering the crude approximations — neglect of the more subtle variations in silicon bond angle for instance — can be regarded as excellent.

The present analysis of the density of simple chain and sheet silicates contains this essential ingredient: chains and sheets forming the silicate anionic lattice can be considered separately from the alkali cations which can be thought of as a separate surface phase in the "molecular composite". The volume of the silicon-oxygen component is altered by oxygen bond angle changes and by changes in conformation associated with the transformation from chains to sheets, say. To some extent this may be treated separately from the effect of the cations which is to increase the interplanar separation between notionally close-packed oxygen ions. The total volume change is a sum of the two contributions. This assumes, of course, that the separate changes in structure are separable in this way and, more importantly, that they are compatible. But compatibility is not necessarily assured. Consider the puckering of the sheets introduced in lithium disilicate - figure 7b,c: the associated elastic strain in the Si-O sublattice may be considerable and Dent-Glasser (30) has argued that the non-existence of certain crystalline silicates containing small, highly charged cations such as Mg^{++} or Ca^{++} may be related to the inability of the silicate sheets to pucker sufficiently to meet the coordination requirements of these ions. For the (existing) crystalline structures considered here we may assume some degree of compatibility or cooperative motion leading to observed volumes in good agreement with experiment.

Silicate Glasses

The close correspondence between the densities of compositionally-equivalent crystals and glasses tempts the thought that essentially similar dilatations occur in vitreous chain and sheet silicates. That is, that we may assume the existence of chains and sheets (distorted, inevitably); that we may assume "interfacial" close-packed oxygens dilated by the interstitial cations according to geometrical rules which are similar to those for the crystalline phase. Finally, that some degree of cooperative dilatation can be assumed when we consider the hypothetical process of stuffing the close-packed regions with, say, Na^+ ions.

The ideas expressed above are essentially different from the random network picture of glasses in which alkali cations occupy suitable cavities within a silica "sponge". In the present picture, cation sites are well-defined with respect to their immediate coordination shell but also in relation to each other and to the silicate chains or sheets. In crystals we think of the cations lying outside the chains or sheets: in glasses we are tempted to continue to think in terms of regions inside and outside the silicon-oxygen sublattice. Just as a cation placed (exceptionally) 'inside' the (corrugated) $Si_2O_5^{--}$ sheet in a crystal would have a different effect on the volume than if placed between the sheets, so we can postulate differences in glasses also.

The greatest difficulty stems from the notion of cooperative dilatation. The geometrical factors representing the dilatation introduced in the close-packed regions by a Na^+ cation, say, were calculated on the assumption that planes of 'interfacial' O^- ions were pushed apart, that is that each Na^+ ion in the structure introduces a similar dilatation so that sheets or chains are separated — correlated displacements, therefore. Even if we assume that the structure of disilicate glasses is composed of $Si_2O_5^{--}$ sheets, bent, torn and distorted perhaps, it is difficult to see how the puckering shown in figure 7c can occur unless,

(a) the sheets and the interfacial M^+ O^- regions are essentially continuous over distances equivalent to perhaps ten $O-O$ interatomic distances;

(b) that over similar distances the sheets are parallel and are not <u>torn</u> (i.e. that the normals to the planes do not switch discontinuously) within regions of the order of, perhaps, 1-2 nm. Otherwise the strain induced in the silicon-oxygen framework in achieving adequate close-packing for the Li^+ ions, say, is likely to be so large that the structure may become 'non-existent'. For glasses, this means separation into two phases where compatibility is more easily achieved — a silica-rich 'skeleton' on the one hand and more bendable chain-like silicates on the other.

In an attempt to examine relative difficulties of 'cooperative' displacements inherent in random or quasi-random structures of interconnected SiO_4 tetrahedra, and the possible incompatibility of randomness in structures with local structural constraints as integral features of the model, Gaskell (31) has built a quasi-random model for glassy metasilicates.

The starting point is the h.c.p. oxygen sublattice which, in the limit of small metal cations, represents the structure of Li and Na metasilicates. In the crystal, tetrahedral interstices are filled in a regular way with Si and Li atoms, to give chains parallel to the z-axis (Fig 8). Each Si-O chain is surrounded by six Li-O chains as shown. In the model for the glass, the h.c.p. oxygen sublattice is retained but the choice of tetrahedral interstice is more random. Specifically, no chains are allowed to be cross-linked — since ^{29}Si NMR data shows essentially only Si atoms connected to two bridging oxygens (32). It is not possible to avoid chain terminations, however. Furthermore, each Si-O chain is surrounded by six Li-O chains. Finally, the overall ratio of Si and Li remains at 1 : 2. Subject to these conditions, the occupancy of tetrahedral interstices is controlled by a random number generator.

A physical model was constructed using polystyrene spheres and the coordinates refined by energy minimization using a modified Keating potential energy function with stretching, V_s, and bending, V_b terms,

$$V = \sum_i V_s + \sum_j V_b \qquad V_s = \frac{3\alpha}{16\,d^2} \sum_{1,i} (r_{1i} \cdot r_{1i} - d^2)^2$$

$$V_b = \frac{3\beta}{8d^2} \sum_{\substack{1,[i,i'] \\ i \neq i'}} (\underline{r}_{1i} \cdot \underline{r}_{1i'} - d^2 \cos^2\theta)^2 \qquad \ldots(2)$$

where the sum is over the i stretching and j bending coordinates, $\underline{r}_{1i}$ is the vector connecting atoms 1 and i, d and θ are the equilibrium bond length and bond angle, respectively. Values for d and θ are chosen separately depending on whether the Si-O bonds are bridging or non-bridging — guidance being sought from the crystalline structures. Similarly, constants for metal-centred polyhedra are chosen so that four bonds to oxygen are shorter than a fifth. Interaction between non-bridging oxygen atoms are also modelled by a Keating stretching term.

Success depends crucially on the choice of realistic values for α, β etc, and this choice was guided by a preliminary calculation in which the h.c.p. oxygen array was regularly 'stuffed' with Si and Li (Na) atoms to produce a topology identical with the metasilicate crystal. Constants were adjusted until the best _geometrical_ fit was obtained with experimental structural data for Li and Na metasilicates.

The crucial parameter is α_{O-O}, that is the potential energy term between non-bridging oxygen ions. There is very little guidance obtainable from the literature on the magnitude of this interaction which was therefore varied over two decades. Calculated radial distribution functions and V_O values were compared with experimental data – the fits for the optimum value of α_{O-O} which also minimizes the strain energy being shown in Fig. 10. Although there is no certainty that this value of α_{O-O} is appropriate for glasses — or even for _crystals_ of different symmetry - it is difficult to see how it can be bettered.

Calculations for the quasi-random model with the 'best fit' crystal value of α_{O-O} produced structures with values of V_O which are significantly smaller than experimentally measured data as Fig. 11 shows: the calculated value for V_O for a-Na_2SiO_3 is about 13% less than the measured value, for glassy lithium metasilicate the discrepancy is about 8%. Note that the calculated density is too high - in the case of Na_2SiO_3 the density for glass is about 10% higher than the _crystal_.

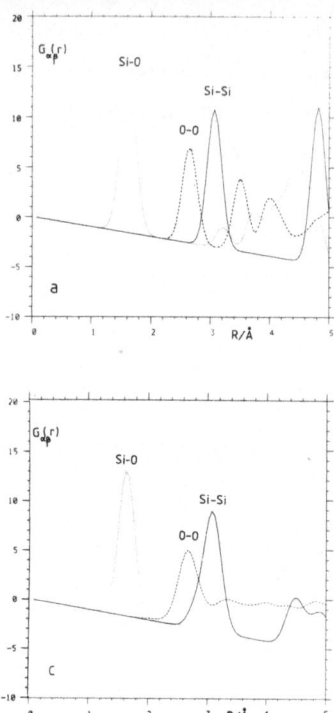

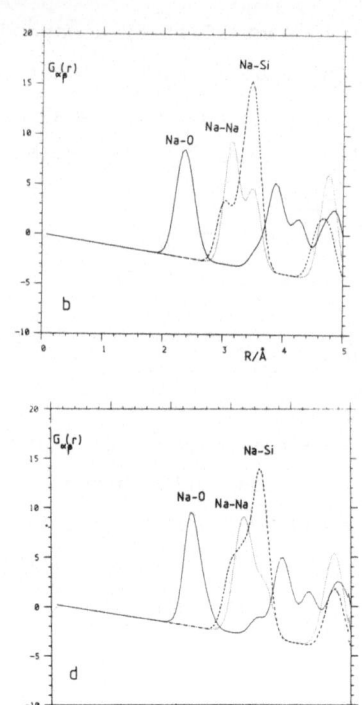

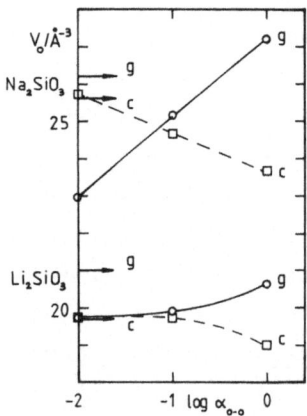

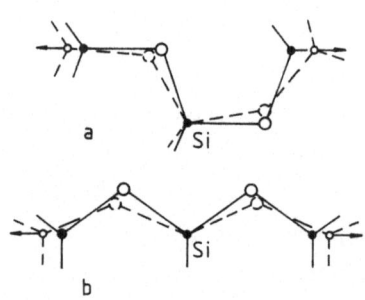

Fig. 10.   a),b) Experimental partial pair distribution functions (ppdfs)
for c-Na$_2$SiO$_3$.   c),d) Computed ppdfs for a model of c-Na$_2$SiO$_3$ obtained by energy-minimization of a starting structure based on the h.c.p.
lattice shown in Fig.8.   Parameters in the interatomic potential
energy function were varied to obtain the fit shown and these parameters were then used in computations for the glasses.

Fig.11 V$_O$ computed for crystalline,
c, and glassy, g, Li$_2$SiO$_3$ and
Na$_2$SiO$_3$ as functions of the non-
bridging O stretching interaction
α_{O-O}. For values of α_{O-O} giving
good agreement with experimental V$_O$
data (arrows) for crystals, values
of V$_O$ for glasses are seriously
in error.

Fig. 12. a) Expansion of Si-Si
distances parallel to the
hexad axis involves bending
at oxygen atoms only since
cooperative rotation of tetrahedra is possible. b) For
chains normal to the hexad
axis, as in a random model,
cooperative motion is not
possible and bending of the
'stiffer' Si angle occurs.

The reason for these discrepancies can easily be found. In the crystalline lattice, dilatation of the h.c.p. lattice by Na, say, is a resultant of elongation of the chains along the hexad axis and an enlarged
interchain distance normal to this axis . As Figure 12 shows, the dilatation, in both directions, which a large cation requires can be obtained,
within limits, without any constraints apart from (possibly) bending at
oxygen, which is relatively easy. For Si-O chains which run normal to
this direction, as they must, at times, in a random lattice, then bending
at silicon is also necessary and the bending force constant at Si is about
ten times that at O. Consequently the cooperative nature of the dilatation is lost and the volume is constrained at an anomalously low value.
Thus although this model satifies the local coordination constraints it
does not lead to satisfactory packing as measured by the density.

It could be argued that a more natural starting point would have been
a dense-random packed oxygen sublattice: Si and Li could then be stuffed
into tetrahedral interstices to give connected chains. The initial close-
packed random structure has a lower packing fraction, 0.63 compared with
0.74 for h.c.p. However, it requires to be proved that such a model would
also satisfy the constraints representing coordination, anionic composition - and packing densities: lithium metasilicate, even as a glass has a
very high packing fraction ~ 0.47 and, at first sight, it would seem that
a model structure would become too <u>open</u> if a starting structure with a
packing fraction, η_O ~ 0.63 is stuffed with alkali ions.

A personal view would be that for a successful model of the density of
these simple glasses it will be necessary to move in the other direction -
that is, towards domain models where chains lie parallel,but not necessarily straight, for distances of perhaps 20Å and then change orientation
passing into other domains where parallel chains exist for another 20Å and
so on. A schematic two-dimensional representation is shown in Figure 13.

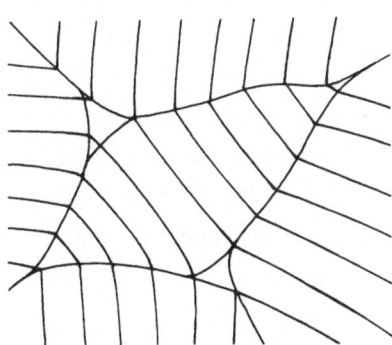

Fig. 13   Schematic domain structure for metasilicate glasses.

The importance of the optical properties of glasses requires no emphasis — yet, despite the evident value of any understanding, we have no
adequate quantitative treatment. The questions that emerge depend on the
energy range.

At high energies, greater than, say, 30 eV, and up to a few keV, transitions involve excitation of core states. While there is some influence
of the structure on the details of the absorption spectrum, giving rise to
chemical shifts, near-edge spectra (Kossell structure) or EXAFS effects,
these features have little or no effect on 'socially- or commercially-
useful' properties.

In the range from the visible (~3eV) to the onset of core-shell excitations, optical transitions involve electronic excitation from states near
the upper edge of the valence band to low-lying conduction band states.
Such processes are usually marked by strong absorption and give rise to
dispersion — the frequency-dependence of the refractive index measured in
the visible. Quantitative knowledge of such processes has considerable
value therefore. Moreover the structure — at least the local structure —
is explicitly involved in the detailed mechanism.

Below the band gap the absorption coefficient is relatively small and
is normally the result of defects, 'colouring' impurities such as ions of
transition metals or rare earths with unfilled 'inner' d or f shells,
respectively, or tail-state processes giving rise to 'Urbach' law absorption. The latter will in general be structure-sensitive, as shown for
instance by the pressure-dependence of the absorption edge of several
chalcogenides (33), as also are the detailed spectra of colouring ions
through their interactions with surrounding 'ligands'.

Vibrational modes occur in the energy range below 1 eV with intense
absorption and reflection (Restrahlen) bands below about 200 meV. Such
features are strongly structure-sensitive — and since the vibrational
spectrum is determined by the details of local and medium-range structure, it is extensively studied as part of any structural investigation.

At very low energies vibrational motions of weakly bound species
such as alkali ions, "floppy" oxygen atoms, interstitials, etc., become
merged with diffusive motion leading to interaction with low frequency
electrical fields or mechanical stresses giving a variety of effects,
dielctric loss, internal friction, tunnelling states, quasi-elastic light
scattering, etc.

In this section we concentrate on the upper reaches of this energy spectrum emphasizing the valence-conduction band transitions and their impact on the visible spectrum — particularly the refractive index.

Refractive indices of silicate glasses

The optical properties are characterized by the real and imaginary parts of a complex refractive index or dielectric constant. Thus:

$$n = n + ik, \qquad \varepsilon = \varepsilon_1 + i\varepsilon_2$$

where $\quad \varepsilon_1 = n^2 - k^2, \quad \varepsilon_2 = 2nk \qquad \qquad \ldots(4)$

The real and imaginary components are not independent and are inter-convertible through Kramers-Kronig dispersion relations. Specifically, the low-frequency dielectric constant, at frequency ω_a, $\varepsilon_1(\omega_a)$ — that is, an extrapolated value of the square of the 'optical' refractive index measured in the visible and neglecting phonon and lower frequency components — is given by

$$\varepsilon_1(\omega_a) = 1 + \frac{2}{\pi} \int_0^\infty \frac{\omega\, \varepsilon_2(\omega)\, d\omega}{\omega^2 - \omega_a^2} \approx 1 + \frac{2}{\pi} \int_{\omega_g}^\infty \frac{\omega\, \varepsilon_2(\omega)\, d\omega}{\omega^2 - \omega_a^2} \qquad \ldots(5)$$

Generally, ε_1 will be required at frequencies ω_a, corresponding to 1-2eV and ε_2 will be vanishingly small below the band gap, $\hbar\omega_g \approx 5$ eV, so that this becomes the lower limit of integration. Given accurate values of $\varepsilon_2(\omega)$ for ω greater than ω_g, the behaviour of $\varepsilon_1(\omega)$ in the visible is completely specified. Such measurements are, however, difficult to obtain with adequate accuracy and to sufficiently high energies. The prospect of acquiring refractive index data to the accuracy required in optical design cannot be considered realistic. Predictive recipes will continue to be based therefore on arbitrary fitting of experimental measurements using 'dispersion' relations and composition-dependent 'fudge' factors. What we might, realistically, hope for would be quantitative relationships where the terms make some sort of physical sense.

Progress towards a semi-quantitative predictive treatment may be made by noting, firstly, that although the ε_2 spectrum will be structure-sensitive in general, it is observed to be only weakly dependent on medium-range structure. That is, once the local structural parameters have been included — coordination number, ligand type, bond length distribution — other, longer-range structural changes have little effect <u>except through macroscopic properties such as the density</u>. This statement may be exemplified by the qualitative similarity of the ε_2 spectra of quartz and

vitreous silica and crystalline and amorphous silicon. Secondly, such
detailed differences as do exist are suppressed in the integration over
the entire ε_2 spectrum, which leads to ε_1 by a Kramers-Kronig transform,
(eq. 5). Thirdly, the $\omega^2 - \omega_a^2$ term in the denominator of equation 5 has
the effect of diminishing the contribution of the higher energy, soft
X-ray, part of the spectrum which is difficult to measure accurately.
Finally, the total absorption is subject to a sum rule; firstly, that
expressing the connection between the plasma-like, oscillations of all the
electrons in the solid at energies above the highest transition, and the
electron density per unit volume N.

$$\frac{2}{\pi} \int_0^\infty \omega \, \varepsilon_2(\omega) \, d\omega = \frac{4\pi \, N \, e^2}{m} \equiv \omega_p^2$$

or

$$\int_0^\infty \omega \, \varepsilon_2(\omega) \, d\omega = 2\pi^2 \, N_a \, N_{eff} \, / \, m \qquad \qquad \ldots(6)$$

Here, N_a is a suitably-defined atomic (or molecular) density and N_{eff} is
the number of electrons per formula unit and m the electron mass.

Also, the integral over the entire $\varepsilon_2(\omega)$ spectrum is related to the
low-frequency or static dielectric constant ε_o by

$$\varepsilon_1(0) - 1 = \frac{2}{\pi} \int_0^\infty \frac{\varepsilon_2(\omega) \, d\omega}{\omega} \qquad \qquad \ldots(7)$$

which follows immediately from equation (5)

Alternatively, the spectrum $\varepsilon_2(\omega)$ can be regarded as a sum over j
discrete oscillators centred at ω_j and of strength $\Delta\varepsilon_j$, where

$$\Delta\varepsilon_j = \frac{2}{\pi} \int_{band} \frac{\varepsilon_2(\omega) \, d\omega}{\omega}$$

Then,

$$\varepsilon(\omega_a) = 1 + \sum_j \frac{\omega_j^2 \, \Delta\varepsilon_j}{\omega_j^2 - \omega_a^2} \qquad \qquad \ldots(8)$$

The quantity $\Delta\varepsilon_j$ has the meaning of an oscillator strength (often
denoted by f_j).

Single oscillator models of Dispersion

The value of approximate methods for calculating $\varepsilon_1(\omega)$ lies in the
fact that the actual spectrum which often has a complex structure, extending over several hundred or thousand electron volts, can be approximated

by an effective spectrum with a simpler functional form or even parameters relating to single effective oscillators. This is the stratagem which underlies the so-called Sellmeier (classical oscillator) formulae, and the effective (single) oscillator models of Penn (34), Phillips (35), Van Vechten (36), Harrison and Pantelides (37), and Wemple and Di Domenico (38) and Wemple (39,40).

As an example, the refractive index of vitreous silica is given in the 'Handbook of Optics' (41) in the form:

$$n^2 - 1 = \frac{0.6961663 \; \lambda^2}{\lambda^2 - 0.0684043^2} + \frac{0.4079426 \; \lambda^2}{\lambda^2 - 0.1162414^2} + \frac{0.897479 \; \lambda^2}{\lambda^2 - 9.896161^2} \qquad \ldots (9)$$

where the units of λ are μm and the last term corresponds to the phonon contribution. (It is assumed that the quoted number of significant figures is realistic.) However, the values of λ_j bear only approximate relations to features of the $\varepsilon_2(\omega)$ spectrum - figure 14, as would be expected since two effective oscilators are employed to represent the entire electronic spectrum.

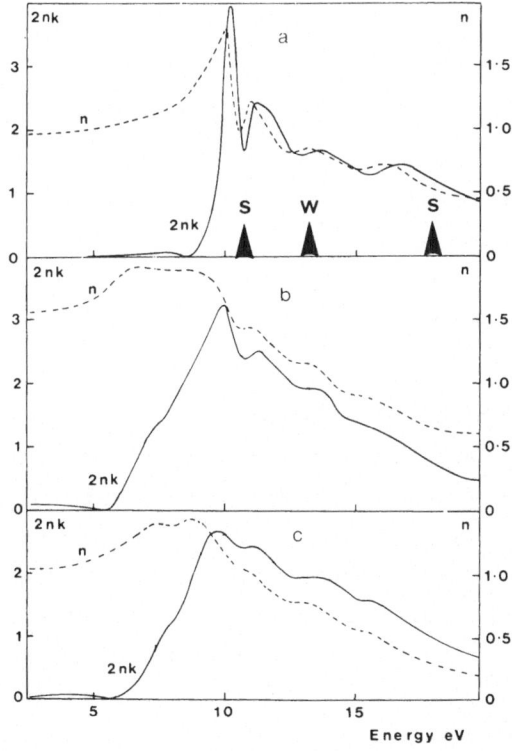

Fig. 14. Refractive index, n, and the imaginary part of the complex dielectric constant, $\varepsilon_2 \equiv 2nk$, for a) vitreous silica, b) sodium silicate glass, $3Na_2O \cdot 7SiO_2$, c) window (Float) glass. Equivalent oscillators are also shown in a) S - Sellmeier eq. 9; W - Wemple.

An even simpler relationship is proposed by Wemple for the case of SiO_2 (39) and for a range of other glasses and numerous crystalline solids (40). Writing equation (8) in terms of energy:

$$\varepsilon_1(E_a) - 1 = \sum_j \frac{E_j^2 \, \Delta\varepsilon_j}{E_j^2 - E_a^2} \qquad \ldots(10)$$

Replacing the entire ε_2 spectrum by a single effective oscillator leads to:

$$\varepsilon_1(E_a) - 1 = \frac{F \, E_0^2}{E_0^2 - E_a^2} = \frac{(\varepsilon_1(0) - 1) \, E_0^2}{E_0^2 - E_a^2} \qquad \ldots(11)$$

where F is an effective oscillator strength, E_0 the effective energy and $\varepsilon_1(0)$ is the extrapolated value of ε_1 to zero frequency, but omitting phonon and other low energy contributions.

Wemple then defines a so-called dispersion energy;

$$E_d^2 = F^2 \, E_0^2 = \left(\int_{E_g}^{\infty} E^{-1} \varepsilon_2(E) \, dE\right)^3 / \int_{E_g}^{\infty} E^{-3} \varepsilon_2(E) \, dE \qquad \ldots(12)$$

In terms of the moments of the $\varepsilon_2(\omega)$ spectrum, E_0 is given by:

$$E_0^2 = \frac{\int_{E_g}^{\infty} E^{-1} \varepsilon_2(E) \, dE}{\int_{E_g}^{\infty} E^{-3} \varepsilon_2(E) \, dE} \qquad \ldots(13)$$

Thus $\varepsilon_1(E) - 1 = E_d \, E_0 / (E_0^2 - E_a^2) \qquad \ldots(14)$

The advantages claimed by Wemple and co-workers for this form of the Sellmeier expression, are that the model parameters, particularly E_d, are relatable to certain structural properties — nearest neighbour coordination number of the cation, N_c, the formal anion valency, Z_a, and the effective number of valence electrons per anion, N_e. Thus, for example, E_d in six-coordinated alkali halides has the value 12.6±1.4 eV, whereas for 8-coordinated halides E_d = 16.2±1 eV, close to the ratio 8/6. E_d in six-coordinated oxides is 24.7±2.8 eV, almost twice that for the salts containing (monovalent) halide ions.

As a useful rule, Wemple and Di Domenico propose that the quantity $\beta \equiv E_d/N_c Z_a N_e$ has an approximately constant value in large groups of crystals containing a single anionic species and that β = 0.26 for ionic

materials and 0.37 for covalent solids. The value of E_d which represents
an effective oscillator strength is found to be relatively independent of
the energy of the band gap which Wemple and Di Domenico interpret in terms
of a model $\varepsilon_2(E)$ function in which the conductivity $\sigma = E\varepsilon_2(E)/4\pi\overline{h}$ is
constant. However, Aslakson (42) has pointed out that this is only one
example of a family of satisfactory functions which includes a series of
δ-function peaks that, in different materials simply scale with the band
gap, E_g. It seems unlikely that the generality of Wemple and Di Domenico's parameters depends on such accidents and, unless insensitivity to
the detailed form of $\varepsilon_2(E)$ lies at the heart of it, the apparent simplicity has not, to my knowledge, been adequately explained.

Returning to glasses, the values of E_d for quartz and vitreous silica
are 18.10 and 14.71 eV — a difference of 19%. Wemple points out that most
of this discrepancy can be accounted for by a 17% difference in the density — which, it is argued, reflects the fact that E_d is a function of the
near-neighbour oxygen shell surrounding silicon, i.e. SiO_4 tetrahedra and
that these are inefficiently packed in the glass.

An alternative formulation of Wemple's rules (40) applied to SiO_2 is
that $f \equiv E_d/(N_e N_a d^3 Z_a)$ is approximately constant for similar materials,
where d is the nearest neighbour (Si-O) distance and N_a is the (oxygen)
anion density, related to the volume per oxygen, V_O by $N_a = 1/V_O$. Inserting values appropriate to $a\text{-}SiO_2$, $N_a = (1/22.6) = 0.044$ A^{-3}, $N_e = 8$,
$d = 1.6$Å, $Z_a = 2$, $E_d = 14.7$ eV, then the constant f, = 5.1, a value also
found when appropriate constants are inserted for α-quartz. Wemple notes
similar values of f for amorphous and crystalline As_2S_3, and points out
that this would be expected since there are no significant bonding changes
"within the basic molecular units (SiO_4 and As_2S_3)". This is in contrast
to the lack of constancy in f observed in amorphous and crystalline forms
of the elemental semiconductors, Ge, Si, Se, Te, As. An explanation is
given by Wemple in terms of "valence electron localization."

For all its simplicity, the interesting constancy of $f \equiv E_d/(N_e N_a d^3 Z_a)$
is not obvious (at least to this author). N_e is equated to 8 representing
the filled outer octet of s,p electrons with appropriate modifications if
d-shell excitation is involved. The quantity $N_a d^3$ represents a dimensionless anion density within a sphere, centred on the cation, of radius equal
to the nearest-neighbour bond length, and thus quantitatively takes care
of differences between close-packed and open structures and, appropriately
for this discussion, emphasizes the central contribution of V_O to the
dispersion of the silica polymorphs since this is the only important
variable, E_O being approximately constant for vitreous silica and α-quartz.

For binary silicates, Wemple's treatment is less convincing. The logic of the local structural argument might suggest that the spectrum should now be represented by two effective oscillators representing the local environment of Si and Na (say) in interstices of the oxygen sub-lattice, a point made also by Robertson (43) in the context of PbO-SiO$_2$ glasses. As shown below, the $\varepsilon_2(\omega)$ spectra for binary silicate glasses can clearly be decomposed into two regions reflecting such local environments. Wemple prefers, however, to average such contributions retaining a single effective oscillator.

'Exact' fits to the $\varepsilon_2(\omega)$ spectrum of silica

A number of experimental measurements of the optical constants of silica have been reported, based on Kramers-Kronig transforms of the normal incidence reflectivity (usually) or occasionally the energy loss function, Im($-1/\varepsilon$), obtained from electron energy loss spectroscopy, EELS. Thus Philipp (50), figure 18; Platzoder (48), Ellis et al (44,45), figure 14, have published $\varepsilon_1, \varepsilon_2$ data derived from normal incidence reflectivity up to about 20 eV and Buechner (51) from EELS data to about 40 eV, figure 15. For fast electrons (~1-10 keV), values for Im($-1/\varepsilon$) can be compared with inelastic mean free path data. An admirably comprehensive review of the experimental data for a-SiO$_2$ and theoretical interpretations has been given by Griscom (52). Figure 16 indicates Griscom's summary of the core-shell and valence band excitations.

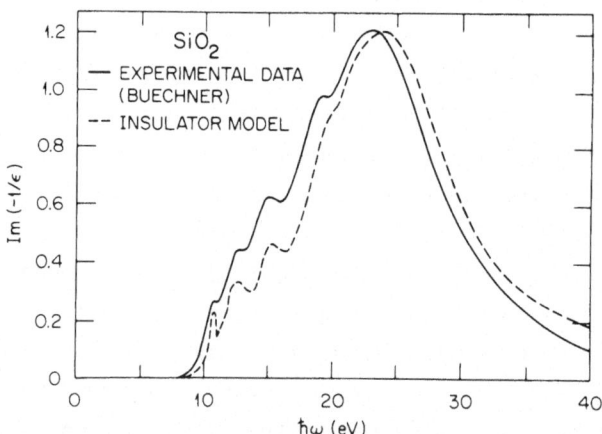

Fig.15. Energy loss function, Im($-1/\varepsilon$), for α-quartz. The solid line is the experimental data of Buechner (51) and the dashed line the function computed by Ashley and Anderson (53).

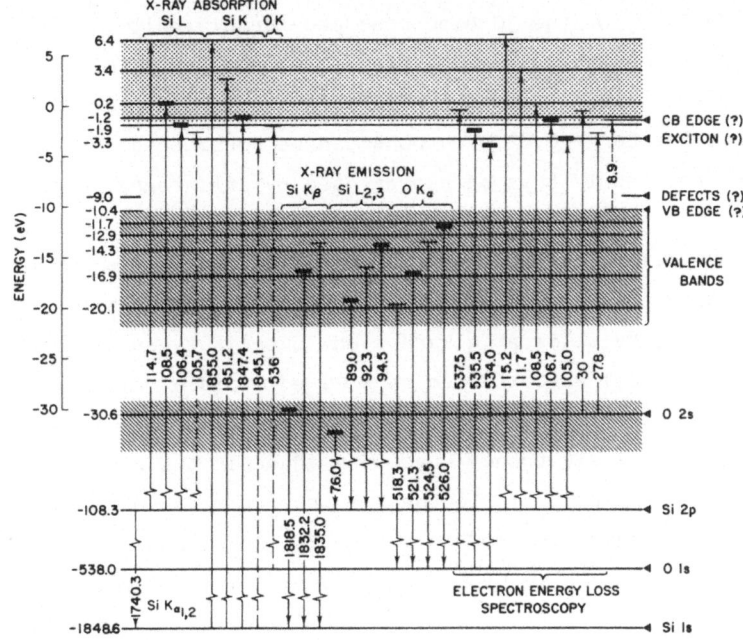

Fig. 16 Empirical energy band diagram for SiO_2 compiled by Griscom (52). The energy zero is the vacuum level.

Some of this data has been used by Ashley and Anderson (53) to provide a quantitative model of the response of the valence electrons. The stratagem has been to account the transitions from the core-shells shown in Table 2 by calculating the effective number of electrons, N_{eff}, participating in transitions from the given core level to the conduction band continuum from values of the generalised oscillation strength, df/dE. Thus

$$N_{eff} = \int_{c}^{\infty} \frac{df}{dE} \, dE$$

Table 2. Binding energy and effective number of electrons participating in excitations from core levels to the continuum.

Element	Shell	Binding Energy/ eV	N_{eff}
Si	$(1s)^2$	1829	1.55
	$(2s)^2$	152	1.31
	$(2p)^6$	108	7.00
O	$(1s)^2$	537	1.66

From Table 2, of the 30 electrons per formula unit (SiO_2), 9.86 silicon electrons and 2 x 1.66 oxygen electrons are involved in transitions above the energy corresponding to the Si L_{23} edge at 109 eV. Thus, 30-9.86 - 2 x 1.66 = 16.82 electrons can be considered to be 'valence electrons' and should thus be represented by an integral over the spectrum of $\omega\varepsilon_2(\omega)$ (equation 6). In fact this integral to 20 eV does not agree with the sum rule: 7.87 effective electrons per SiO_2 are observed from Philipps' data for α-quartz, rather than 16.82. Values for vitreous silica obtained by Ellis et al are even lower. Oxygen L_1 transitions (from the 2s level) are expected at ~ 29 eV, with a calculated strength of 2.28 electrons. This transition, and the deficient spectral intensity equivalent to 6.67 electrons, have been added in the region between 20 and 35 eV to produce a good fit to the spectrum from 20-40 eV derived from Buechner's EELS data, (scaled to agree with Philipp's data at 20 eV).

Results are shown in figures 17 and 18. The effective number of electrons participating in transitions to energy $\hbar\omega$ is shown in figure 17 and clearly indicates the contributions of valence and core-shell transitions. Fitted data for $\varepsilon_2(\omega)$ has been transformed to give $\varepsilon_1(\omega)$ (equation 5) and the resulting data is compared with experiments in figure 18. Values for the loss function, Im $(-1/\varepsilon)$ are computed from fitted ε_1 , ε_2 data and compared with Buechner's experimental data in figure 15.

Such careful experimental work and computational simulation is likely to prove invaluable in future in trying to answer structure-property relationships in the necessary detail.

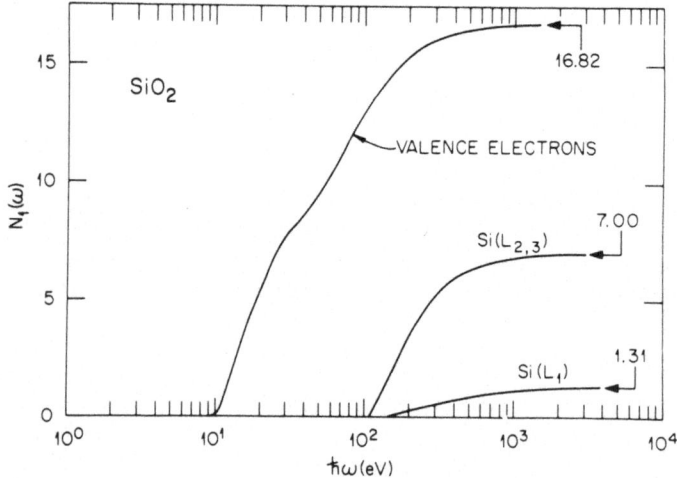

Fig. 17. Effective number of electrons, $N_1(\omega)$, involved in transitions up to energy $\hbar\omega$. Data from Ashley and Anderson (53).

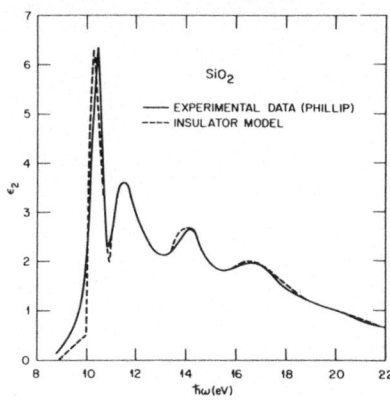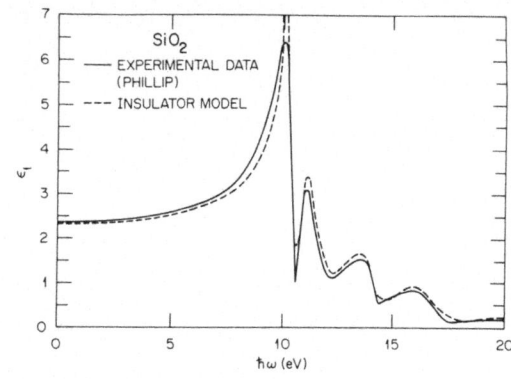

Fig. 18. a) Calculated (53) (dashed line) and experimental values (50) (full line)of ε_2 for quartz. b) Similar data for ε_1.

Experimental $\varepsilon_2(\omega)$ data for Binary Silicates

A number of workers have studied the reflectivity of binary silicates and in some cases, data has been transformed to give optical constants. Thus Ellis et al have reported results for a series of Na_2O- and PbO-SiO_2 glasses (44,45), as shown in figures 14 and 19. Spectra for some lead silicate glasses have also been reported by Bourdillon et al (46). More recently, vacuum ultra-violet spectra for several alkali silicates, borates and phosphates has been published by Hirota et al (47). Data for commercial silicate glasses have also been reported by Platzoder (48) and Sigel (49).

These spectra display a number of characteristic features. In the sodium silicates, addition of sodium induces absorption at energies below the band gap of SiO_2, marked by a shoulder in $\varepsilon_2(\omega)$ in the region 6-9 eV. Structure in the silica spectrum above 10 eV is still recognisable in the alkali silicate glasses also.

The onset of low energy absorption is even more pronounced in silicates containing lead oxide. In glassy, lead disilicate, the data of Ellis et al shows intense peaks in $\varepsilon_2(\omega)$ with distinguishable peaks at 4.4, 5.3 and 6.7 eV, together with a diminished contribution from features observed in the spectrum of a-SiO_2. As the lead oxide content increases further, all the low energy features begin to dominate the spectrum. The result of intense, low energy features on the extrapolated dielectric constant $\varepsilon_1(0)$ is very pronounced.

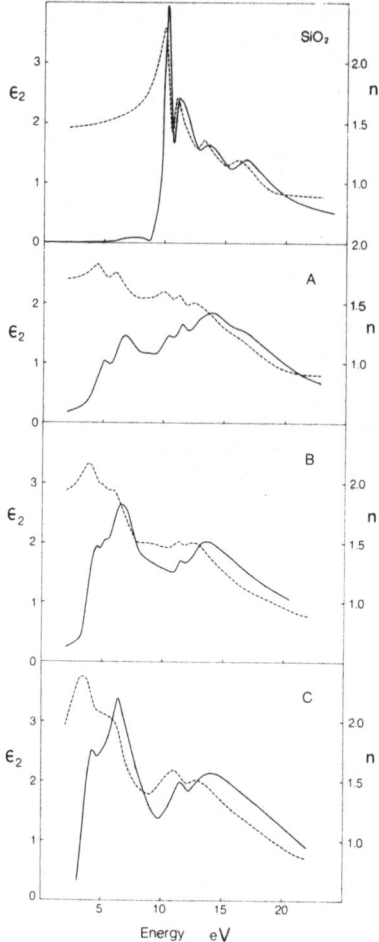

Fig. 19 Refractive index, n (dashed line) and ε_2 (full line) values
for silica and three lead silicate glasses (45). Approximate
compositions: A — PbO.2SiO$_2$, B — 3PbO.2SiO$_2$, C — 10 PbO.SiO$_2$.

Modifying equation 7

$$\varepsilon_1'(0) = 1 + \frac{2}{\pi} \int_{\omega'}^{\infty} \frac{\varepsilon_2(\omega)}{\omega} d\omega$$

it is possible to calculate the contribution $\varepsilon_1'(\omega)$ to $\varepsilon_1(0)$ from
absorption processes occurring at frequencies greater than ω'. Values of
$\varepsilon_1(0) \equiv n_0^2$ for the four glasses shown in figure 20 are presented in
figure 21. Note the large contribution to n_0^2 from features in $\varepsilon_2(\omega)$ in
the region of 4-9 eV. For the equimolar metasilicate composition
PbO.SiO$_2$, about 40% of $\varepsilon_1(0)$ arises from transitions between 4 and 9 eV.
This emphasizes the importance of transitions just above the band gap in

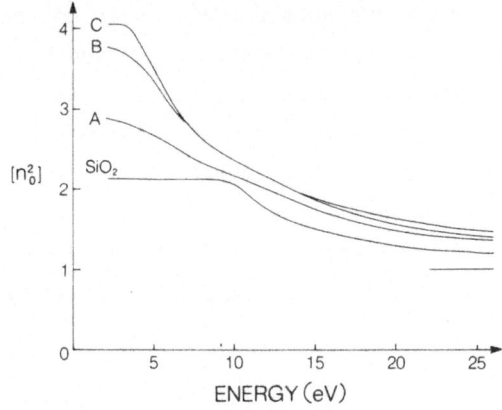

Fig. 20. Energy-dependence of the contribution $\varepsilon_1'(\omega)$ from absorption processes at energies greater than ω', eq.15. for silica and lead-containing glasses, A,B,C, (Fig. 19).

dispersion calculations, a point already discussed in terms of the effective oscillator formalism.

Interpretation of the optical transitions in Silica

Several workers have reported band structure calculations for crystalline silica; for brevity, no attempt is made to review each approach. Rather the work of Chelikowsky and Schluter (54) on α-quartz is quoted as an example of models for crystalline SiO_2 and, by extension, vitreous silica. A self-consistent pseudopotential calculation of the valence band states is shown in figure 21 with experimental photoemission data.

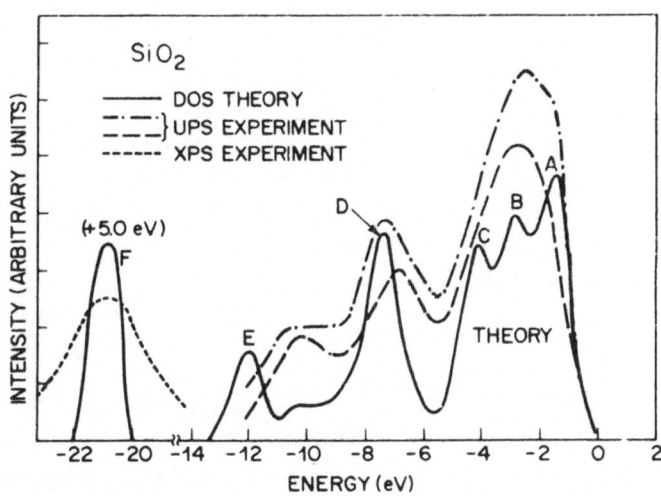

Fig. 21. Calculated density of states for α-quartz (full line),(54) compared with X-ray (XPS) and ultraviolet photoemission data, (UPS).

The upper valence band represents predominantly non-bonding, 'lone-pair' oxygen p orbitals, the charge densities being shown in figure 22. The upper valence band level shows charge localised at oxygen but not along the Si-O bond in contrast to the peak at -7eV (D) which clearly represents a bonding state. The lowest state shown in figure 21, (F), corresponds to oxygen 2s: note only small hybridization of the oxygen s and p states.

Figure 23 shows experimental data and calculated estimates of $\varepsilon_2(\omega)$. The first strong peak at ~10eV is thought to be excitonic. The lowest

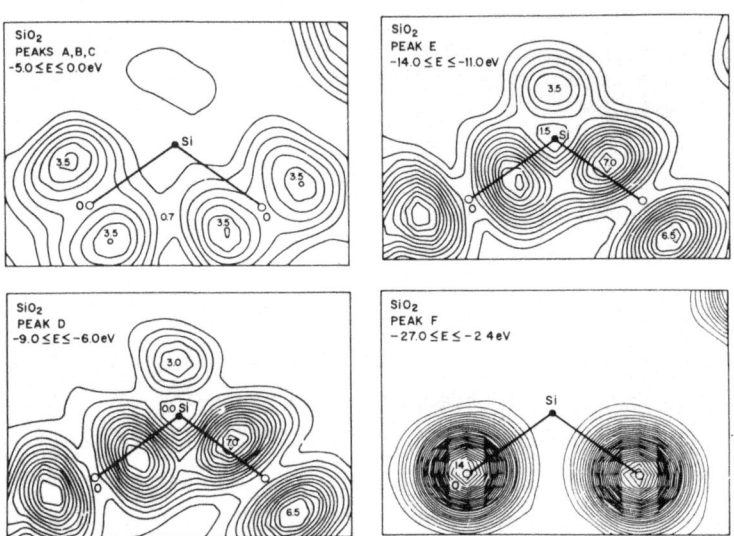

Fig. 22. Valence pseudocharge densities for features labelled in Fig. Charge densities are plotted on the O-Si—O bonding plane (54).

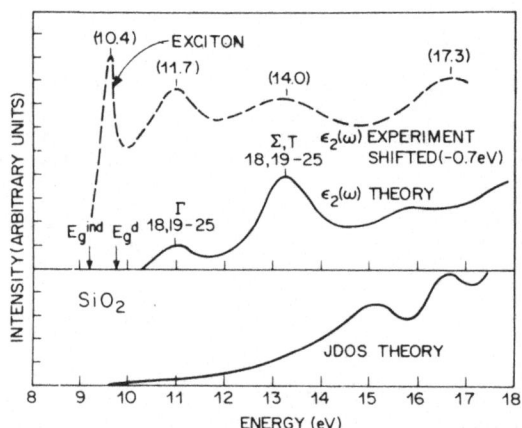

Fig. 23. Calculated joint density of states (54) for direct transitions (bottom) and $\varepsilon_2(\omega)$ after inclusion of matrix elements. Comparison is with an experimental ε_2 spectrum.

direct valence-conduction gap is dipole-forbidden and the first allowed
transition occurs just over 1 eV higher involving transitions from O 2p-
like valence states to conduction band states which are largely O 3s in
character. Pseudocharge densities for the two states indicate the
pronounced localized nature of the transitions — centred on the oxygen
ions — hence the importance of the similarity of local structure in
amorphous and crystalline SiO_2.

Interpretation of transitions in Alkali Silicates

Ellis et al (44) describe extended Huckel calculations of the band
structure of small clusters, modelling the local structure of sodium
silicates. Calculated values of joint densities of state (JDOS) functions, representing (approximately) conduction — valence band transition
probabilities (but neglecting dipole selection rules) are shown in figure
24 for SiO_2 and sodium silicate. Alkali ions cause pronounced features at
low energies which are due to transitions from O 2p levels, as in SiO_2, to
conduction-band states which are largely Na 3s and Ca 4s. The number of
electrons involved in such transitions is very small (~ 0.9) which is in
marked contrast to the relatively large effect on $\varepsilon_1(0)$, again a result of
the low energy of the transition.

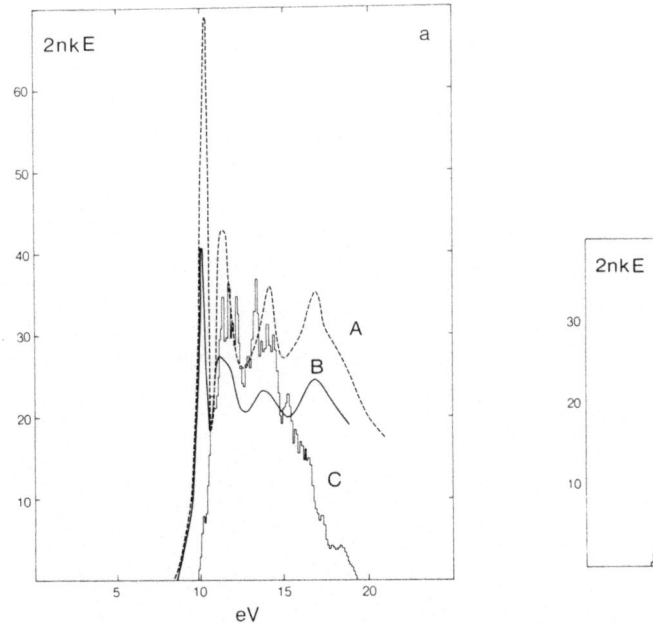
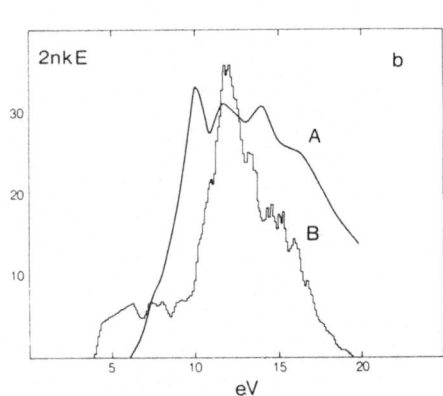

Fig. 24.   JDOS (histograms) for a) vitreous SiO_2 and b) sodium silicate
together with a) experimental $E\varepsilon_2$ data for quartz (dashed line) and
vitreous SiO_2 (full line) and b) $E\varepsilon_2$ for a $3Na_2O.7SiO_2$ glass (44).

The ε_2 spectra for various vitreous lead silicates have been reported by Bourdillon et al (46), Ellis et al (45) and Robertson (55). The minimum energy gap (~4eV) is found to be indirect, with peaks in the joint density of states at 6 and 11.5 eV. The lower feature corresponds to Pb 6s → 6p transitions while the higher energy peak corresponds to O 2p → Pb 6p. Robertson has calculated the band structure of an idealized model of lead metasilicate: structures were chosen in which the Pb atoms were sufficiently close-packed to produce the distinct Pb-Pb first neighbour peak seen in the radial distribution function at 3.9 A (56). Two different Pb-O coordination polyhedra were modelled - an <u>asymmetric</u> unit with 8-coordinated Pb atoms, but with the Pb atoms displaced forming a square pyramid with the four nearest oxygens and a <u>symmetric</u> unit with the Pb atom at the centre. Calculated JDOS and ε_2 data are shown in figure 26 where it can be seen that the optical gap is smaller for the symmetric than the asymmetric site. The symmetric site is rejected by Robertson for this reason. The asymmetry is explained in terms of stabilization resulting from a Peierls distortion - suggesting a local structure essentially equivalent to tetragonal PbO. It is not clear to what extent such distortions require correlation of nearest-neighbor Pb cations as observed in these glasses.

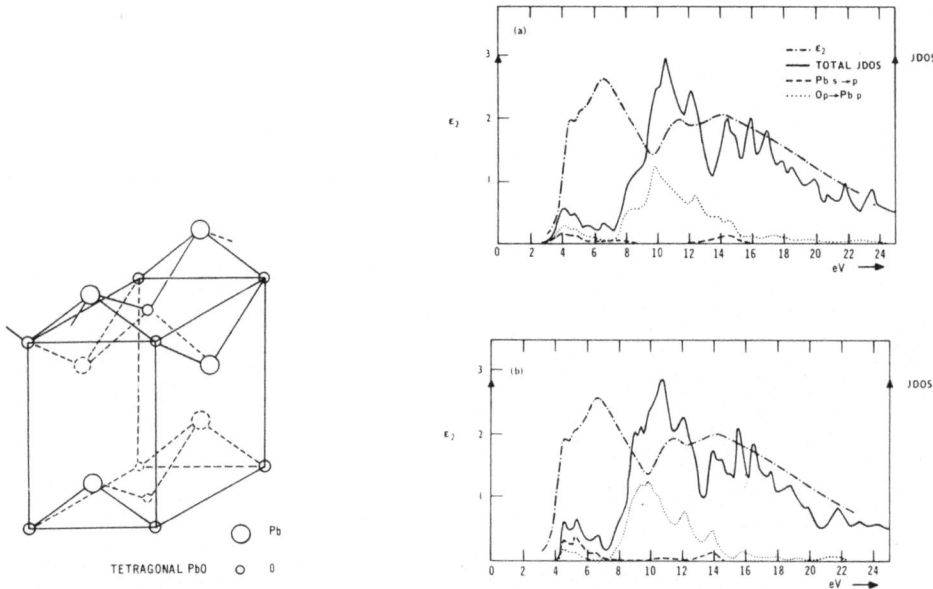

Fig. 25. a) The local coordination in tetragonal PbO b) JDOS calculated for symmetric (upper diagram) and asymmetric Pb sites (lower), in models for lead silicates, compared with ε_2 data for lead silicate glass $3PbO.2SiO_2$ (45). From Robertson (55).

Summary

The foregoing discussion shows that the sum total of knowledge of the optical behaviour of silicate glasses is rather impressive. Optical constants for a number of key materials have been measured — often over huge energy ranges. In addition to the work of Ashley et al on SiO_2 (53) quoted earlier, Ellis et al (57) have reported the optical constants for SiO_2 and a commercial soda-lime-silica glass from the vacuum ultraviolet to the microwave region - figure 26. Broadly speaking, the transitions have been catalogued and related to fairly detailed local structural information. Qualitatively, at least, it is possible to explain the high and low values of refractive index observed in 'light' and 'dense' glasses.

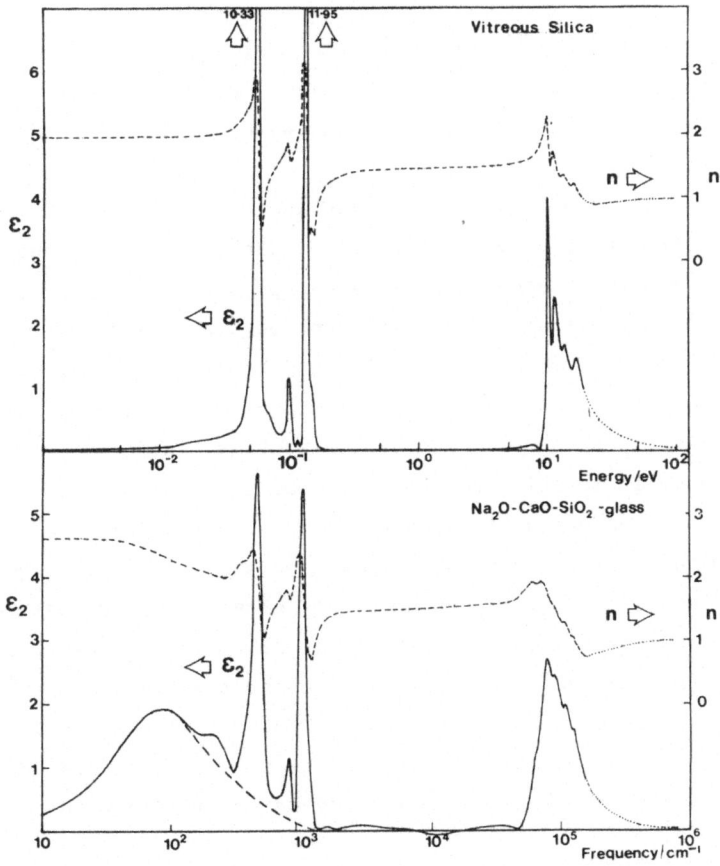

Fig. 26. Refractive index and $\varepsilon_2(\omega)$ data for vitreous silica and a commercial soda-lime silica window glass (Pilkington 'Float' glass) over an extended energy range from microwave frequencies to the vacuum ultraviolet or soft X-ray region. Low frequency results are from Birch et al (61). Note the pronounced, broad peak near 0.01 eV which appears to be the signature of the alkali and alkaline earth cations.

Nor are the achievements of approximate models to be minimized. In addition to the empirical relations between dispersion and coordination number, atomic volume etc. mentioned earlier, it has proved possible to calculate more discriminating derivative properties, such as the temp-erature- (58) and pressure-dependence of n (33) using a modified Sellmeier equation and a Penn model respectively. Wemple (39) has also been able to give a guide to the refractive index - dispersion relations observed in glasses by showing that the reciprocal dispersive power or Abbe number, $V_d = (n_D -1/(n_F -n_C)$ (where D, F and C are standard wavelengths) depends on n in an obvious way:

$$V_d = 0.72 \ E_o^2 \ (n_D/(n_D -1) = 0.72 \ E_d^2 \ n_D \ (n_D -1)^2 \ /(n_D +1)$$

Since the factor $n_D/(n_D + 1)$ is almost constant, V_d depends princip-ally on E_O (or E_d). Values of V for the likely range of E_d in glass-forming systems show a reasonable agreement with the observed range of optical glasses.

This type of treatment has also proved useful in the design of optical fibres. Apart from precise refractive index data, it is also necessary to minimize the temporal dispersion of a signal pulse by arranging that the minimum dispersion shall approximately coincide with the signal carrier frequency. Di Domenico (59), Wemple (60) and Nassau (61) have calculated the variation of this quantity with composition for a number of model binary and multicomponent oxide and halide systems - with some success.

For all this, it is still possible to feel distinctly uneasy. In this author's opinion such unease is likely to be dispersed only when the 'exact' approach exemplified by Ashley et al's analysis of vitreous silica has been performed on a sufficient number of glasses that we can feel confident that <u>accurate</u> predictions of material optical properties are available over a wide range of material types so that the 'empirical' constants are no longer purely empirical. That is, they are rigorously derivable from optical constants, they relate directly to local structural features and are thereby coupled implicitly or explicitly with chemical constitution.

Many of these desiderata are merely only a little way over the horizon - but they <u>are</u> over the horizon and only as the horizon is extended will it be possible, genuinely, unambiguously, and with total intellectual honesty, to displace the empirical composition-dependent constants which have proved to be the foundation of much of the optical lens designer's art.

ACKNOWLEDGEMENTS

The generous support of Pilkington Brothers PLC is gratefully acknowledged. The author is also grateful for discussions with Mr. J. C. Savill and for help with the diagrams from Mr. K. Papworth.

References

1. Zachariasen, W. H. J.Am.Chem.Soc. (1932) 54, 3841.
2. Gaskell, P. H.; Tarrant, I. D. Phil.Mag.B (1980) 42, 265.
3. Wright, A. C.; Sinclair, R. N. "Physics of SiO_2 and its interfaces" (ed. S. T. Pantelides); Pergamon:Oxford. 1977 p. 133.
4. Griscom, D. L.; Friebele, E. T.; Sigel, G. H.; Ginther, R. J. "Structure of Non-Crystalline Solids" (ed. P. H. Gaskell); Taylor and Francis:London 1977 p. 113.
5. Dupree, R.; Pettifer, R. F. Nature (1984) 308, 523.
6. Gladden, L. F.; Carpenter, T. A.; Elliott, S. R. Phil.Mag. (1986) in press.
7. Mozzi, R. L.; Warren, B. E. J. Appl. Crystallogr. (1969) 2 164 ; Evans, K. M.; Gaskell, P. H.; Nex, C. M. M. "Structure of Non-Crystalline Materials II" (eds. P. H. Gaskell, E. A. Davis and J. M. Parker); Taylor and Francis:London p. 426.
8. Lamparter, P.; Sperl, W.; Steeb, S.; Bletry, J.; Z.Naturforsch. (1982) 37a, 1223.
9. Lamparter, P.; Steeb, S.; "Rapidly Quenched Metals" (eds. S. Steeb and H. Warlimont); North-Holland:Amsterdam (1985) Vol. 1; p. 459.
10. Pannisod, P.; Bakonyi, I.; Hasegawa, R.; Phys.Rev. B. (1983) 28, 624.
11. Dubois, J. M.; Gaskell, P. H.; Le Caer, G. Proc.Roy.Soc. A (1985) 402, 323.
12. Orwell, G. "Animal Farm — A Fairy Story" Secker and Warburg: London (1945) p28.
13. Turnbull, D. Contemp. Phys. (1969) 10, 473.
14. Davies, H. A. Physics.Chem.Glasses. (1976) 17, 159.
15. Uhlmann, D. R. J.Non-Cryst.Solids. (1977) 25, 42.
16. Sadoc, J. F.; Mosseri, R. Phil.Mag. (1982) B 45, 467.
17. Nelson. D. R. Phys Rev B (1983) B28, 5515.
18. Frank, F. C. Proc. Roy. Soc. (1952) A215 43.
19. Phillips, J. C. "Solid State Physics" (eds. H. Ehrenreich, F. Seitz and D. Turnbull) Academic Press:New York (1982) 37, 93.
20. Weber, M. J. J.Non-Cryst.Solids (1985) 73, 351.
21. Gaskell, P. H. Acta Metall. (1981) 29, 1203.
22. Johnson, W. L.; Williams, A. R. Phys.Rev.B. (1979) 20, 1640.
23. Turnbull, D. Scripta Metall. (1977) 11, 1131.
24. Bennett, L. H.; Watson, R. E. Scripta Metall (1982) 16, 1379.
25. Gaskell, P. H. J.de Physique (1982) 43, C9, 101.
26. Huppert, H. E.; Sparks, R. S. J. Ann. Rev. Earth Plan. Sci. (1984) 12 11.
27. Huggins, M. L. J.Opt.Soc.Amer. (1940) 30, 420; Stevels, J. M. "Handbuch der Physik" (ed. S. Flugge); Springer-Verlag:Berlin. (1962) 13, 510.
28. Hesse, K-F. Acta Cryst. (1977) B33, 901.
29. McDonald, W. S.; Cruickshank, D.W.J. Acta Cryst. (1967) 22, 37.
30 Dent-Glasser, L. S. Z.Krist. (1979) 149, 291.
31. Gaskell, P. H. J.de Physique (1985) Colloque C8, 3.
32. Dupree, R.; Holland, D.; McMillan, P. W. Pettifer, R. F. J.Non-Cryst.Solids. (1984) 68, 399.

33. Weinstein, B. A.; Zallen, R.; Slade, M. L.; Mikkelson, J. C.Jr. Phys.Rev.B (1982) 25, 781.
34. Penn, D. R. Phys.Rev. (1962) 128, 2093.
35. Phillips, J. C. Phys.Rev.Lett. (1968) 20, 550.
36. Van Vechten, J. A. Phys.Rev. (1969) 182, 891.
37. Harrison, W. A.; Pantelides, S. T. Phys.Rev. B (1976) 14, 691.
38. Wemple, S. H.; Di Domenico, M. Jnr. Phys.Rev. B (1971) 3, 1338.
39. Wemple, S. H. Phys.Rev. B (1973) 7, 3767.
40. Wemple, S. H. J.Chem.Physics (1977) 67, 2151.
41. Wolfe, W. L. "Handbook of Optics" (eds. W. A. Driscoll, W. Vaughan) McGraw-Hill:New York (1978) 7-85.
42. Aslaksen, E. W. Phys.Rev.Lett. (1970) 24, 767.
43. Robertson, J. J.Non-Cryst.Solids (1980) 42, 381.
44. Ellis, E.; Johnson, D. W.; Breeze, A.; Magee, P. M.;Perkins, P. G. Phil.Mag. (1979) B40, 105.
45. ibid p. 125.
46. Bourdillon, A. J.; Khumala, F.; Bordas, J. Phil.Mag. B (1978), 731.
47. Hirota, S. T.; Izumitani, T.; Onaka, R. J.Non-Cryst.Solids (1985) 72, 39.
48. Platzoder, K. Phys.Stat.Sol. (1968) 29, K63.
49. Sigel, A. H. J.Phys.Chem.Solids. (1971) 32, 3273.
50. Philipp, H. R. J.Phys.Chem.Solids. (1971) 32, 1935.
51. Buechner, V. J.Phys.C. (1975) 8, 155.
52. Griscom, D. L. J.Non-Cryst.Solids (1977) 24, 155.
53. Ashley, J.C.; Anderson, V.E., J.El.Spec. (1981) 24, 127.
54. Chelikowsky, J. F.; Schluter, M. Phys.Rev B. (1977) 15, 4020.
55. Robertson, J. Phil.Mag. B (1981) 43, 497.
56. Mydlar, M. F.; Kreidl, N. J.; Hendron, J. K.; Clayton, G. T. Physics.Chem.Glasses (1970) 11, 196.
57. Ellis, E.; Gaskell, P. H.; Johnson, D. W. in Physics of Non-Crystalline Solids (ed.G. H. Frischat) Trans Tech Publ: Aedersmanndorf (1977) p312.
58. Goldschmidt, D. Phys.Rev (1983) B28, 7177.
59. Di Domenico, M.Jr. Appl.Opt. (1972) 11, 652.
60. Wemple, S. H. Appl.Opt. (1979) 18, 31.
61. Nassau, K. Bell.Syst.Tech.J. (1981) 60, 327.
62. Birch, J. R.; Cook, R. J.; Harding, A.F.; Jones, R.G.; Price, G. D. J.Phys.D: Appl. Phys., (1975) 8 1353.

INFRARED AND RAMAN STUDIES OF

Si-CHALCOGENIDE GLASSES

M. Tenhover, R. S. Henderson, M. A. Hazle,
D. Lukco, and R. K. Grasselli

Standard Oil Research and Development
The Standard Oil Company
4440 Warrensville Center Road
Warrensville Heights, Ohio 44139

ABSTRACT

The atomic structure of Si_xX_{1-x} (X=S,Se) glasses is investigated
with the combined use of Infrared and high resolution Raman spectroscopy.
These measurements permit a careful analysis of the various vibrational
modes and the evaluation of structural models for the glasses. Glassy
SiX_2 alloys are composed of both chains of edge-sharing tetrahedral units
and corner-sharing units in which the Si atoms are four-fold coordinated
to X atoms and the X atoms are two-fold coordinated to Si atoms. These
edge-sharing and corner-sharing tetrahedral units are also present
throughout the glass forming range. In the entire composition range
studied (0.05<x<0.40) heteropolar bonding is strongly favored. In X-rich
glasses (x<1/3), the tetrahedral $Si(X_{1/2})_4$ units coexists with S(Se)
rings and chains. A new molecular unit is present along with the Si
tetrahedral units in the Si-rich glasses (x>1/3) and is identified as
ethane-like $X_3-Si-Si-X_3$ units. The combined use of infrared and Raman
spectroscopies provides the strongest evidence to date for the presence
of extensive short and medium range order in the Si-chalcogenide glasses.

In contrast to the considerable number of investigations that have been conducted on the structure and properties of SiO_2 and related glasses, glasses of the series Si-S, Si-Se, and Si-Te have received relatively little attention. Much of the early interest in chalcogenide glasses focussed on device applications and this topic is still of considerable interest (1-6). Glasses containing Si along with S,Se,and Te such as Si-Ge-As-S have been developed for memory and switching applications. In addition to this, studies of the binary and pseudo-binary Si-chalcogenides glasses include X-ray diffraction (Si-Te(7)), infrared absortion (Si-Te(8)), Raman scattering (Si-S(9),Si-Se(10-13)), EXAFS and neutron diffraction (Si-Se(13)), optical absorption (Si-S, Si-Se(14)), and thermal properties (Si-Se(13-14)). Renewed interest in glasses of this type has resulted from their application as Li-ion conductors(15-16) and fundamental questions concerning the nature of glasses. The concept of optimum coordination numbers for the mechanical and chemical stability of glasses developed by Ovshinsky(17), Phillips(18), Dohler et al(19), and Thorpe(20) have been extensively examined in chalcogenide glasses, especially in the Ge- and As- based chalcogenide glasses.

Like the Ge-based chalcogenide glasses (such as Ge-S and Ge-Se), the description of the atomic stucture of the Si-S and Si-Se glasses is still evolving. The first Raman studies of these materials focussed on the question of medium range order in the glasses SiS_2 and $SiSe_2$(21). Based on the similarity between the Raman spectra of the glassy and crystalline forms of these materials it was concluded that the glasses were composed of bitetrahedral units. Assignment of the prominent Raman lines in the crystalline and glassy forms were based on the expected four lines of $Si(X_{1/2})_4$ (X=S,Se) tetrahedral units and a Raman line associated with the stretching mode of the X atoms in extended edge-sharing tetrahedral chains. More detailed and rigorious assignments of the Raman spectrum of crystalline $SiSe_2$(11) and the Raman and IR spectra of crystalline and glassy $SiSe_2$(12) provided additional support for extensive medium range order in Si-X glasses. Griffiths et al (11) investigated glasses of Si_xSe_{1-x} (x>0.333) (Si-rich glasses) and developed a structural model for glassy $SiSe_2$ and the Si-rich glasses. They proposed that in contrast to crystalline $SiSe_2$ in which all the Si atoms are in extended edge-sharing tetrahedral chains, the glassy form contains both edge-sharing and corner sharing tetrahedra.

In the present work, new results of infrared (IR) and Raman experiments are reported on the Si_xS_{1-x} and Si_xSe_{1-x} glassy systems. In particular, IR measurements have been made on both the Si-S and Si-Se systems as a function of Si content. The IR measurements are an important complement to the previous Raman measurements and permit an evaluation of the various proposals concerning the structure of the Si-based glasses. These measurements provide additional insight into the atomic structure of the Si-based glasses.

EXPERIMENTAL

All the samples of this study were prepared as bulk glasses. The starting materials were chunks of vacuum degassed Si and Se and resublimed S. Powders of the starting materials were not used due to concerns about oxygen contamination of the glasses and to insure that the desired compositions were not changed by these oxygen impurites leading to second phases. The chunks of Si, S and Se were vacuum sealed in fused silica ampoules that had been throughly degassed at high temperatures ($1000^{o}C$) The temperatures required to synthesis these alloys ($>1100^{o}C$) are near the safe limits of fused silica. One problem encountered in the early work was the formation (via crystallization) of cristobalite on the surface of the fused silica tubes which reduces the transparency and mechanical strength of the tubes. This problem was eliminated by the following treatment. After sealing the samples in the silica tubes, the outsides of the tubes were throughly washed in acetone, followed by ethanol, and then rinsed is distilled water. After this treatment, the tubes were handled exclusively with forceps or gloves. All the water quenching was performed in room temperature distilled water. This procedure resulted in clear transparent sample tubes and facilitated the determination of the homogenity of the samples. This is an important consideration since the Si_xS_{1-x} and Si_xSe_{1-x} glasses are difficult to prepare as homogeneous glasses, especially in the Si-rich region ($x>0.34$). The samples were vertically suspended in a furnace and slowly heated over a period of 6 hours to $1100^{o}C$. There were repeatedly quenched in distilled water and flipped over allowing the sample to flow back upon itself. The quenching/flipping process was performed 4 times/day. The homogeneity of the samples was gauged by visual inspection and by recording Raman Spectra along the length of the samples while they were still sealed in the fused silica ampoules. They were judged to be homogeneous when they appeared uniform in color and when the Raman spectrum didn't change along the length of the tube. The average time for a sample to meet the homogenity criteria appeared to increase with increasing Si-content. The typical time for $SiSe_2$ was 6-8 days while a sample such as $Si_{0.38}Se_{0.62}$ required alloying over a period of 21 days. All the measurements reported here are on samples determined to be homogeneous using the above criteria. At high Si concentrations (beyond $Si_{0.38}Se_{0.62}$), samples could not be made homogeneous by the method described above. Two other references however claimed to have made homogeneous samples up to and including $Si_{0.40}Se_{0.60}$ (11,13). Reference 11 further claims that phase-separation occurs in the Si-Se system in the region ($0.338<x<0.345$). The present work in agreement with reference 13 finds no indication of phase separation; all the samples appeared from the visual and Raman measurements to be single phase on a macroscopic scale.

The Raman measurements were performed as described in reference 10. All Raman measurements were done with the samples still sealed in their silica ampoules. The IR measurements were performed in a somewhat unconventional manner compared to what has been traditionally done in Ge- and As- chalcogen glasses. The concern in the case of the Si- glasses is that they are air sensitive, with one of the products of the reaction being SiO_2 (a strong IR absorber). This problem prohibits the usual IR reflectance measurements. The glassy samples were removed from the fused silica tubes in an ultrapure Ar dry box (less than 1 ppm H_2O). The glasses were powdered and mixed with dry CsI, pressed into a pellet, and mounted in an IR cell having polyethlene windows. The IR measurements were made in the transmission mode with a Nicolet 2005X Vacuum Bench Fourier Transform IR Spectrometer in the region 180 to 900 cm^{-1}. To test this system and method of sample preparation, several Ge_xS_{1-x} (x= 0.0, 0.10, 0.25) glasses were examined and the IR spectra are shown in figure 1. These spectra are in excellent agreement with those obtained in the reflection mode by Lucovsky et al(22). The disadvantage of this method lies in the inability to determined the true IR thickness of the sample and therefore make quantitative comparisons between samples. This is a serious restriction and limits the useful information obtainable from the IR measurements. In reference 22 for example, careful quanitative measurements in Ge_xS_{1-x} glasses allowed the investigators to test several models to determine the role excess S plays in S-rich glasses.

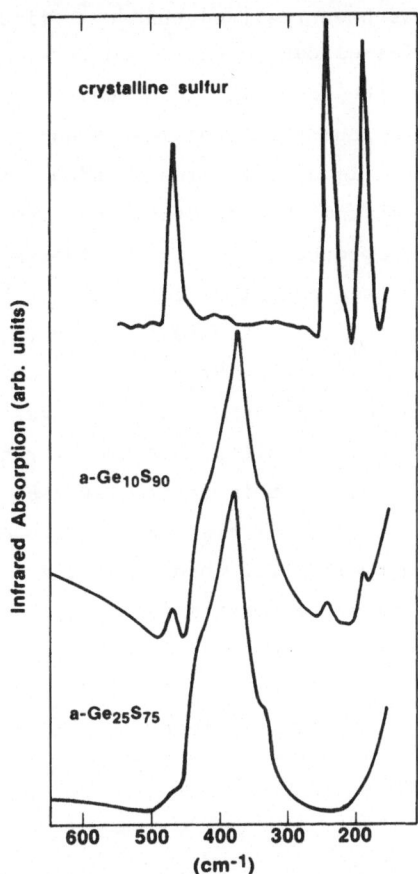

Figure 1.   Infrared spectra of two Ge-S glasses and crystalline sulfur. The measurements were made in the transmission mode.

RESULTS AND DISCUSSION

Ge-Based Chalcogenide Glasses

It was mentioned in the introduction that the description of the atomic structure of Ge-S and Ge-Se glasses is still evolving and is the topic of some controversy. It is useful at this point to give a brief perspective on the status on the understanding of the atomic structure of Ge-glasses. Since the early work of Tronc et al (Ge-Se(23)) and Lucovsky et al (Ge-S(22)) it has been realized that to a first approximation the structure of the Ge glasses can be described by a chemically ordered network model in which the Ge and X atoms obey the 8-N rule. In such a model, the Ge atoms are four-fold coordinated and the X atoms are two-fold coordinated. There is a very strong tendency for heteropolar bonding in these systems and this tendency has important consequences for the optical and thermal properties of these glasses(18). The Raman and IR spectra of the Ge-X glasses contain modes that can be associated with the four normal modes of $Ge(X_{1/2})_4$ tetrahedral units (E, F_2, A_1, F_2). Like the isolated molecules $GeCl_4$ and $GeBr_4$, all four of these modes are Raman active and the F_2 modes are IR active. The observation that some modes are both IR and Raman active implies that the vibrational units do not have a center of inversion. In this particular case it also means that the tetrahedral units in the Ge-X glasses are vibrationally isolated from the rest of the network (i.e. corner sharing tetrahedra). The experimently observed mode frequencies for $Ge(X_{1/2})_4$ tetrahedral units in the glasses along with the corresponding Ge-halogen compounds are listed in Table I. The good agreement between the modes in the glasses and the Ge-halogen compounds conclusively illustrates the presence of tetrahedral units in the glasses.

Table I.   Experimently observed vibrational modes of Ge-halogen and Ge-chalcogen tetrahedral units. Both Raman (R) and infrared (IR) active modes are listed.

Mode	$GeCl_4$ (cm^{-1})	$Ge(S_{1/2})_4$ (cm^{-1})	$GeBr_4$ (cm^{-1})	$Ge(Se_{1/2})_4$ (cm^{-1})
$F_2(R,IR)$	459	375	332	258
$A_1(R)$	397	345	236	198
$F_2(R,IR)$	171	147	111	115
$E(R)$	125	114	75	82

One feature of the raman spectra of GeS_2 and $GeSe_2$ that has attracted considerable attention is the so-called A_1-companion mode. In $GeSe_2$, this mode occurs at about 213 cm^{-1} and has no analog in the raman spectrum of polycrystalline $GeSe_2$. Early on, Nemanich and Solin (24) suggested that this feature indicates the presence of rings of atoms in the glassy structure. Bridenbaugh et al (25) proposed that the A_1 companion modes were the signatures of layer-like units in the glassy network. These units are proposed to be reconstructed fragments of the high temperature crystalline phase of $GeSe_2$. This model has been further developed and used to explain a variety of the properties of the glasses such as photo-induced quasicrystallization (26) and Mossbauer substituion experiments (27). The work of Kawamoto and Kawashima (28) on GeS_2 also stressed the similar nature of the amorphous and crystalline forms of GeS_2. Their comparison of the IR and raman spectra of glassy and polycrystalline GeS_2 led them to conclude that glassy GeS_2 is composed of tetrahedral GeS_4 units similar in structure to those found in the high temperature form of crystalline GeS_2. Another important conclusion of their work was that the two dimensional character of the crystal was to some extent retained in the glass. Recent work on the structure of GeS_2 and $GeSe_2$ has suggested that the A_1-companion line can be accounted for by the presence of small rings of atoms in the glassy network (30,31). Like the high temperature forms of $GeSe_2$ and GeS_2, the glass is imagined to consists of both edge-sharing and corner-sharing tetrahedra. A careful x-ray diffraction study has been recently reported on glassy GeS_2 and $GeSe_2$(32). The experimently derived radial distribution functions (RDF) were compared to several model RDF's. The conclusion of this process was that the structure of the two Ge-glasses are similar and that it could be modeled by a three dimensional connected network of $Ge(X_{1/2})_4$ units 1/4 of which are edge-sharing and 3/4 of which are corner sharing.

Although the detailed nature of glassy GeS_2 and $GeSe_2$ is not universally agreed upon, it appears clear that they are not random covalent networks. Most measurements and intrepretations have pointed to the existence of medium range order related to that which occurs in the crystalline forms of these materials. In both the crystalline and glassy forms, the $GeS_2(Se_2)$ materials exhibit edge-sharing and corner-sharing tetrahedral units. Despite the uncertainities concerning the exact nature of the atomic structure of the stoichiometric GeS_2 and $GeSe_2$ glasses, some of the changes that occur when the chalcogenide content is varied are easy to understand. For S or Se rich glasses (x<1/3), properties such as glass transition temperature, crystallization temperature, glass forming ability, and optical absorption are in large part determined by the presence of homopolar X-X bonds (X=S,Se) (26,33,34). These materials can be described as molecularly phase separated glasses (22,23). for the X-rich glasses, the Ge atoms are coordinated to four X neighbors and act to cross link the Se or S units. the signature of X-X bonds and $Ge(X_{1/2})_4$ tetrahedral units are clearly seen in the Raman and IR spectra. For very X-rich Ge glasses, the Ge atoms exists in isolated $Ge(X_{1/2})_4$ tetrahedral units. In this case, the Raman and IR spectra are consistent with the vibrations of GeX_4 molecules and X rings and/or chains. In the Ge-rich (x>1/3) region, it has been proposed that a new molecular unit appears in the glasses having an ethane-like structure $X_3-Ge-Ge-X_3$. One key feature here is that Ge-Ge bonds now occur in the glasses. Recent Mossbauer experiments on Sn-doped Ge-X glasses have suggested that the situation is more complex, but retain the idea that Ge-Ge bonds start forming for x>1/3 (35).

CRYSTALLINE SiS_2 AND $SiSe_2$

Figure 2 shows the Raman and IR spectra of crystalline SiS_2. The spectra of the various forms of $SiSe_2$ have been shown in references 11 and 12. Early analysis of the Raman spectra of these materials emphasized the similarity between amorphous and crystalline $SiS_2(SiSe_2)$ and identified the modes at approximately 121(68), 174(110), 427(242), and 625(463) as the normal modes (E, F_2, A_1, F_2) of $Si(X_{1/2})_4$ tetrahedral units (10,21). The high resolution Raman measurements of Griffiths et al (11) and the IR measurements reported in reference 12 show that this was not the best description of the vibrational modes of these materials. One key observation is that the F_2 modes of vibrationally isolated $Si(X_{1/2})_4$ tetrahedra should be both IR and Raman active. As figure 2 clearly shows, there are no prominent vibrational modes in crystalline SiS_2 that are both IR and Raman active; a similar result for $SiSe_2$ was illustrated in reference 12. The results of a closer analysis of the expected vibrational spectra of crystalline SiS_2 and $SiSe_2$ was given in reference 12 (a different analysis leading to a similar description appeared in reference 11). The details are presented here to illustrate the method and to provide additional insight into the meaning of the Raman and IR spectra of the glasses.

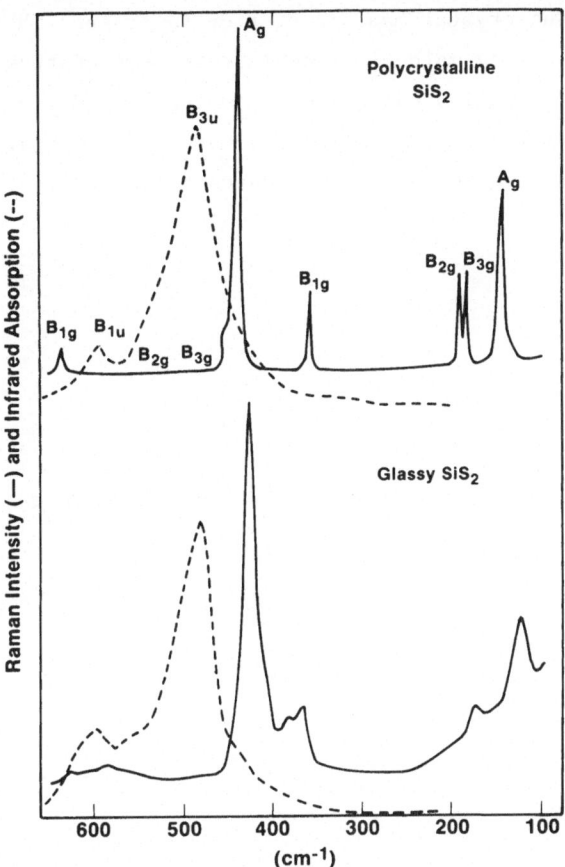

Figure 2.   Raman and infrared (IR) spectra of polycrystalline and

glassy SiS_2. The IR measurements were made in the

transmission mode. The relative intensities for all four

spectra are arbitrary.

Crystalline SiS_2 and $SiSe_2$ are orthorhombic (Ibam, C42-type, D_{2h}^{26}
with four $Si(X_{1/2})_4$ molecules in the unit cell(36). Using the
correlation method (37) it is a straight forward task to determine the
number and symmetry types of the expected modes. There are nine possible
Raman active (gerade) modes ($2A_g$, $3B_{1g}$, $2B_{2g}$, $2B_{3g}$), five possible IR
active (gerade) modes ($1B_{1u}$, $2B_{2u}$, $2B_{3u}$), and four modes that are neither
Raman nor IR active (A_u, B_{1u}, B_{2u}, B_{3u}). In the crystalline compounds,
the fact that no modes are observed that are both Raman and IR active is

expected since the crystal has a center of inversion. The next step in the analysis is to use molecular models to make tentative assignments and for this case, the well-studied model compounds Al_2Cl_6 and Al_2Br_6 (nonplanar bridging X_2Y_6 $-D_{2h}$ molecules) are appropriate. They have similar structure, masses and force constants to the materials of present

Table II. Raman and IR modes of the compounds Al_2Cl_6 and Al_2Br_6.

Mode	Al_2Cl_6 (cm^{-1})	Al_2Br_6 (cm^{-1})
$n_1(A_g)$	501	410
$n_2(A_g)$	336	212
$n_3(A_g)$	217	142
$n_4(A_g)$	115	70
$n_5(A_u)$	--	--
$n_6(B_{1g})$	438	354
$n_7(B_{1g})$	168	82
$n_8(B_{1u})$	625	500
$n_9(B_{1u})$	---	---
$n_{10}(B_{1u})$	---	---
$n_{11}(B_{2g})$	608	491
$n_{12}(B_{2g})$	---	116
$n_{13}(B_{2u})$	420	341
$n_{14}(B_{2u})$	---	90
$n_{15}(B_{3g})$	---	---
$n_{16}(B_{3u})$	484	373
$n_{17}(B_{3u})$	---	198
$n_{18}(B_{3u})$	---	110

interest. The model compounds also have 18 possible vibrational modes of which nine are Raman modes and nine are IR modes. The observed modes of Al_2Cl_6 (analog to SiS_2) and Al_2Br_6 (analog to $SiSe_2$) are listed in Table

Table III. Assignments of the vibrational modes of crystalline SiS_2 and $SiSe_2$.

Mode	Crystalline SiS_2 (cm^{-1})	Crystalline $SiSe_2$ (cm^{-1})
$n_1(A_g)$	430	248
$n_2(A_g)$	138	82
$n_3(A_u)$	IA[*]	IA[*]
$n_4(B_{1g})$	625	512
$n_5(B_{1g})$	351	204
$n_6(B_{1g})$	NO[**]	NO[**]
$n_7(B_{1u})$	590	492
$n_8(B_{1u})$	IA	IA
$n_9(B_{2g})$	NO	355
$n_{10}(B_{2g})$	181	122
$n_{11}(B_{2u})$	402	397
$n_{12}(B_{2u})$	NO	NO
$n_{13}(B_{2u})$	IA	IA
$n_{14}(B_{3g})$	NO	345
$n_{15}(B_{3g})$	175	114
$n_{16}(B_{3u})$	487	468
$n_{17}(B_{3u})$	NO	NO
$n_{18}(B_{3u})$	IA	IA

---

[*] IA= inactive mode

[**] NO= mode not observed.

II. These assignments along with the vibrational modes of simple tetrahedral molecules such as $SiCl_4$ and $SiBr_4$ motivates the assignments show in Table III for SiS_2 and $SiSe_2$. A final simple check on this tentative assignment can be obtained by looking at the ratios of the respective modes in the SiS_2 and $SiSe_2$ cases. Some of these ratios are easy to interpret. For example, those in the 1.7-1.78 range indicate vibrational modes in which the chalcogen atoms are in motion with respect to fixed Si atoms. At present no detailed valence force field calculations have been used to confirm this assignment.

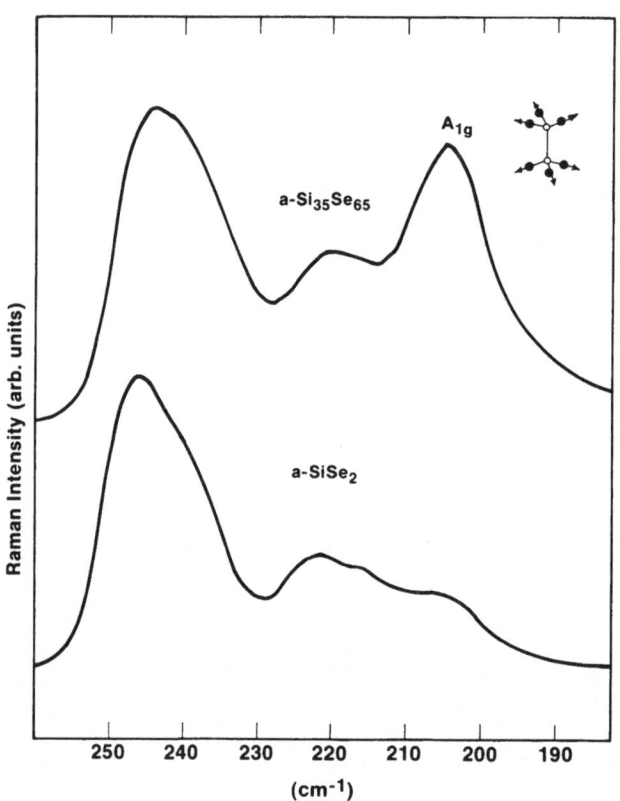

Figure 3.   Raman spectra of two Si-Se glasses. The insert (top right) illustrates the A_{1g} mode of an ethane-like molecule.

The Raman and IR spectra of glassy SiS$_2$ are shown in figure 2 and the Raman spectrum of glassy SiSe$_2$ is shown in figure 3. Comparing the previous Raman and IR spectra of glassy SiSe$_2$ with glassy SiS$_2$ (figure 2) it can be seen that they have similar spectra. Of special interest is the observation that like the crystalline compounds, there are no dominant modes which are both Raman and IR active. This is evidence that the basic unit in the glass has a center of inversion and supports the claim that the fundamental structural unit in the Si-glasses is not a tetrahedra, but an edge-sharing tetrahedral chain. The earlier evidence for this claim rested on the similarity between the Raman spectra of the crystalline and glassy forms of these alloys. Earlier work (11,12) stressing the similar nature of the glasses and polycrystalline materials assigned the Raman and IR modes in the glasses using the same format as that used in the crystalline case. This is misleading and will be avoided in the present work since the Raman probes are basically sensitive to the local structure of the glasses. In figure 2, only minor differences can be seen between the spectra of polycrystalline and glassy forms of SiS$_2$ such as the broadening of the Raman and IR lines. In the present work, care will be taken to not over-interpret the Raman and IR spectra of these glassy alloys and to attempt to explain the spectral results with a minimum of assumptions and speculations. At this time, with both the IR and high resolution Raman results available, it is appropriate to focus on the differences between the spectra of the crystalline and glassy materials. One very important difference is the observation that there are some weak lines in glassy SiS$_2$ and SiSe$_2$ that are both Raman and IR active. In particular, the lines at about 465 cm^{-1} in glassy SiSe$_2$ and at about 597 cm^{-1} in glassy SiS$_2$ are both Raman and IR active. In the bond stretching region, the glassy samples exhibit structure not found in the spectra of the crystalline samples. Griffiths et al (11) were the first to point out that in glassy SiSe$_2$ the Raman features at approximately 243 and 215 cm^{-1} contain additional structure. The Raman line at about 242 cm^{-1} has a shoulder at about 240 cm^{-1} and the line at 215 cm^{-1} is composed of at least two contributions located at 219 and 211 cm^{-1}. These are vibrational modes in the bond-stretching region and can be seen in figure 3. As figures 2 and 4 show, the same features can be identified in the Si-S glasses. In particular, the splitting of

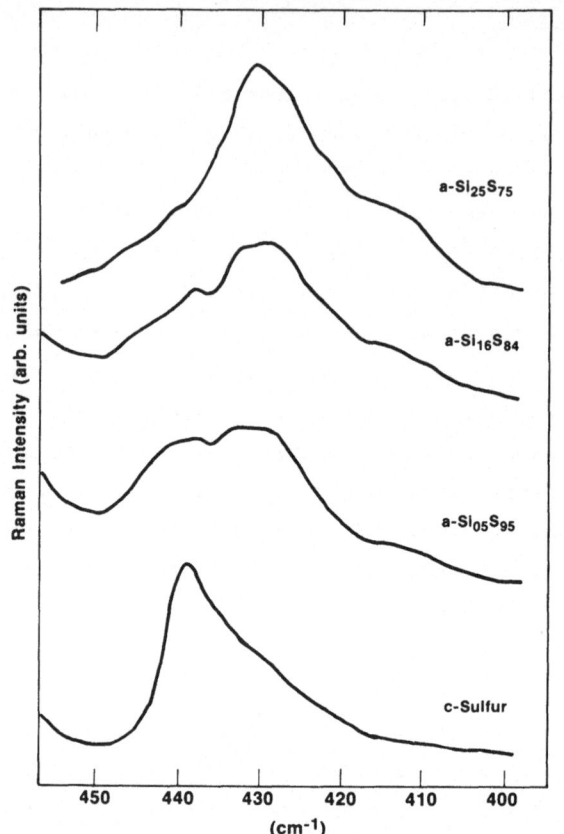

Figure 4.   Raman spectra of several Si_xS_{1-x} glasses and crystalline
sulfur in the region 400–430 cm^{-1}.

the peak near 380 cm^{-1} is clearly seen and a shoulder (415 cm^{-1}) is found
on the central peak at 427 cm^{-1}. Table IV lists the four peaks of
interest here for both SiS_2 and $SiSe_2$. In column three the ratio of the
modes is given and this ratio is precisely what one would expect using a
model in which only the chalcogenide atoms are involved in the
vibrational motion (21). Griffiths et al (11) proposed that in glassy
$SiSe_2$, the modes at 236 and 218 cm^{-1} should be associated with corner-
sharing tetrahedral units. All four of the Raman modes in the bond
stretching region are highly polarized and IR inactive. Another
important difference between the spectra of crystalline and glassy
materials is concerned with the B_{1g} mode at about 204 cm^{-1} in c-$SiSe_2$ and
351 cm^{-1} in c-SiS_2. In these materials, this mode corresponds to a
scissor-like mode with S(Se) atoms stretching about fixed Si atoms in the

342

Table IV. Raman modes in the bond stretching region for SiS_2 and $SiSe_2$ glasses.

SiS_2 (cm^{-1})	$SiSe_2$ (cm^{-1})	Ratio
427	242	1.76
415	237	1.75
381	219	1.74
367	211	1.74

tetrahedral chains. Earlier studies (9,11,12,21) associated either one of the peaks or all of the broad Raman scattering in the range 211-219 cm^{-1} in glassy $SiSe_2$ or 367-381 cm^{-1} in glassy SiS_2 with this feature in the crystalline spectra. From experiments on small single crystals of SiS_2, this B_{1g} mode has been found to be unpolarized. This is of course consistent with its microscopic nature. However, the fact that the nearby modes in the glass are highly polarized coupled with the rather large shift between the B_{1g} line in crystalline SiS_2 and what has been assumed to be the related lines in the glass (shift= 16-30 cm^{-1}) makes this previous identification suspect. In contrast to the large shift of these modes between the crystalline and glassy spectra, the most intense mode in glassy SiS_2 corresponds nicely with the symmetric stretching mode of the Si-tetrahedra in the edge-sharing chains of the crystalline structure (A_g mode).

The above observations concerning the IR/Raman activity of some of the modes, the polarization of lines in the glasses, and the shift of the lines can be explained by a simple model involving two structural units in the glasses: edge-sharing tetrahedral chains and corner-sharing tetrahedral units. The edge-sharing units have a center of inversion, while the corner sharing units do not. The expected modes of the $Si(X_{1/2})_4$ corner-sharing tetrahedra can be estimated through the use of the molecular cluster model which has been reported in references 9 and 10. In Table V, the modes of the model compounds $SiCl_4(Br_4)$, the results of the molecular cluster calculations, and the appropriate modes of the spectra of the glasses are listed. Note that in addition to the excellent agreement in terms of the mode frequencies, the various modes have the correct Raman/IR activities and polarizations. For the modes associated with the edge-sharing tetrahedral units, the best models are the crystalline forms of the materials. The modes of the glasses that best correspond to those found in the crystalline samples are shown in Table VI. Again, in addition to the agreement between the mode frequencies, the modes have the correct Raman/IR activities and

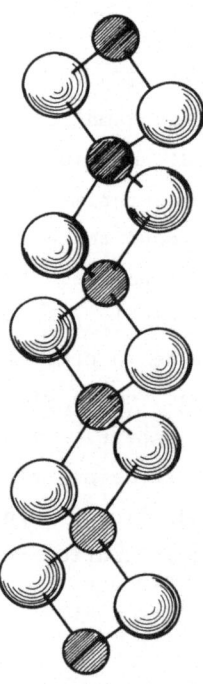

Figure 5.   Edge-sharing tetrahedral chains of crystalline SiS_2 and $SiSe_2$ in which the small closed circles represent Si atoms and the large open circles represent S/Se atoms.

Table V. Vibrational modes in glassy $SiS_2(Se_2)$, molecular $Si(X_{1/2})_4$ clusters, and Si-halogen compounds. R–Raman active, IR– infrared active, P– polarized, and U– unpolarized.

Mode	Glassy SiS_2 (cm^{-1})	Molecular Cluster $Si(S_{1/2})_4$ (cm^{-1})	$SiCl_4$ (cm^{-1})	Glassy $SiSe_2$ (cm^{-1})	Molecular cluster $Si(Se_{1/2})_4$ (cm^{-1})	$SiBr_4$ (cm^{-1})
E	121 (U/R)	121 (U/R)	150 (U/R)	68 (U/R)	68 (U/R)	90 (U/R)
F_2	----	142 (U/R,IR)	221 (U/R,IR)	----	85 (U/R,IR)	137 (U/R,IR)
A_1	381 (P/R)	380 (P/R)	423 (P/R)	219 (P/R)	224 (P/R)	249 (P/R)
F_2	590 (U/R,IR)	587 (U/R,IR)	608 (U/R,IR)	465 (U/R,IR)	460 (U/R,IR)	487 (U/R,IR)

Table VI. Identification of some of the modes in glassy $SiSe_2(S_2)$ with the corresponding modes in crystalline $SiSe_2(S_2)$.

Mode	$SiSe_2$		SiS_2	
	crystal (cm^{-1})	glass (cm^{-1})	crystal (cm^{-1})	glass (cm^{-1})
B_{1g}	512	512 (U)	625	625 (U)
B_{1u}	492	490	590	597
B_{3u}	468	460	487	481
B_{2g}	355	350 (U)	NO*	NO
B_{3g}	345	350 (U)	NO	NO
A_g	248	243 (P)	430	427 (P)
B_{2g}	122	115 (U)	181	177 (U)
B_{3g}	114	110 (U)	175	174 (U)

* – mode not observed.

polarizations. Note that no modes in the glasses have been assigned to the mid-frequency B_{1g} mode found in the crystalline materials. The two remaining bond stretching modes in Table IV for glassy SiS_2 (415, 367 cm^{-1}) and glassy $SiSe_2$ (237, 211 cm^{-1}) have not been identified by either the corner or edge-sharing tetrahedral units. They likely represent Si-tetrahedral sites which differ slightly from those described above.

The description of the structure of SiX_2 glasses resulting from the above analysis is that it consists of predominantly edge-sharing tetrahedral units with a smaller number of corner-sharing tetrahedral units. As discussed in the original work on SiX_2 glasses (21), the length of these edge-sharing tetrahedral chains (pictured in figure 5) can in principle range from isolated bitetrahedral units to extended semi-infinite chains (as in crystalline SiX_2). Other complementary measurements such as X-ray diffraction or Si-NMR measurements are needed to determine the chain lengths.

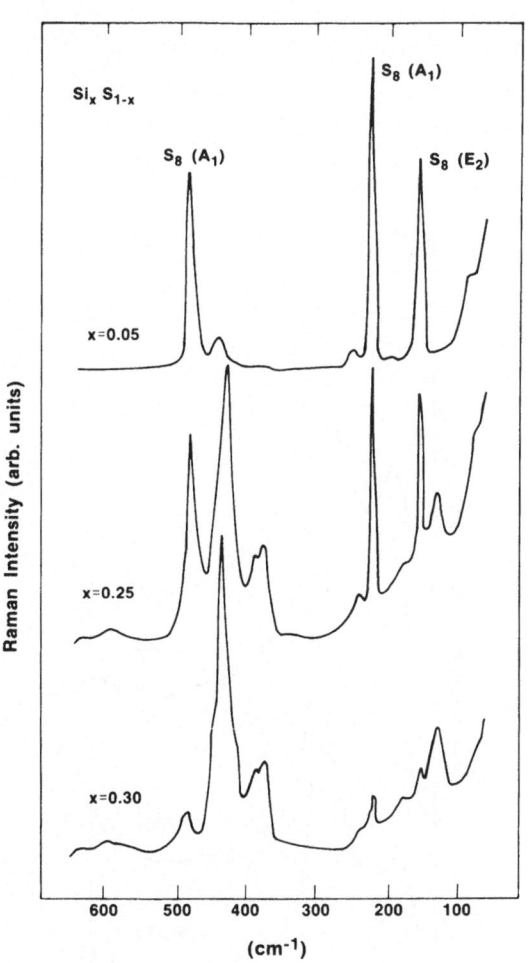

Figure 6.   Room temperature Raman spectra of S-rich Si_xS_{1-x} glasses.

The Raman and IR spectra of the sulfur rich Si_xS_{1-x} glasses are
shown in figures 4,6,and 7. For low Si concentrations, the three
dominant lines (A_1,A_1,E_2) of S_8 units are clearly seen in the Raman
spectra and with lower relative intensity in the IR spectra. The
intensity of the IR lines is related to the dipole moment of the
vibrational mode. This fact means that the IR spectra are dominated by
the vibrational modes of heteropolar bonds such as Si-S and Si-Se. In
reference 9, an unusual compositional dependence of the Raman mode

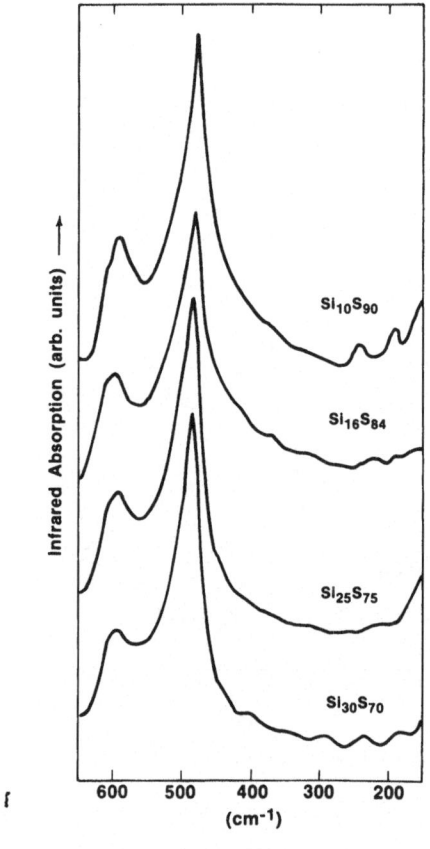

Figure 7.   Infrared spectra of several Si_xS_{1-x} glasses. The
measurements were made in the transmission mode at room
temperature.

associated with the extended chains (about 375 cm^{-1}) relative to the mode associated with Si in tetrahedral units (about 420 cm^{-1}) was observed. This behavior was proposed to result from the emergence of isolated Si(S$_{1/2}$)$_4$ units in the glasses as the sulfur content is increased. The description of the local environments about Si atoms was that at and near SiX$_2$, all the Si atoms are in bitetrahedral units and that as the S content is increased some Si tetrahedral sites appear that are not part of bitetrahedral chains. This proposal is similar to that frequently used in the Ge-X glasses in the X-rich region. In the case of the Si-S glasses this proposal now appears to be incorrect. The isolated Si(S$_{1/2}$)$_4$ tetrahedral units (i.e. those not part of bitetrahedral chains) should have F$_2$ modes that are both IR and Raman active. This does occur in the case for the Ge-X glasses where corner sharing tetrahedra are present in the glasses (18,22,23,26,28,32). As figure 7 shows, the IR spectra of the Si$_x$S$_{1-x}$ glasses do not change with composition and do not exhibit any new features that can be associated with the change in the proposed nature of the Si environments. In all the Si-S glasses studied, no changes are observed between the relative intensities of the modes which are both Raman and IR active and those which are exclusively Raman active. These results prompted a closer inspection of the previous Raman measurements. In reference 9 , the ratio of the area of the peaks associated with SiS$_2$ bitetrahedral units (Raman modes at 427 and 415 cm^{-1}) to Si in tetrahedral units (Raman modes at 381 and 367 cm^{-1}) was observed to decrease with increasing S-content (x<0.25). This behavior can be seen in figure 6 , as a change in the relative intensity of the Raman modes. Higher resolution Raman spectra are shown in figure 4, and it shows that as the sulfur content is increased, the Raman peak at about 427 cm^{-1} is enhanced by a contribution from the weak Raman mode in sulfur at about 440 cm^{-1}. In fact, if this contribution is substracted from the spectra, it is found that the ratio of Si atoms in bitetrahedral units to those in tetrahedral units is a constant independent of sulfur content for x<1/3. The IR results along with the higher resolution Raman measurements indicate that the best description of the atomic structure of the S-rich glasses is that (like the S-rich Ge-S glasses) they are molecularly phase separated. The microstructure consists of SiS$_2$ edge and corner-sharing units and S$_8$ rings for x<0.25, and S rings as well as fragments of rings and chains as x approaches 1/3. This is in contrast to the S-rich Ge-S glasses where the Ge atoms act to cross link the S rings and chains. No information can be obtained on the length of the SiS$_2$ bitetrahedral chains from this compositional study. Based on the Raman and IR measurements it would appear that the basic chain structure and perhaps length do not depend on the sulfur content.

No results have been obtained by us on the Raman spectra of Se- rich glasses. Several IR spectra are shown in figure 8 of Si_xSe_{1-x} glasses. Unlike the Si_xS_{1-x} glasses, the IR spectra appear to be changing with Se-content. Interpretation of the IR spectra in this case must await the completion of the Raman measurements of the Si_xSe_{1-x} glasses. One additional piece of indirect evidence concerning the structure of the Se-rich glasses is available from the glass transition measurements from reference 14. Here the T_g values of the Si_xSe_{1-x} ($x<1/3$) closely follow those of the Se-rich Ge_xSe_{1-x} glasses. This simply means that in both of these systems, the homopolar Se-Se bonds are responsible for the values of T_g (14,34).

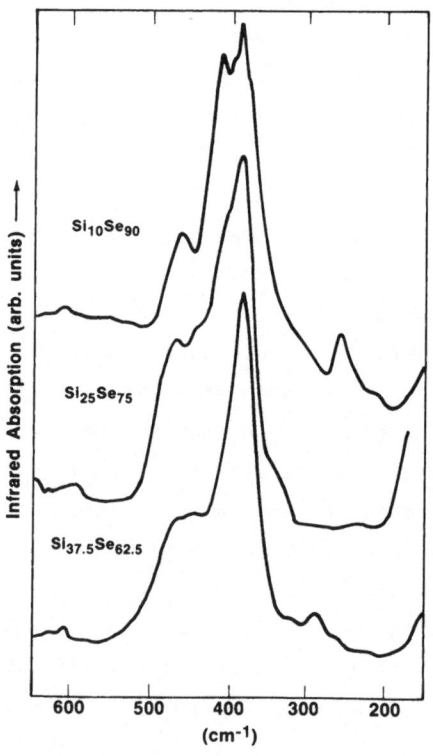

Figure 8.   Infrared spectra of several Si_xSe_{1-x} in the transmision mode.

The Raman spectrum of glassy $Si_{0.35}Se_{0.65}$ and the IR spectrum of glassy $Si_{0.375}Se_{0.625}$ are shown in figures 3 and 8. Previous investigations (10) have discussed the sudden appearance of Si-Si bonds and the description of the Si-glasses in terms of the chemically ordered network model (CONM). The identification of the Si-rich molecular unit is based on compositional studies (which indicate that the composition of this cluster is Si_2Se_3) and comparisons with the halogen model compounds Cl_3-Si-Si-Cl_3 (Br_3-Si-Si-Br_3). The vibrational modes of the these compounds along with the modes observed in the Si-rich glasse are listed in Table VII. In accordance with the 8-N rule, the Si atoms stay 4-fold coordinated and the Se atoms stay 2-fold coordinated in this composition range. The Se atoms are all bonded to two Si near neighbors and there are two distinct Si tetrahedral sites. The two sites are Si with four Se neighbors (in the tetrahedral units) and Si atoms (in the ethane-like units) having 3 Se neighbors and 1 Si near neighbor. The IR measurements provide little additional information on the nature of the glasses in this composition range. The IR spectrum of $Si_{37.5}Se_{62.5}$ is very similar to that for crystalline $SiSe_2$ and glassy $SiSe_2$ (12). This indicates that the Si-rich molecular units are not strongly IR active and that the tetrahedral units persist in this compositional region. Some additional

Table VII.A comparison of the Raman modes of the ethane-like molecules Si_2Cl_6 and Si_2Br_6 with the modes observed in Si-rich Si-X glasses.

Mode	Cl_3-Si-Si-Cl_3 (cm^{-1})	S_3-Si-Si-S_3 (cm^{-1})	Br_3-Si-Si-Br_3 (cm^{-1})	Se_3-Si-Si-Se_3 (cm^{-1})
$n_1(A_{1g})$	351	335	223	206
$n_{11}(E_g)$	211	205	130	125
$n_2(A_g)$	127	110	80	68

indirect information about the structure of the Si-rich glasses can be obtained from the recent work on the optical absorption and thermal properties of these glasses (14). In this work, it was found that for x>1/3 the optical bandgap decreases with increasing Si content (see figure 9) while the glass transition increases with increasing Si content. This is just the opposite of what is commonly observed in chalcogenide glasses. For example, in the Ge-Se glasses the emergence of Ge-Ge bonds results in a drop in both the T_g and optical band gap (33,34). In simple terms, the drop in optical bandgap indicates that weaker bonds are being formed in the glassy network and this usually leads to a drop in the glass transition values. In the Si-Se system, T_g continues to increase with composition despite the presence of Si-Si bonds in the glasses. In reference 14 it was suggested that this behavior indicates that the Si-rich units (ethane-like units) act to cross-link the $SiSe_2$ chainlike units. This idea is similar to the role proposed by Griffiths et al (11) for the Si-rich clusters in Si-Se glasses based on Raman measurements. In the Si-rich glasses, the overall

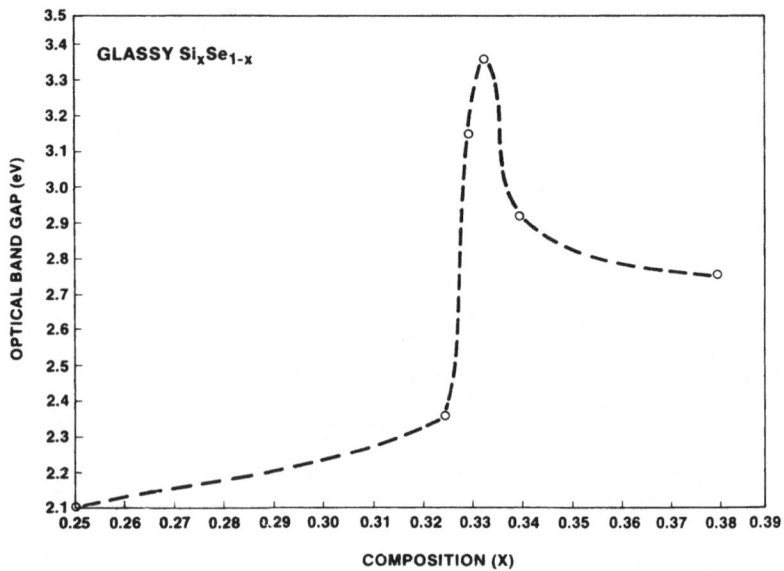

Figure 9. Compositional dependence of the optical band gap of Si_xSe_{1-x} glasses.

rigidity appears to be increasing despite the addition of weaker bonds (Si-Si versus Si-Se) in the glassy network. One further connection that can be made is between the time required to synthesize the glasses and the cross-linking of the

SiSe$_2$ chains. In the experimental section it was pointed out that the time needed to prepare homogeneous glasses appears to increase with increasing Si-concentrations. Since the samples were all prepared in the same manner and given the same treatments this effect must result from a change in the nature of the melts. One proposal is that the viscosity of the molten alloys is increasing with Si-content which slows down the actual alloying process. The connection between this observation and the cross-linking suggestion is that in addition to increasing the T_g of the glasses the Si-rich units are expected to increase the viscosity of the molten alloys. Further measurements of the viscosity as a function of composition would be very interesting in this system.

The discussions above have assumed that the Si, Se, and S atoms all obey the 8-N coordination rule throughout the glass forming region. As an alternative explanation of the spectral and physical data, the consideration of Si-rich glasses which do not strictly adhere to the 8-N rule can be given. In such systems, four-fold coordination of the sulfur atoms in a pseudo-square pyramidal arrangement may be occuring. Sulfur atoms coordinated in this fashion would exhibit stretching frequencies in the area currently designated for the Si-rich molecular units. Invoking 4-fold coordination for some chalcogenide atoms can explain some of the physical data in these glasses. In the Si-Se system, a increase in T_g with increasing Si content indicates an increasingly cross-linked network of chains caused by an increase in the number of 4-fold coordinated sulfur atoms. As a direct consequence of this cross linking, a concurrent increase in viscosity would be expected. Further, not only would this increased viscosity be responsible for the longer reaction preparation times, but alloying would also become more difficult with increasing Si content as increased regularity is needed between chains to allow such coordination. At present, no direct evidence exists for such bonding arrangements in the Si-rich glasses and Se-NMR experiments are planned to directly probe the Se environments in the Si-Se glasses. These experiments should provide the first direct determination of the local structure about the chalcogenide sites in the Si-based glasses.

CONCLUSION

Combined infrared and high resolution Raman measurements have been used to probe the atomic structure of Si_xX_{1-x} (X=S,Se) glasses. The glasses can be described by the chemically ordered network model (CONM) in which the atoms obey the 8-N rule and there is a strong tendency for heteropolar bonding. Throughout the glass forming region ($0.05<x<0.38$), the IR and Raman measurements indicate the presence of SiX_2 edge-sharing and corner-sharing tetrahedral units. For $x<1/3$ (X-rich glasses), homopolar X-X bonds are detected and the X rings/chains may coexist with the SiX_2 tetrahedral units. In the Si-rich glasses ($x<1/3$) ethane-like Si_2X_3 molecular units apear along with the SiX_2 tetrahedral chains.

REFERENCES

1. Bicerano, Josef; Ovshinsky, S. R. <u>Jour. of Non-Cryst Solids</u> (1985) <u>74</u> 75.

2. Ovshinsky, S. R. <u>J. de Phys. C4</u> (1981) suppl. 10, <u>42</u> C4-1095.

3. von Gutfeld; R. J. and Chaudhari, P. <u>J. Appl. Physics</u> (1972) <u>43</u> 4688.

4. Feinleib, J.; deNeufville, J.; Moss, S. C.; Ovshinsky, S. R. <u>Appl. Phys. Lett.</u> (1971) <u>18</u> 254.

5. Rajchman, J. A. <u>J. Appl. Phys.</u> (1970) <u>41</u> 1376.

6. Fritzsche, H.; Ovshinsky, S. R. <u>Jour. of Non-Cryst Solids</u> (1970) <u>2</u> 148.

7. Schoening, F. R. L. <u>Jour. of Materials Science</u> (1979) <u>14</u> 2397.

8. Petersen, U. E.; Birkolz, U.; Adler, D. <u>Phys. Rev. B</u> (1973) <u>8</u> 1453.

9. Tenhover, M.; Hazle, M. A.; Grasselli, R. K. <u>Phys. Rev. B</u> (1984) <u>29</u> 6732.

10. Tenhover, M.; Hazle, M. A. ; Grasselli, R. K. ; Tompson, C. W. <u>Phys. Rev. B</u> (1983) <u>28</u> 4608

11. Griffiths, J. E.; Malyj, M.; Espinosa, G. P.; Remeika, J. P. <u>Phys. Rev. B</u> (1984) <u>30</u> 6978.

13. Johnson,R.W.; Susman,S.; McMillan,J.; Volin, K.J. <u>Material Research Bulletin</u> (1986) <u>21</u> 41. Johnson, R. W.; Susman, S.; Price, D.L. <u>Journal of Non-Cryst. Solids</u> (1985) <u>75</u> 57.

12. Tenhover, M.; Henderson, R. S.; Lukco, D.; Hazle, M. A.; and Grasselli, R. K. <u>Solid State Comm.</u> (1984) <u>51</u> 455

14. Harris, J. H.; Tenhover, M. accepted for publication <u>Jour. of Non-Cryst Solids</u> (1985).

15. Ribes, M.; Barrau, B. ; Souquet, J. <u>Jour. of Non-Cryst Solids</u> (1980) <u>38+39</u> 271.

16. Akridge, James R. U.S. Patent 4,465,746 (1984).

17. Ovshinsky, S. R. <u>Phys. Rev. Lett.</u> (1968) <u>21</u> 1450.

18. Phillips, J. C. <u>Jour. of Non-Cryst. Solids</u> (1980) <u>35+36</u> 1157.

19. Dohler,G. H.; Dandoloff, R.; Bilz, H. <u>Journal of Non-Cryst. Solids</u> (1980) <u>42</u> 87.

20. Thorpe, M. F. <u>Jour. of Non-Cryst. Solids</u> (1983) <u>57</u> 355.

21. Tenhover, M.; Hazle,M. A.; Grasselli, R. K. <u>Phys. Rev. Lett.</u> (1983) <u>51</u> 404.

22. Lucovsky, G.; Galeener, F. L.; Keezer, R. C.; Geils, R. H.; Six, H. A. <u>Phys. Rev. B</u> (1974) <u>10</u> 5134.

23. Tronc, P.; Bensoussan, M.; Brenac, A.; Sebenne, C. Phys. Rev. B (1973) 8 5947.

24. Nemanich, R. J.; Solin, S. A. Solid State Comm. (1977) 21 273.

25. Bridenbaugh, P. M.; Espinosa, G. P.; Griffiths, J. E.; Phillips, J. C.; Remeika,J. P. Phys. Rev. B (1979) 20 4140.

26. Griffiths, J. E.; Espinosa, G. P.; Remeika,J. P. ;Phillips, J. C. Phys. Rev. B (1982) 25 1272.

27. Boolchand, P.; Grothaus, J.; Brésser, W. J.; Suranyi, P. Phys. Rev. B (1982) 25 2975.

28. Kawamoto, Yoji; Kawashima, Chizu Mat. Res. Bull. (1982) 17 1511.

29. Nemanich, R.; Galeener,F. L.; Mikkelson, Jr., J. C.; Connell,G. A. N.; Etherington, G.; Wright, A. C.; Sinclair, R. N. Physica (Utrecht) (1983) 117+118B 959.

30. Nemanich, R. J. Jour. of Non-Cryst. Solids (1983) 59+60 551.

31. Lucovsky, G.; Wong, C. K.; Pollard, W. B. Jour. of Non-Cryst. Solids (1983) 59+60 839.

32. Feltz, A.; Pohle, M.; Steil, H.; Herns, G. Jour. of Non-Cryst. Solids (1985) 69 271.

33. Nemanich, R. J.; Connell, G. A. N.; Hayes, T. M.; Street, R. A. Phys. Rev. B (1978) 18 6900.

34. Sarrach, D. J.; deNeufville, J. P.; Haworth, W. A. Jour. of Non-Cryst. Solids (1976) 22 245.

35. Boolchand, P.; Grothaus, J,; Tenhover,M.; Hazle, M. A.; Grasselli, R. K. accepted for publication Phys. Rev. B (1985).

36. Peters, J.; Krebs, B. Acta. Cryst. (1982) B38 1270.

37. Fateley, W. G.; Dollish, F. R.; McDevitt, N. T.; Bentley, F. F. "Infrared and Raman Selection Rules for Molecular and Lattice Vibrations: The Correlation Method". Wiley-Interscience: New York, 1972.

COMPOUND INDEX

AlB_{12}	189
$Al_{1.05}B_{0.7}N$	189
$Al\ Be_{0.7}B_{22}$	189
Al_2Br_6	338
$Al(CH_3)_3$	101
Al_2Cl_6	338
$AlMgB_{12}$	189
Al-Mn alloy	294
AlN	189
$Al(NO_3)_3$	48
$Al(OC_3H_7)_3$	19
$Al(OC_4H_9)_3$	17
Al_2O_3	21, 48, 101, 174, 185, 276, 277, 285
$2\ Al_2O_3\ 2\ MgO\ 5\ SiO_2$ (cordierite)	188
Al_2O_3-SiC	184
$3\ Al_2O_3\ 2\ SiO_2$ (mullite)	188
Al_2TiO_5	185
Al_2O_3-ZrO_2	175
As_2S_3	314
As_2Se_3	198, 199
B_4C	179
B_2H_6	101, 113, 114, 115
BCl_3	97
BN	179
$BaTiO_3$	188
BeC_2B_{12}	189
BeC_2B_2	189
$BeSiN_2$	189
CH_4	82, 95
C_2H_6	101
$C_4H_9(OH)$	21
$CH_3CH_2=CHSiCl_2$	92
cyclo-$[(CH_3)_2(CH_2=CH)SiNH]_3$	85
$[(CH_3)_2(CH_2=CH)Si]_2NK$	85
$(CH_3)_2(CN)CN=NC(CN)(CH_3)_2$	92
CH_3I	86, 91
CH_3SiCl_3	83
$(CH_3)_2SiCl_2$	82
$(CH_3SiH)_{0.2}(CH_3Si)_{0.8}$	83
CH_3SiH_3	95
CH_3SiHCl_2	83, 84, 85, 86
CO_2	230, 254
CaB_6	189
$CaWO_4$	184, 185
$CeCr_2B_6$	189
Cl_2SiH_3	101
Co_3GaN	189
Co_3GeN	189
Co_3InN	189
Co_3SnN	189
Co_3ZnN	189
Cr_2AlC	189
Cr_2O_3	185
Cr_3C_2	189
CrN	189
$Dy_3AlC_{0.7}$	189
$Er_5Si_3C_2$	189
Fe_3AlC	189
Fe_3AlN	189

$(Fe(CO)_3)_3CON_3P_3Cl_4$	76
$(Fe(CO)_4)_2N_3P_3Cl_4$	76
Fe_3GaN	189
Fe_3GeC	189
Fe_3GeN	189
Fe_3InN	189
Fe_3MgN	189
$Fe_{2-3}N$	144
Fe_4N	144
Fe-Ni-B alloy	297
$FeRu(CO)_3(CP)_2N_3P_3F_3$	76
Fe_3SnC	189
Fe_3SnN	189
Fe_3ZnC	189
Fe_3ZnN	189
$Ga_{.5}Al_{.5}N$	189
GaAlAs	287
GaAs	138, 155, 156, 177, 197, 276, 277, 281, 282, 287
$Ga_{.28}B_{.85}N$	189
$GeBr_4$	334
$GeCl_4$	334
$Gd_3Fe_5O_{12}$	276, 277
Ge-S glasses	332, 333, 334, 335
GeS_2	335
Ge-Se glasses	333, 334, 335
$GeSe_2$	335
H_2	111
H_3	113, 115
HMn_2O_4	126
H_3PO_3	129
H_3PO_4	129
$H_2Ti_4O_9 \cdot H_2O$	125
HfC	189
Hf_2InN	189
Hf_2SnN	189
InP	281
KCl	100
KH	86, 87, 90
$K_2Ti_4O_9$	125
LaBC	189
LaB_2C_2	189
LaB_2C_4	189
$LiAlB_{14}$	189
Li_2AlN_2	189
$LiAl_7SiO_4$	185
$LiBN_2$	189
Li_2CO_3	126
Li_9CrN_5	189
Li_3GaN_2	189
Li_5GeN_3	189
LiMgN	189
Li_7MnN_4	189
$LiMn_2O_4$	126
$LiNbO_3$	184, 185
$LiNbO_3$	276, 277, 279-282
$LiNO_3$	48
$c-Li_2SiO_3$	302
Li_2SiO_5	304
$LiTiN_3$	189
LiZnN	189

Li_3ZnN	189
$Mg_5Ni_{20}B_6$	189
$MgAl_2O_4$	174
$MgFeTaO_5$	185
$MgSiN_2$	189
$MgSiO_3$	300
$MgSnB_2O_6$	185
$MgTi_2O_5$	185
$(Mn_{0.89}Mn_{0.11}O_{1.69})_2OH \cdot H_2O$	126
MnO_2	126
Mn_2O_3	126
$MnSO_4$	126
Mn_3ZnN	189
Mo_2BC	189
Mo-Ru-B alloy	297
Mo-Ru-Si alloy	297
$(NPCl_2)_n$	74
$(NPF_2)_n$	74
$[NP(OC_6H_5)_2]_n$	74
$[NP(OCH_2CF_3)_2]_n$	74
$[NP(OCH_2CH_2OCH_2CH_2OCH_3)_2]_n$	76
NH_3	79, 82, 95
NH_4Br	90
Na_2CO_3	126
$Na_xH_{2-x}Zr(PO_4)_2 \cdot yH_2O$	123
$Na_4Mn_{14}O_{27}$	126
$Na_4Mn_{14}O_{27} \cdot 9H_2O$	126
Na_2O-SiO_2 glasses	319
$3Na_2O \cdot 7\ SiO_2$	313
Na_2SiO_3	302
$Na_2(SrOH)_2Ti_9O_{20}$	126
$Na_2Ti_6O_{13}$	126
$Na_2Ti_nO_{2n+1}$	125
$Na_4Ti_9O_{20}$	126
$Na_4Ti_nO_{2n+2}$	125
$Na_{2n+1}Ti_nO_{3n+1}$	125
$NaZr_2(PO_4)_3$	124, 127
$Na_{1+x}Zr_2Si_xP_{3-x}O_{12}$ (Nasicon)	127
$Na_{2.88}Zr_{1.68}Si_{1.84}P_{1.16}O_{12}$	127
$Na_{3.0}Zr_{1.93}Si_{1.9}P_{1.1}O_{11.91}$	127
$Na_{3.2}Zr_{1.68}P_{1.84}P_{1.16}O_{11.54}$	127
$Na_{3.3}Zr_{1.65}Si_{1.9}P_{1.1}O_{11.5}$	127
$Na_4Zr_2Si_{2.25}P_{1.8}O_{15}$	127
Nb	276, 277
Nb_2AlC	189
NbC	189
Nb_5Ge_3B	189
NbN	189
(Nb,Ta)N	189
Nb_5ZnN_x	189
Ni_3Al	259-261, 268-271
Ni_3AlN	189
Ni-B alloy	293
a-Ni_4B alloy	293
Ni_3InN	189
Ni-P alloy	293
Ni_3ZnN	189
organocyalosiloxanes	71
$[OSi(CH_3)_2]_4$	71
$Pb(Fe,Nb,W)O_3$	285
Pb_5GeO_{11}	279

$3PbO \cdot 2SO_2$	324
$PbO-SiO_2$ glasses	316, 319, 320
$PbTiO_3-PbZrO_3$	188
$Pb(Zr,Ti)O_3$	276, 277
Pd-Cu-P alloy	297
Pd-Cu-Si alloy	297
polyphosphazenes	72
polysilanes	71
polysilazanes	71
R_2SiOR_2	21
R_3SiOR	21
$RSi(OR)_3$	21
$[Ru(bipy)_3]^{2+}$	132
RuO_2	285
SF_6	97
Si	276, 277
$Si_3Al_4N_4C_3$	189
$Si_3Al_5N_5C_3$	189
$SiBr_4$	344, 345
Si_2Br_6	351
SiC	79, 82, 86, 93, 95, 104, 105, 106, 115, 162, 178, 179, 184, 185
$[Si(CH_3)_2]_6$	71
$[Si(CH_3)(H)-CH_2]_n$	71
$SiCl_4$	344, 345
Si_2Cl_6	351
SiHx	200
SiH_2	111
SiH_4	95, 97, 110, 113
silicic acid	52
Si_3N_4	79, 82, 86, 93, 95, 104, 105, 106, 115, 179, 184, 190
$Si(OC_2H_5)_4$	17, 24, 26, 28, 29, 30, 31, 39, 40
SiO_2	31, 48, 185, 292, 300, 314, 315, 316, 319, 321, 330, 332
$Si_2O(OH)_6$	18
$Si_3O_2(OH)_8$	18
$Si_6O_6(OH)_{12}$	18
$Si_{10}O_{11}(OH)_{18}$	18
$Si_{18}O_{22}(OH)_{28}$	18
Si_xS_{1-x}	329, 330, 331, 342, 347, 348, 350, 351, 354
$Si(S_{1/2})_4$	329, 330, 336, 337, 344, 345, 349
SiS_2	329, 330, 336, 337, 339-346, 349, 354
Si_2S_6	351
Si_xSe_{1-x}	329, 330, 331, 350, 351, 352, 354
$Si(Se_{1/2})_4$	329, 330, 336, 337, 344, 345
$SiSe_2$	329, 332, 336, 337, 339-346, 351-354
Si_2Se_6	351
Si-Te glasses	330
$SnMn_3C$	189
$SrCl_2$	126
$SrTiO_3$	126
$Sr_2Ti_9O_{20}$	126
Ta_2AlC	189
TaC	189
TaC HfC	188, 189
TaC ZrC	188, 189
TaN	189
Ta_2VC_2	189
Ti_2AlC	189

Ti_2AlN	189
TiB_2	101, 189
TiC	189
$TiCl_4$	101
Ti_2GaN	189
Ti_2InN	189
Ti_3InN	189
TiN	162, 189
TiO_2	126, 185, 189
TiO_4	101
$Ti(OC_2H_5)_4$	17, 21
Ti_3SiC	189
Ti_3TlN	189
$Ti_4Zn_2N_x$	189
V_2AlC	189
V_2GaN	189
WC	189
$WCo\ B$	189
$W_{10}Co_4C_4$	189
YB_2C_2	189
YB_2C	189
$YCr\ B4$	189
$Y_2Re\ B_6$	189
ZnS	190, 281
$ZnSe$	177, 190
ZrB_2	189
ZrC	189
$Zr(HPO_3)_2$	130
$Zr(HPO_4)_2 \cdot H_2O$	122
$Zr(HPO_4)_2$	130
$[Zr(HPO_4)_{1.3}(O_3P\text{-}ph\text{-}ph\text{-}ph\text{-}PO_3)_{4.33}]$	129
$ZrIr_3B_4$	189
ZrN	189
$Zr(NaPO_4)_2$	126
$Zr(NaPO_4)_2 \cdot 3H_2O$	123
ZrO_2	21, 41, 43, 44, 128, 174
$Zr(OC_4H_9)_3$	17
$Zr(O_3PC_6H_4\text{-}C_6H_4PO_3)_{0.5}(HPO_4)$	130
$Zr(O_3PC_6H_5SO_3H)(HPO_4)$	132
$Zr[O_3PO(CH_2CH_2O)_4H]_2$	132
$ZrTlN$	189
$Zr_4Zn_2N_x$	189

INDEX

a-Si:H	208, 209, 212, 213
acoustic arrays	276, 279
activated reactive experiment	141, 142, 160
aerogels	46, 49, 61
alcogel	40
alkali silicate glasses	291
alkosysilanes	13, 21, 22
amorphous	72, 148, 295
chalcogenides	199,
germanium	149
phosphorus	137
silicon	138, 149, 151
materials	2, 103, 136
oxides	61
semiconductor	195, 199, 207, 217, 227
silicon	112, 113, 114, 116, 195, 196, 197, 198, 199, 201, 208, 209, 217, 218, 221
antireflective coating	59
ARPS	9
austenitizing region	246, 251
bandtail states	209, 210, 211
battery	76
biocompatible	72
bloating	53, 55
bonding	257
boriding	144, 145
bubble domains	277, 280, 281, 283
bulk synthesis	97
capacitive coupling	141
carbide dispersion strengthening	251
carbon-carbon composites	81
carburizing	144, 145
catalysis	
design	1, 267
hetergeneous	67
catenation	70
cellular	
dendritic growth	248
segregation patterns	242, 245
CEMS	9
ceramic	67
material	80
ceramic powders	98, 99
characteristics	100, 105
ceramicization	72
ceramics	2, 13, 14, 19, 22, 170
economic potential	169
multilayer	277
precursors	13
chalcogenides	198, 199, 331, 332, 333, 341
charge couple devices	283
chemical vapor deposition	176, 177, 178
chemical vapor infiltration	185
chemistry	
macromolecules	67
small molecules	67
solids	67
coatings	22, 180, 181

coefficient of expansion	124
components of materials design	3
computation	3, 4
experiment	3, 4, 8
modeling	3, 4, 6
statistics	3, 4
theory	3, 4, 5
composite arrays	281
composition dependence	294, 295
compressive stress	154, 155, 156, 160
computers	264
Cray-1s	264
VAX	264
conductivity	199
cooling rate	239
coordination rule	353
copper phthalocyanine	75
coprecipitation	122
corrosion resistance	144
crystallography	275
CVD process	98
dangling bond	196, 197, 199, 201
data base	5
deep level spectroscopy	207
defects	173
degree of dissociation	136
dendritic growth	248
densification	54, 55
density	294, 295, 301
density of states	197, 201, 206
density-composition relationships	291
deposition on surfaces	97
design	
catalyst	1
materials	1, 3
microstructure	1
tools	138, 142
diamond	258
dielectric breakdown	96
dielectric constant	276
diffusion	26
diffusion length	210
dilation volume	299
directed energy processing	254
discrete components	281
diversity	202
DOE-BES	263
DOE-ECUT	263
domain divided devices	279
domain model	309
drying	46
EELS	9
electric properties	197
electrical conductivity	69, 70
electroactive	67, 72, 74, 75
electrochemical methods	122
electrolytic conduction	70, 72
electromagnetic beam deflection	232
electron beam	
glazing	229, 232, 234, 248, 249, 252
oscillating	233

technique	232
electron energy loss spectroscopy, EELS	316, 318
electron spin resonance	113, 201, 206
electron transfer	265
electrostatic precipitation	100
engineering	
crystallographic	275, 276, 279, 288, 289
domain	279
etching	137
etching of surfaces	97
eutectic alloys	248
EXAFS	9, 257
fast ion conductors	60
fiber diffraction	74
fictive temperature	56, 57
field effect transistors	208, 209
fluid catalytic cracking, fcc	267
fluid flow effects	240
foaming	53
FTIR	257
full scale integration	281
functionality	257
gas pyrolysis	
inert	82
reactive	82
Ge-chalcogenide glasses	331, 333
gelling	22
gels	
acid catalyzed	45, 49, 50, 51, 52, 61
base catalyzed	45, 49, 50, 51, 52, 61
colloidal	45
polymer	45, 52
germanium	198, 199, 197
glass transition temperature	72
glow discharge	137, 142, 143, 163
graft polymer	86, 87
graphite	258
Gunn effect	197
Hall effect	203
Hall-Shockley experiment	203
hard coatings	160
He pycnometry	106
heat flow models	
Cline and Anthony	239
Greenwald	237, 240
hexachlorocyclotriphosphazene	72
hierarchy of models	7, 261
high pressure synthesis	122
HREM	262
hybrid circuits	281
hydriding	147
hydrolytic polycondensation	13, 34
hydrolysis	15, 19, 39, 40
hydrophone	276
hydrothermal methods	122
inductively coupled discharge	140, 142
infrared spectroscopy	257, 330, 331, 332, 333, 341
inorganic	14
integrated ceramic packaging	284
integrated optics	277, 281
intercalation	122, 129, 131, 132

intercalation compounds   2

ion
    exchange   122, 123, 126, 129
    implantation   146, 218, 223
    plating   141, 142, 160
isotope separation   107
Josephson devices   277, 284
Keating potential energy function   307
kinetics
    crystallization   293
    hydrolysis   19
    polymerization   19, 34
laser
    CO_2 230
    diagnostics   98, 102
    glazing   229, 234, 245, 248, 249, 252
    high intensity   96
    IR   96, 98
    manufacturing cost analysis   106
    primary energy sources   96
    process modeling   101
    synthesis condition   105
    UV   96
    visible   96
laser initiated CUD   98, 108, 109
laser initiated gas phase synthesis   95, 97
LEED   9
line source melting   232
lithia-alumina-silica system   48
lithium metasilicate   301, 303, 306, 307
local structure   291, 293, 318
macromolecular science   80
macromolecules   67, 68
    properties   69
manganese dioxide   122
    MnO_2 materials design   121, 128, 133
martensitic strengthening   249
materials design   1, 3, 15, 36, 143, 170, 190, 257, 262, 263, 266, 311
    advancement areas   4, 5
    advances   5
    computation   263, 265
    definition   261
    economic benefits   266
    theory   263, 265
materials modification   14, 19
materials science   67
matrix composites
    ceramic   80
    glass   80
    metal   80
    polymer   80
maximum melt depth   237
medium range structure   291, 293
melt spinning   82, 83
melt spraying   183
melt zone geometry   234, 236, 237
mercury porosimetry   50
metal alkoxides   13, 14, 16, 19, 21
metallic glasses   293, 295
metallurgy   67

microcrystalline	294
microporous materials	39
microstructure	45, 49, 54, 61, 178
design	1
microwave semiconductor	277, 282
mobilities	198, 209
mobility of carriers	203, 204, 205
molecular computers	288
monoliths	39
moving point melting	232
multiphoton absorption	96
nanocrystalline materials	136, 151, 152, 153, 154
nasicon	60, 127, 128
network glasses	292
network structure	19, 29
neutron diffraction	197
nitriding	144, 145, 147, 148
nitrogen sorption analysis	50
non-destructive evaluation, NDE	190, 191, 202
nonisothermal	137, 138, 143, 146, 163
nuclear magnetic resonance, NMR	22, 24, 30, 83, 85, 87, 88, 89, 90, 93, 306
magic-angle spinning	9
olefin hydrosilylation	91
optical	
absorption	208
constants	294, 325
fibers	326
properties	310, 326
wave guides	157
order of reactions	19
oxide glasses	14, 295, 300
oxide materials	79
p-carboxyphenylenephosphate	131
p-n junctions	199
packing	172
partial chemical equilibrium	149, 151
partial integration	281
particle size	172
phospham	74
photoadsorption	205, 206
photoconductivity	138
photoconductor	199, 200
photoemission experiments	207
photoluminescence experiments	207
photolysis	122
photovoltaic applications	208
physical vapor deposition	183
piezoelectrics	276, 279, 286
plasma	
assisted CVD	136
assisted Mo-CVD	162
chemistry	136, 138
column	141
epitaxy	156
induced CVD	136, 148, 157, 158
low pressure	136, 143, 160
parameters	142
spraying	137, 160, 174, 183
suppression	230
thermal	136, 163

treatment	136
poly(dichlorophosphazene)	72
poly(dimethylsiloxanes)	81
poly(dimethylsilymethylenes)	83
poly(ethylene oxide)	70
poly(sulfur nitride)	70
polyacetylene	70
polymeric sulfur	72
polymerization	15, 39, 40
polymers	2, 67, 137, 178
inorganic	67, 68
semiconductor	67, 75
polyphosphazene	71, 74
pore size	129, 130
porosity	51, 52
post-mixing	19
potential fluctuations	196
powder	
ceramic	98, 99, 100, 105
purity	174
preceramic polymer	79, 80, 81, 82
preceramic polysilazanes	86
precipitation	249
processing	
ceramics	170
CUD	176
densification	172
film-fed	175
hot isostatic pressing	172
hot pressing	171, 172
melt	174, 175
melting	14
plasma	171
powder	171
pyrolysis	179, 180
sintering	14
skull melted	175
proton	122
pyrolysis	122
quartz	314
quasi-random model	306, 307
radial distribution function	292
radial flow parallel plate reactor	138
Raman scattering	154
Raman spectroscopy	218, 219, 220, 223, 224, 226, 329, 331, 332, 333, 341
random network model	217, 218, 293, 294
rapid solidification	229, 230, 234, 254
melt quenching	229
refractive indices	310, 311, 313, 325
residual stresses	178
resistivities	200
scanning tunneling microscope, STM	8, 262
Sellmier formulae (model)	313, 325
semiconductors	2, 70, 72, 360
microelectronics	277, 282
thin films	98
Si-chalcogenide glasses	330, 332, 341
sialons	190
silane decomposition	149
silica	79

silicate   39, 79, 300
  binary   319
  chains   302, 303, 304, 305
  glasses   305
  lead   293, 295
  sheets   304, 305
silicon   198, 199, 200, 197
silicon carbide   79
silicon nitride   79
silylamide-catalyzed reaction   85, 86
single oscillator models   312
sintering   171, 191
size-exclusion liquid chromatography, SEC   22, 23, 24, 28, 30
sodium metasilicate   301, 303, 306, 307, 323
sodium titanates   122
sol-gel   71, 183
  process   39, 122
  transition   40
solid electrolytes   75
solidification parameters   234, 239
solution viscosity   41
specific heat   294
spinnability   45
sputtering   183
  DC   141
  RF   141
Staebler-Wronski effect   112, 113, 114, 116
statistics   3, 4
structural factors
  neutron structure   294
  x-ray structure   294
structural field map   6
structural materials   67
structure   257
structure-property relationships   258, 263, 276, 291, 293, 318
substrate bias   138, 141, 142, 149, 154, 155
sulfur   333
superconductor microelectronics   284
surface
  alloying   234, 253
  modification   144, 229, 230, 234, 252, 254
  segregation   8
surface phase   305
  cooperative   306
  dilatation   304, 309
surface voltage method   211
tailoring   68
tempering region   246
tetracyanoquindimethene   74
tetraethyl silicate   71
tetrahedral packing   293
texture   49, 52
thermal expansion coefficient   294
thermal protection   182
thermogravimetric analysis, TGA   87
thermopower   113
time to gel   40, 41, 43, 44, 45
transition metal - metalloid alloys   291, 295
Ultrasonic transducers   277
ultraviolet photoelectron spectroscopy   321
uniformity   178, 183

Urbach energy	205
Urbach law	310
vacuum ultra-violet spectroscopy	319
valence electron localization	315
vapor phase transport	122
viscoelastic behavior	40
viscosimeter	54, 56
viscosity	44, 294
behavior	40, 44
intrinsic	44
reduced	44
relative	44
solution	45
specific	44
vitreous silica	315
Voronoi polyhedron	296
waveguides	59
wear	182
wear resistance	144
Wemple oscillator model	313, 314, 315, 316
x-ray diffraction	93, 197
x-ray lithography	286
x-ray photoelectron spectroscopy, XPS	9, 321
xerogels	46, 49
xerography	208, 210
yield strength	270
Zachariasen	292
zirconia-silica system	41, 45
zirconium phosphate	122, 123, 125
zirconium phosphonates	131